ANNALS OF
THE NEW YORK ACADEMY
OF SCIENCES

Volume 490

EDITORIAL STAFF
Executive Editor
BILL BOLAND
Managing Editor
JUSTINE CULLINAN
Associate Editor
STEFAN MALMOLI

The New York Academy of Sciences
2 East 63rd Street
New York, New York 10021

SECOND ASPEN WINTER
PARTICLE PHYSICS CONFERENCE

ANNALS OF THE NEW YORK ACADEMY OF SCIENCES
Volume 490

SECOND ASPEN WINTER PARTICLE PHYSICS CONFERENCE

Edited by Loyal Durand

The New York Academy of Sciences
New York, New York
1987

Library of Congress Cataloging-in-Publication Data

Main entry under title:

Aspen Winter Particle Physics Conference (2nd: 1986)
    Second Aspen Winter Particle Physics Conference.

    (Annals of the New York Academy of Sciences,
ISSN 0077-8923; v. 490)
    Papers presented at the Second Aspen Winter
Particle Physics Conference series, held at the
Aspen Center for Physics in Aspen, Colo., Jan. 12–18,
1986.
    Bibliography: p.
    Includes index.
    1. Particles (Nuclear physics)—Congresses.
I. Durand, Loyal, 1931– . II. Title.
III. Series.
Q11.N5 vol. 490 [QC793]   500 s   [539.7′21]   87-5671
ISBN 0-89766-375-6
ISBN 0-89766-376-4 (pbk.)

SP
Printed in the United States of America
ISBN 0–89766–375–6 (cloth)
ISBN 0–89766–376–4 (paper)
ISSN 0077–8923

ANNALS OF THE NEW YORK ACADEMY OF SCIENCES

Volume 490
April 30, 1987

SECOND ASPEN WINTER PARTICLE PHYSICS CONFERENCE[a]

Editor and Conference Chairman
LOYAL DURAND

Honorary Chairman
MARTIN M. BLOCK

Organizing Committee
M. M. BLOCK, L. DURAND, S. MESHKOV, D. PINES, M. RUDERMAN,
D. SCHRAMM, L. M. SIMMONS, JR., and F. ZACHARIASEN

Scientific Advisory Committee
J. ADOUZE, J. D. BJORKEN, M. DAVIS, V. FITCH, W. FOWLER,
M. GELL-MANN, J. PEEBLES, M. REES, C. RUBBIA, W. SARGENT,
S. TING, and S. WOJCICKI

CONTENTS

[a]The papers in this volume were presented at the Second Aspen Winter Particle Physics Conference Series, which was held at the Aspen Center for Physics in Aspen, Colorado, on January 12–18, 1986. The conference was sponsored by the Aspen Center for Physics, with support from the Aspen Foundation, the Department of Energy, and the National Science Foundation.

Introductory Remarks

LOYAL DURAND

Department of Physics
University of Wisconsin–Madison
Madison, Wisconsin 53706

The Aspen Winter Physics Conferences were inaugurated in 1985 with two week-long conferences on different aspects of elementary particle physics. The 1985 conferences were the first of a continuing series sponsored by the Aspen Center for Physics. The 1986 program included three conferences, with successive weeks dealing with astrophysics, particle physics, and condensed matter physics. The present volume contains the proceedings of the week on particle physics.

The 1986 Aspen Winter Conference on Particle Physics was held January 12–18, 1986, with 69 participants from around the world. The principal subjects of the conference were e^+e^- and hadronic collider physics, QCD and meson spectroscopy, experiments on heavy flavor production, accelerator neutrino physics, nonaccelerator experiments, and the new theoretical ideas connected with superstrings. The conference ended with a public lecture for the Aspen community by W. F. Fry (University of Wisconsin–Madison) on "The Secrets of Stradivari's Violins: Physics and the Art of the Master Fiddlemakers."

I would like to thank the many people who contributed to the success of the conference, in particular, the members of the Scientific Advisory Committee for Particle Physics (J. D. Bjorken, V. Fitch, M. Gell-Mann, C. Rubbia, S. Ting, and S. Wojcicki) for their cogent suggestions; the members of the particle physics organizing committee that I had the pleasure of chairing (M. Block, L. Durand, S. Meshkov, L. M. Simmons, Jr., and F. Zachariasen) for their work on the many aspects of the conference; and the speakers in the scientific program, most of whose talks are summarized here. The style of the conference had been set in 1985 by Martin Block, who also laid the groundwork for the 1986 conference and remained as overall Honorary Chairman. Martin and Bea Block and Nick and Maggie de Wolf again entertained the entire group in their inimitable styles. The operation of the facilities at the Aspen Center for Physics, and other local arrangements, were handled by Sally H. Mencimer and her staff with their usual expertise and efficiency.

The Conference on Particle Physics was supported by grants from the Aspen Foundation, the U.S. Department of Energy, and the National Science Foundation.

We would like to thank the New York Academy of Sciences for editing and publishing these proceedings.

Theoretical Survey
of Electron-Positron Physics[a]

FREDERICK J. GILMAN

Stanford Linear Accelerator Center
Stanford University
Stanford, California 94305

INTRODUCTION

Everyone is aware of the success of the standard model. However, we are not satisfied with it as a final theory of Nature. It is incomplete. Many parameters, such as quark masses and weak mixing angles, are arbitrary and are simply put in from outside the standard model. In addition, the reason for quark and lepton families or generations is not explained, let alone a connection between quarks and leptons. And then we have the hierarchy problem—how is a scale as "small" as the weak scale determined if we start at a grand unified scale or at the Planck scale?

Therefore, we are constantly probing—checking and then rechecking with more accuracy the predictions of the standard model. We are looking for a discrepancy—a crack that will provide us with an insight as to what lies beyond the standard model.

With this in mind, this survey will be somewhat larger in scope than just electron-positron physics. First, we will review some of what has been checked of the standard model's assumptions or predictions. Then, we will review some of the parameters that are inputs to the standard model and, in particular, the latest values for the magnitudes of the Kobayashi-Maskawa matrix elements. After that, we will proceed to two areas where there may be problems: CP violation and tau lepton decay. Finally, we will examine a couple of places in which new phenomena could show up: heavy neutral leptons and extra vector bosons associated with a larger gauge group.

STANDARD MODEL PARAMETERS

The standard model has been subjected to extensive reviews.[1] We will not attempt another comprehensive review here, but instead will concentrate on a few salient points and new analyses.

There are a number of very impressive measurements over the past few years that make us quite confident about the relevance of the standard model as a description of Nature. None is so direct as the discovery[2] of the W and then[3] of the Z by the UA1 and UA2 collaborations at CERN. Their masses are in accord with theoretical predictions. Moreover, if the masses are converted to a value of $\sin^2\theta_W$ and compared with other determinations from neutral current experiments in both purely leptonic (such as

[a]This work was supported by the Department of Energy, Contract No. DE-AC03-76SF00515.

1

$e^+e^- \rightarrow \mu^+\mu^-$ and neutrino-electron elastic scattering) and semileptonic (such as deep inelastic neutrino-nucleon and polarized electron-deuteron inelastic scattering) processes, there is very impressive agreement among the various experiments.[4] These results are not yet quite at the level of testing the electroweak radiative corrections at the one loop level, but the time for doing so is coming soon.

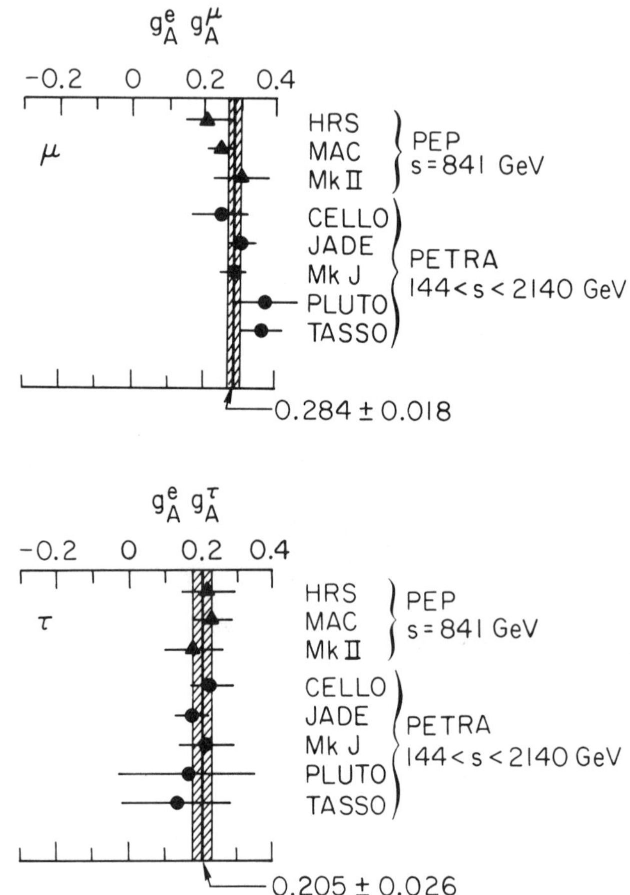

FIGURE 1. The $g_A^e g_A^\mu$ and $g_A^e g_A^\tau$ values derived from experiments measuring the respective $\mu^+\mu^-$ and $\tau^+\tau^-$ front-back asymmetries at PEP and PETRA. The world average values are indicated, which are to be compared to the standard model expectation for both quantities of $+0.25$. This figure is from reference 5.

The axial-vector couplings of the μ and τ leptons and c and b quarks to the Z have been determined from the front-back asymmetry in electron-positron annihilation. The lepton couplings are now determined to roughly 10% from the combined PEP and PETRA measurements[5] shown in FIGURE 1 and are in agreement with the standard model within two standard deviations. Similar results hold for the c and b quarks, with larger errors. Said in a different form, if our choice of the difference of weak isospins, $I_{3R} - I_{3L}$ (which is proportional to g_A), is restricted to half-integral values, then the data unambiguously favor the values that result if one makes the standard model assignments of weak isospin for these quarks and leptons.

Essentially the same measurements put limits on an effective four-Fermi interaction arising from compositeness.[6] Writing the coefficient of this interaction as g^2/Λ^2 and taking $g^2/4\pi = 1$, the recent measurements[7] of $e^+e^- \rightarrow e^+e^-$ and $e^+e^- \rightarrow \mu^+\mu^-$ put a lower limit on Λ of about 2 TeV.

One of the sets of parameters that we are unable to predict from inside the standard model is the relation between the mass eigenstates and weak eigenstates for quarks. The relation between them is expressed by a matrix known as the Kobayashi-Maskawa[8] matrix, as they exhibited an explicit parametrization in the six quark case in 1973. We define this unitary matrix as

$$
\begin{pmatrix} d' \\ s' \\ b' \end{pmatrix} = \begin{pmatrix} V_{ud} & V_{us} & V_{ub} \\ V_{cd} & V_{cs} & V_{cb} \\ V_{td} & V_{ts} & V_{tb} \end{pmatrix} \begin{pmatrix} d \\ s \\ b \end{pmatrix}, \tag{1}
$$

where we have followed the usual convention of leaving the charge $2e/3$ quarks unmixed, and have expressed the relation by a rotation acting on the mass eigenstates to obtain the weak eigenstates of the charge $-e/3$ quarks.

Because the matrix elements come from outside the standard model, as do the quark masses, they are a potential window on the physics beyond: best of all, if we knew what to do with this information, it would be something that we would have in our hands today. At a somewhat less ambitious level, there may be a connection between the quark masses and the matrix elements[9]—at a minimum reducing the number of free parameters, and perhaps giving us insight into a symmetry or dynamics that relates one generation to another.

In addition, knowledge of these matrix elements is useful for more mundane engineering purposes in calculating the effects of loops involving virtual heavy quarks (as in the next section). Last, but not necessarily so in importance, checking on unitarity of the matrix, assuming there are three generations, could, by its failure, provide evidence for a fourth generation. (We immediately note that there is no evidence for this at the present time. See below.)

The value of each of the matrix elements can, in principle, be determined from weak decays of the relevant quarks, or, in some cases, from deep inelastic neutrino scattering. Our knowledge of the various matrix elements at this time comes from the following sources:[10]

(1) $|V_{ud}| = 0.9729 \pm 0.0012$ from nucleon beta decay when compared to muon decay, after incorporating newly refined corrections from Marciano and Sirlin[11] for the order α structure dependent terms and leading log radiative corrections, which are summed using the renormalization group.

(2) $|V_{us}| = 0.221 \pm 0.002$ from K_{e3} and hyperon decays, with a new analysis by Leutwyler and Roos[12] that takes account of isospin violation between charged

and neutral K decays and brings the values extracted for $|V_{us}|$ from the two decays into agreement at the 1% level.

(3) $|V_{cd}| = 0.24 \pm 0.03$ from neutrino production of charm[13] and subsequent semileptonic decays.

(4) $|V_{cs}| > 0.66$ from the width for $D \rightarrow Ke\nu$ and the (very conservative) assumption that the relevant form factor occurring in D_{e3} decay, $f_+^D(0)$, is less than 1.

(5) $|V_{ub}/V_{cb}| < 0.19$ from the newer (and less stringent) limit on $b \rightarrow u/b \rightarrow c$ of <0.08.[14]

(6) $0.037 < |V_{cb}| < 0.053$ from the present world average b lifetime.[14]

Putting these constraints together with unitarity and the assumption of three generations gives the following limits on the Kobayashi-Maskawa matrix elements:

$$\begin{pmatrix} 0.9742 \text{ to } 0.9756 & 0.219 \text{ to } 0.225 & 0 \text{ to } 0.008 \\ 0.219 \text{ to } 0.225 & 0.973 \text{ to } 0.975 & 0.037 \text{ to } 0.053 \\ 0.002 \text{ to } 0.018 & 0.036 \text{ to } 0.052 & 0.9986 \text{ to } 0.9993 \end{pmatrix}. \tag{2}$$

The data are consistent with there being just three generations: neither do they preclude there being more than three generations. If we assume there are four or more generations, then the constraints from unitarity are, of course, much less effective:

$$\begin{pmatrix} 0.9710 \text{ to } 0.9748 & 0.218 \text{ to } 0.224 & 0 \text{ to } 0.01 & \cdots \ \cdots \ \cdots \\ 0.192 \text{ to } 0.288 & 0.66 \text{ to } 0.98 & 0.037 \text{ to } 0.053 & \cdots \ \cdots \ \cdots \\ 0 \text{ to } 0.14 & 0 \text{ to } 0.72 & 0 \text{ to } 0.999 & \\ \vdots & \vdots & \vdots & \cdots \ \cdots \ \cdots \end{pmatrix}. \tag{3}$$

The known matrix elements may be used together with unitarity to restrict potential matrix elements to additional quarks; for example, we must have $|V_{ub'}| < 0.088$ from the known elements in the first row. The magnitudes of the matrix elements given above may be used to obtain the angles involved in any particular parametrization, such as that of Kobayashi-Maskawa[8] or that of Maiani.[15]

CP VIOLATION

More than thirty years after its discovery, the only place where CP violation has been definitively observed is still the neutral K system and, in particular, in the decay $K \rightarrow \pi\pi$. This amplitude may be split into parts, A_0 and A_2, that involve final $\pi\pi$ isospin 0 and 2, respectively. Experimentally, it is known that $|A_0|$ is about twenty times $|A_2|$. This is one manifestation of the $\Delta I = \frac{1}{2}$ rule for strangeness-changing nonleptonic decays. This rule states that transitions that result in an isospin change of $\frac{1}{2}$ (such as from a K meson to two pions with total isospin 0) have amplitudes that are much bigger than transitions that result in an isospin change of $\frac{3}{2}$ (such as from a K meson to two pions with total isospin 2).

The usual convention is to choose phases so that the larger amplitude, A_0, is real; then the parameter ϵ that involves CP violation in the mass matrix is given by

$$\epsilon \approx \frac{1}{2\sqrt{2}} e^{i\pi/4} \frac{\mathrm{Im}\, M_{12}}{\mathrm{Re}\, M_{12}}, \tag{4}$$

where Re M_{12}, the real part of the off-diagonal element of the $K^0 - \bar{K}^0$ mass matrix, is equal to one-half of the mass difference between K_S and K_L, ΔM_K.

The other parameter for the neutral K system, ϵ', which is connected to CP violation in the decay amplitude, is in this convention,

$$\epsilon' = \frac{i}{\sqrt{2}} e^{i(\delta_2 - \delta_0)} \frac{\mathrm{Im}\, A_2}{A_0}, \tag{5}$$

with δ_0 (δ_2) being the $\pi\pi$ phase-shift in the isospin 0 (2) channel.

We calculate Im M_{12} from the box diagram. This should be a reliable short-distance calculation because it is momentum scales between m_c and m_t that contribute an imaginary part to M_{12} in the standard model. However, if we do the calculation in the usual quark basis, we must remember that A_0 generally has a phase, $e^{i\xi}$, due to virtual heavy quarks in loops contributing to the decay amplitude. To get back to the usual convention where A_0 is real, we redefine the K^0 and $\bar{K}^0$ states so as to eliminate this phase. This induces phase changes elsewhere so that to lowest order in the small quantity ξ,

$$\mathrm{Im}\, M_{12}^{sd} \rightarrow \mathrm{Im}\, M_{12}^{sd} + 2\xi \mathrm{Re}\, M_{12}^{sd}, \tag{6}$$

and, therefore, in the basis where A_0 is real,

$$\epsilon = \frac{1}{\sqrt{2}} e^{i\pi/4} \left(\frac{\mathrm{Im}\, M_{12}^{sd}}{\Delta M_K} + 2\xi \frac{\mathrm{Re}\, M_{12}^{sd}}{\Delta M_K} \right). \tag{7}$$

Here, the superscript sd indicates the short-distance contribution to the particular mass matrix element. Because the amplitude A_2 is real in the quark basis, it now picks up a phase,

$$A_2 \rightarrow A_2 e^{-i\xi}, \tag{8}$$

and, correspondingly, we have

$$\epsilon' \approx \frac{1}{\sqrt{2}} e^{i\pi/4} \frac{A_2}{A_0} (-\xi), \tag{9}$$

where we have used the experimental $\pi\pi$ phase-shifts to write (approximately) the phase of ϵ'.

In the expression for ϵ, we drop the term involving ξ. This is good to 20% or better in light of the recent measurements of ϵ' (see below). Similarly, because of the small measured limit on ϵ', long-distance contributions to ϵ (which should be of order $20\epsilon'$) are also very likely negligible.[16] Taking the short-distance contribution corresponding to the box diagram and $m_q^2 \ll M_W^2$, we then arrive at the expression,

$$\epsilon \approx \frac{e^{i\pi/4}}{\sqrt{2}} \frac{BG_F^2 f_K^2 m_K}{6\pi^2 \Delta M_K} \times s_1^2 s_2 s_3 s_\delta [-\eta_1 m_c^2 + \eta_2 s_2 (s_2 + s_3 c_\delta) m_t^2 + \eta_3 m_c^2 \ln(m_t^2/m_c^2)], \quad (10)$$

where the s_i are the sines of the Kobayashi-Maskawa angles θ_i, $i = 1, 2, 3$. These are known to be small, so the approximation $c_i = \cos \theta_i = 1$ has been used in equation 10. A nonzero value of the angle δ is indicative of CP violation in the Kobayashi-Maskawa parametrization.[8] The factors η_1, η_2, and η_3 are due to strong interaction (QCD) corrections and have the values 0.7, 0.6, and 0.4, respectively, with usual quark and W boson masses.[17] The infamous parameter B is the ratio of the actual value of the matrix element between K^0 and $\overline{K}^0$ states of the operator composed of the product of two $V - A$ neutral, strangeness-changing currents divided by the value of the same matrix element obtained by inserting the vacuum between the two currents.

One can see that there can be a potential problem in getting the right-hand side of equation 10 to reproduce the experimental value of $|\epsilon| = 2.27 \times 10^{-3}$ if the combination of Kobayashi-Maskawa angles, m_t, and B is not large enough. To get some idea of where the experimental situation places us, take $B = \frac{1}{3}$, a b quark lifetime of 1 picosecond (near the present world average[14]), and (as it was a year ago) $b \to u/b \to c < 0.04$. Then for $m_t \lesssim 60$ GeV, we would have trouble satisfying equation 10. For $B \approx 1$, there is no problem satisfying equation 10 for any so far unexcluded value of m_t, just as long as $b \to u/b \to c \gtrsim 0.02$. Smaller values of $b \to u/b \to c$ get us into trouble by lowering the upper bound on s_3.

The story from the past year or so in this regard is one of retreat from a confrontation with the standard model. On the experimental side, the upper limit on $b \to u/b \to c$ has become less stringent as it was realized that the theoretically motivated electron spectra used to fit B meson decay are not a good fit to the much improved measured spectra from CESR.[18] As a result, the upper limit has risen to a more conservative 0.08 rather than the previous 0.04. Thus, s_3 can be larger, relieving some of the pressure on the right-hand side of equation 10.

On the theoretical side, an argument[19] that $B \approx \frac{1}{3}$, based on using $SU(3)$ to relate B to the amplitude A_2, was found to have potential 100% corrections from the next order in chiral $SU(3)$ breaking and is therefore unreliable.[20] Since that time, there has been a flurry of papers on the subject,[21] with values of B ranging from 0.3 to 1.5 or so. In my opinion, the subject is not conclusively settled, and as long as B can be near unity, the standard model is far from the danger of being excluded as having a credible origin for CP violation through the phase in the Kobayashi-Maskawa matrix.

The situation for ϵ' has also undergone a similar relaxation. From equation 9, we derive (by inserting experimentally measured quantities) that[22]

$$\epsilon'/\epsilon = -15.6\xi = 6.0 \, s_2 s_3 s_\delta \left(\frac{\text{Im } \tilde{C}_6}{-0.1}\right)\left(\frac{\langle \pi\pi | Q_6 | K^0 \rangle}{1.0 \, \text{GeV}^3}\right), \quad (11)$$

where Q_6 is the "penguin" operator in the short-distance expansion of the strangeness-changing weak Hamiltonian responsible for K decay,[23] and Im $\tilde{C}_6$ is the imaginary part of the corresponding Wilson coefficient with the Kobayashi-Maskawa factor taken out.

The value of -0.1 for this last quantity is relatively stable from calculation to calculation because the imaginary part depends on momentum scales from m_c to m_t, where the short-distance expansion is well justified. The value of the matrix element of Q_6 is much less certain. If it is large enough to explain the magnitude of A_0 (which is a long-standing puzzle), then combined with the value of $s_2 s_3 s_\delta$ needed to fit ϵ (see above), it yields the prediction of $\epsilon'/\epsilon \approx +10^{-2}$. This was basically the original observation in reference 23: if the "penguin" operator is to be an explanation of the $\Delta I = \frac{1}{2}$ rule and the magnitude of A_0, then ϵ'/ϵ should be at the 1% level.

The most recent experiments, on the other hand, obtain[24]

$$\epsilon'/\epsilon = (-0.46 \pm 0.53 \pm 0.24) \times 10^{-2} \tag{12a}$$

and[25]

$$\epsilon'/\epsilon = (+0.17 \pm 0.82) \times 10^{-2}. \tag{12b}$$

At the same time, theoretical calculations of the matrix element of Q_6 have changed dramatically. Early calculations gave numbers of the order of 1 GeV^3, giving hope that one could explain the magnitude of A_0 on the basis of "penguins" and that ϵ'/ϵ is of the order of 1%. In the last couple of years, however, calculations incorporating current algebra constraints in a correct manner give much smaller numbers.[26] These predict values for ϵ'/ϵ of the order 2×10^{-3}. These calculations, while not unassailable, have as impeccable theoretical credentials as any others; they do not, though, allow one to understand the magnitude of the overall $K \to \pi\pi$ amplitude. It is certainly possible to take the attitude that the newer calculations of the matrix element are the correct ones and the origin of the $\Delta I = \frac{1}{2}$ rule lies elsewhere. It is also possible to still claim that "penguins" are the primary explanation of the $\Delta I = \frac{1}{2}$ rule, and either that this is not decisively excluded by the present round of experiments or that the origin of CP violation lies elsewhere than the standard model.[27]

The future in the calculational realm probably belongs to the lattice gauge calculations of matrix elements.[28] While still in their infancy, there are some hopeful signs, but the calculations are done on small lattices and involve approximations or extrapolations that do not allow definitive conclusions. Still, they hold the promise of eventually providing a reliable calculation of these quantities. Experimentally, the next round should push the error bars down to the 10^{-3} level. That should allow a real conclusion to be drawn as to whether the "penguin" operator plays a significant role in K decay if it is assumed that CP violation has its origin in the standard model.

TAU DECAY

All the properties of the tau are consistent with its being a third generation lepton, that is, just another copy of the electron and muon, albeit much heavier. The front-back asymmetry measurements referred to earlier[5] and the momentum spectrum of the final charged lepton in its purely leptonic decay assure us that its assignment to a left-handed weak doublet is correct. The other member of this doublet, the tau neutrino, must be distinct from the electron and muon neutrinos. The upper bound on the tau neutrino mass has recently been lowered to 70 MeV, which is below the mass of the charged lepton of the previous generation.[29]

The one problem of consequence comes in comparing the sum of the exclusive decay mode measurements with the inclusive charged-prong multiplicity measurements. In particular, it is difficult to account for the origin of all the decays of the tau that result in one charged prong.[30] This arises as follows: Let us normalize all the theoretical calculations of decay branching ratios to that for $\tau \rightarrow \nu_\tau e \bar{\nu}_e$, which we take to be 17.9% (the world average value[29] is $17.9 \pm 0.3\%$). Then, the present experimental situation is summarized in TABLE 1 for decays of the tau involving three charged prongs. First of all, we see that the sum of the exclusive modes (almost entirely $\tau^- \rightarrow \nu_\tau \pi^- \pi^- \pi^+$ and $\tau^- \rightarrow \nu_\tau \pi^- \pi^- \pi^+ \pi^0$) is in agreement with the three charged-prong inclusive branching ratio. Second, where there is a theoretical prediction, it is in good agreement with experiment.[29,30]

The agreement between theory (where there is a prediction of some accuracy) and experiment (where there is a definite measurement) is also very good in the case of tau decays involving one charged prong (see TABLE 2). In particular, note the decays $\tau \rightarrow \nu_\tau \pi$ and $\tau \rightarrow \nu_\tau K$, whose rates follow from those for $\pi \rightarrow \mu \nu$ and $K \rightarrow \mu \nu$, respectively. There is also the major decay $\tau^- \rightarrow \nu_\tau \pi^- \pi^0$, whose rate follows from using CVC to relate it to an integral over $e^+ e^- \rightarrow \pi^+ \pi^-$ cross sections, with a result that is in excellent agreement with experiment. While there is not a precise experimental number with which to compare it, the theoretical prediction for $\tau^- \rightarrow \nu_\tau \pi^- 3\pi^0$ follows in a similar way from an integral over the well-measured cross section for $e^+ e^- \rightarrow 2\pi^+ 2\pi^-$, and one has every reason to have confidence in the prediction.[30]

The theoretical upper bounds on the modes in TABLE 2 follow from using isotopic spin invariance to bound one charge combination of the final hadrons in terms of another combination that has been measured; for example, $\tau^- \rightarrow \nu_\tau \pi^- 4\pi^0$ is bounded from the measured decay $\tau^- \rightarrow \nu_\tau 3\pi^- 2\pi^+$. In particular, the major decay mode $\tau^- \rightarrow \nu_\tau \pi^- 2\pi^0$ has a branching ratio that must be less than that of the well-measured mode $\tau^- \rightarrow \nu_\tau 2\pi^- \pi^+$. From the properties of this latter mode,[29] which is dominated by the A_1 resonance decaying into $\rho\pi$, the inequality should be an equality. The 2% limit on the last three modes is a fairly generous upper limit.

The sum of the branching ratios from theory is $\leq 81.7\%$ in TABLE 2, while the measured one charged-prong inclusive branching ratio is $86.6 \pm 0.3\%$.[29] We have

TABLE 1. Three Charged-Prong Decays of the τ

	Branching Ratio (%)	
Decay Mode	Theory[a]	Experiment[b]
$\tau^- \rightarrow \nu_\tau 2\pi^- \pi^+$		8.1 ± 0.6
$\tau^- \rightarrow \nu_\tau 2\pi^- \pi^+ \pi^0$	4.9	5.0 ± 0.6
$\tau^- \rightarrow \nu_\tau (K_s \pi)^-$	0.3	$0.3 \pm 0.1 \pm 0.1$
$\tau^- \rightarrow \nu_\tau K^- \pi^- \pi^+$		0.22 ± 0.14
$\tau^- \rightarrow \nu_\tau K^- K^+ \pi^-$		0.22 ± 0.14
$\tau^- \rightarrow \nu_\tau 2\pi^- \pi^+ 2\pi^0$		
$\tau^- \rightarrow \nu_\tau 2\pi^- \pi^+ 2\pi^0$	<0.4	
TOTAL		13.2 ± 0.3

[a]All theoretical branching rates are normalized to that for $\tau \rightarrow \nu_\tau e \bar{\nu}_e$, taken as 17.9%. Calculations are from reference 30.

[b]Experimental values are taken from reference 29.

TABLE 2. One Charged-Prong Decays of the τ

Decay Mode	Branching Ratio (%)	
	Theory[a]	Experiment[b]
$\tau^- \rightarrow \nu_\tau e^- \bar{\nu}_e$	17.9 (input)	17.5 ± 0.7
$\tau^- \rightarrow \nu_\tau \mu^- \bar{\nu}_\mu$	17.4	18.1 ± 0.6
$\tau^- \rightarrow \nu_\tau \pi^-$	10.9	10.3 ± 1.2
$\tau^- \rightarrow \nu_\tau \pi^- \pi^0$	22.0	22.0 ± 2.1
$\tau^- \rightarrow \nu_\tau \pi^- 2\pi^0$	≤8.1 ± 0.6	
$\tau^- \rightarrow \nu_\tau \pi^- 3\pi^0$	1.0	
$\tau^- \rightarrow \nu_\tau \pi^- 4\pi^0$	<0.1	
$\tau^- \rightarrow \nu_\tau \pi^- 5\pi^0$	<0.1	
$\tau^- \rightarrow \nu_\tau K^-$	0.7	0.6 ± 0.2
$\tau^- \rightarrow \nu_\tau (K\pi)^-$	0.9	0.9 ± 0.3 ± 0.3
$\tau^- \rightarrow \nu_\tau (K\pi\pi)^-$ $\tau^- \rightarrow \nu_\tau (K\bar{K}\pi)^-$ $\tau^- \rightarrow \nu_\tau (K\bar{K})^-$	<2	
TOTAL	<81.7	86.6 ± 0.3

[a]All theoretical branching rates are normalized to that for $\tau^- \rightarrow \nu_\tau e\bar{\nu}_e$, taken as 17.9%. Calculations are from reference 30.

[b]Experimental values are taken from reference 29.

accounted for all the purely leptonic modes and all the modes of the form $\tau^- \rightarrow \nu_\tau (n\pi)^-$ of any consequence, as well as Cabibbo suppressed modes. Therefore, where are the remaining 5% of one prong decays?

One possibility is that the branching ratio for $\tau \rightarrow \nu_\tau e\bar{\nu}_e$, which we took to be 17.9% (and to which we normalized all our theoretical predictions), should be ≈19%. This would scale up all the predicted branching ratios by ≈6%, making the agreement of experiment and theory worse in TABLE 2. It would also put the τ lifetime measurements[14] about one standard deviation away from their predicted standard model value. While the data over the past year have tended to make the discrepancy more significant, it is still possible that most of the problem could be eliminated by the branching ratios moving upward.[31]

A second possibility is that there are other conventional decay modes that we have neglected as very likely small, but that, in fact, have branching ratios amounting to several percent. A candidate for such a role is $\tau \rightarrow \nu_\tau \eta\pi\pi$. There is no particular argument that would demand a sizable branching ratio for this channel, but there is no direct evidence that limits it to the few tenths of a percent level at which one might have naively estimated it.

Finally, there is the possibility of new physics. There cannot be a new elementary charged particle into which the τ decays because it would have been pair-produced in electron-positron annihilation and it would have made the value of R disagree with experiment. Thus, we need a new neutral particle produced along with conventional ones and/or a distortion of the expected branching ratios due to the effects of a heavy virtual particle. It is hard to find a scenario for this situation that is not very contrived (especially if it is not to be in conflict with other existing experiments). Perhaps the whole problem will just creep away a few percent at a time.

HEAVY NEUTRAL LEPTONS

The most straightforward extension of the presently known neutral leptons, that is, the electron, muon, and tau neutrinos, is to add a fourth, so-called "sequential" neutrino as a part of a fourth generation consisting of a charge $2e/3$ quark, a charge $-e/3$ quark, a charge $-e$ lepton, and a neutrino. The left-handed quarks and leptons reside in weak isospin doublets, while all right-handed fermions are singlets, as for the first three generations. Moreover, in the canonical version of the standard model, there is no right-handed neutrino field at all and simply no way to obtain a massive neutrino: the neutrino is left-handed and massless. More generally, going slightly beyond the standard model, it is entirely possible to supply a right-handed singlet field and make such a neutrino a massive fermion. Then, just as for the quark sector, the weak and mass eigenstates will not coincide. This is conveniently expressed in terms of a unitary matrix U that "rotates" the neutrino mass eigenstates (the column index, labeled by the generation number) to the weak eigenstates (the row index, labeled by the corresponding charged lepton). For example, the neutrino that is in a left-handed doublet with the electron, ν_e, is the superposition of mass eigenstates, ν_j, given by

$$\nu_e = \sum_{j=1}^{4} U_{ej}\, \nu_j \tag{13}$$

in the case of four generations. Here, because all the neutral leptons have the same value of weak isospin, there are no lepton flavor-changing neutral currents.[32] This generalizes to adding an arbitrary number of generations. The unitary property of the matrix U implies that the Z couples to each pair of neutrino mass eigenstates with the same universal strength, but there are no couplings of the Z (at tree level) to different mass eigenstates.

Neutral leptons are also predicted that behave as singlets under the weak isospin of the standard model. Such is the case, for example, in some grand unified models such as $O(10)$, where the 16-dimensional representation includes all the usual quarks and leptons of one generation plus a right-handed singlet, neutral lepton.[33] In left-right symmetric theories as well, there are often heavy neutral leptons.[34] In particular, they arise as the partners of the usual charged leptons under the additional $SU(2)$ gauge group with right-handed couplings, but are singlets under the usual $SU(2)$ gauge group with left-handed couplings of the standard model.

A particularly attractive reason for having further singlet neutral leptons is the so-called "seesaw" mechanism[35] for generating neutrino masses. With each left-handed neutrino, one associates a right-handed partner, N, as in the 16-dimensional representation of $O(10)$, so the mass matrix involving ν and N looks like

$$M = \begin{pmatrix} 0 & m_D \\ m_D & M \end{pmatrix}. \tag{14}$$

Here, m_D is a Dirac mass, presumably comparable to a quark or charged lepton mass of that generation, and M is a large Majorana mass. The eigenstates of this matrix will have masses of m_D^2/M and M approximately. Thus, we have light neutrinos as well as heavy ones. The mixing matrix elements, $U_{\varrho N}$, between the light neutrinos (which are members of the usual left-handed isodoublets with corresponding charged leptons, ℓ) and the heavy neutral singlets are of the order of m_D/M. We can also arrange matters so that the light neutrino remains massless, but still mixes with a singlet Dirac heavy neutral lepton.[36,37]

Another possibility is the existence of mirror neutrinos, that is, neutral leptons in right-handed doublets under the usual $SU(2)$, together with corresponding mirror charged leptons.[38] Such neutral leptons are left-handed singlets; hence, the name "mirror fermions" because they are just the mirror image of the usual fermions in the standard model. A recent case in point is provided by the $O(18)$ grand unified model[39] that predicts a fourth generation with the usual left-handed weak interaction couplings as well as four(!) generations of quarks and leptons with right-handed couplings. Each of these generations contains a neutral lepton with a mass below about 40 GeV.

At the same time that theoretical interest has been increasing,[40–42] the scope and sensitivity of experimental techniques have also been undergoing a dramatic change.[43] High flux hadron beams at both low and high energy now permit very low limits being placed on the emission of neutral leptons in leptonic or semileptonic hadron decays. At electron-positron colliders, center-of-mass energies and integrated luminosities have reached the level where weak interaction cross sections are capable of being responsible for production of detectable numbers of new neutral leptons. This, along with the concurrent development of precision vertex detectors at such machines,[44] has made it possible to extend the mass range under investigation by over an order of magnitude. While previously one was limited by the mass of the decaying hadron, in many cases the kinematic limit now is simply the center-of-mass energy of the electron-positron collision. The near future should see this line of analysis take another large jump in sensitivity through the study of Z decays into neutral lepton pairs. This will permit a high sensitivity sweep of the mass range from zero to half the Z mass.

A variety of techniques have been used over the years in high energy physics experiments to search for neutral heavy leptons. Before the searches at electron-positron colliders, most of the previous limits on mixing of a heavy neutrino with the electron or muon neutrino came from hadron leptonic or semileptonic decays. A direct and powerful technique[45] is to study pion or kaon decays at rest, searching for additional monochromatic peaks in the electron or muon momentum spectra in the leptonic decays $\pi \rightarrow e\nu$, $\pi \rightarrow \mu\nu$, $K \rightarrow e\nu$, or $K \rightarrow \mu\nu$, as appropriate. Upper limits on the mixing with a fourth generation neutrino extracted using this technique are shown[46–49] as the limiting curves for $|U_{e4}|^2$ labeled (1), (2), and (3) in FIGURE 2 and the curves for $|U_{\mu4}|^2$ labeled (3) and (12) in FIGURE 3. At the present time, limits on the square of these mixing matrix elements are in the region of 10^{-6} from such experimental searches.

A different, but related technique in that it relies on the decay in flight of the same hadrons, namely, π and K mesons, arose when it was realized[50] that the heavy neutral leptons so produced would themselves decay downstream. It was therefore possible to use existing neutrino detectors or modifications thereof to conduct searches for the subsequent decay of the heavy neutral leptons produced in high intensity π and K beams. In the potentially observed experimental rate, the square of a mixing matrix element enters twice: first in the production where the neutral lepton is born in π or K decay through mixing with the electron or muon neutrino, and then again when the heavy lepton decays weakly into ordinary leptons or leptons and quarks. Depending on

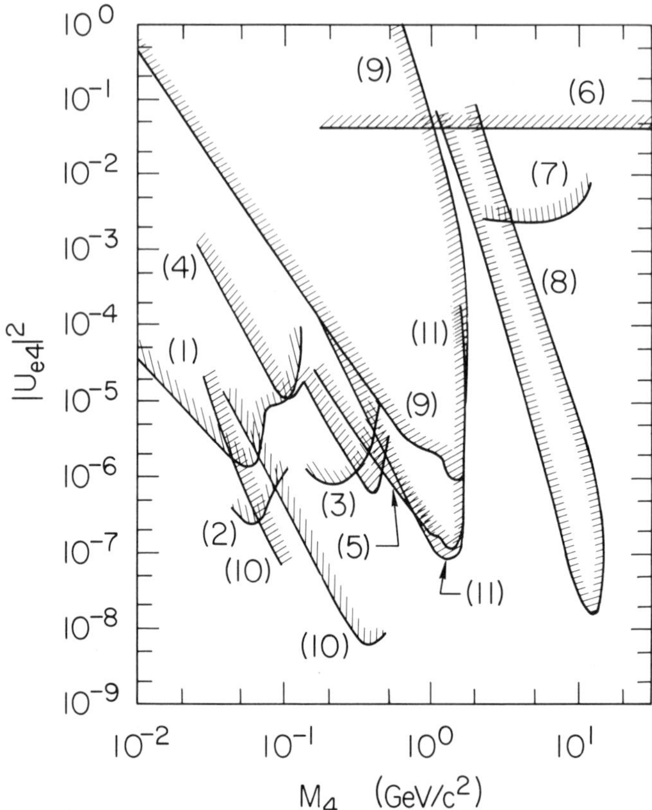

FIGURE 2. Limits on $|U_{e4}|^2$ as a function of the mass M_4 of a sequential, fourth generation neutrino as obtained from (1) TRIUMF $\pi \rightarrow e\nu$,[46] (2) SIN $\pi \rightarrow e\nu$,[47] (3) KEK $K \rightarrow e\nu$,[48] (4) CHARM experiment at CERN with a wideband beam,[51] (5) CHARM experiment at CERN using a beam dump,[53] (6) Universality,[63] (7) Monojet searches at PEP,[62] (8) Mark II secondary vertex search at PEP,[64] (9) Beam dump experiment at Fermilab,[54] (10) Wideband beam experiment at the PS at CERN,[52] and (11) BEBC experiment at CERN using a beam dump.

the beam and the particular sensitivities of the detector, these two mixing matrix elements could be the same or different. Limits obtained from π and K beams using this technique[51,52] are given by curves (4) and (10) in FIGURE 2, where it is seen that for neutral lepton masses below the K mass, they now provide the best upper limits on $|U_{e4}|^2$, going down to below 10^{-8}.

Once we are utilizing hadron decays in flight, we need not restrict our attention to π and K decays, as we are limited by their masses. Charmed meson decays extend the range of investigation considerably. To enhance their proportion of the neutrino flux, one employs a beam dump to absorb longer-lived hadrons before they can decay weakly. This leaves short-lived mesons (and baryons) that have "prompt" decays that are typically a small fraction of an absorption length. With the increased mass range opened up by the charmed meson masses, both purely leptonic and semileptonic decays are potential sources of new neutral leptons in an interesting mass range. Some of the recent limits[53-55] obtained in this manner are given by curves (5), (9), and (11) in FIGURES 2 and 3 for $|U_{e4}|^2$ and $|U_{\mu4}|^2$, respectively. It is also possible to get limits on $|U_{e4}U_{\mu4}|$ using hadron decays in flight (either with beams or in beam dumps) by, for example, producing the heavy neutrino in association with a muon and detecting its charged current decay involving an electron.[51,52]

There is no reason to stop at charm with this technique. Hadrons containing bottom and even top quarks can be employed; one is only limited by their production cross sections in hadron-hadron collisions. Indeed, it has been pointed out[56] that beam dump experiments utilizing B meson decays could extend the limits to masses of roughly 2.5 GeV/c^2 at the 10^{-6} level. One should also note that all the limits we have discussed up to now involve charged current weak interactions to produce the heavy neutral lepton

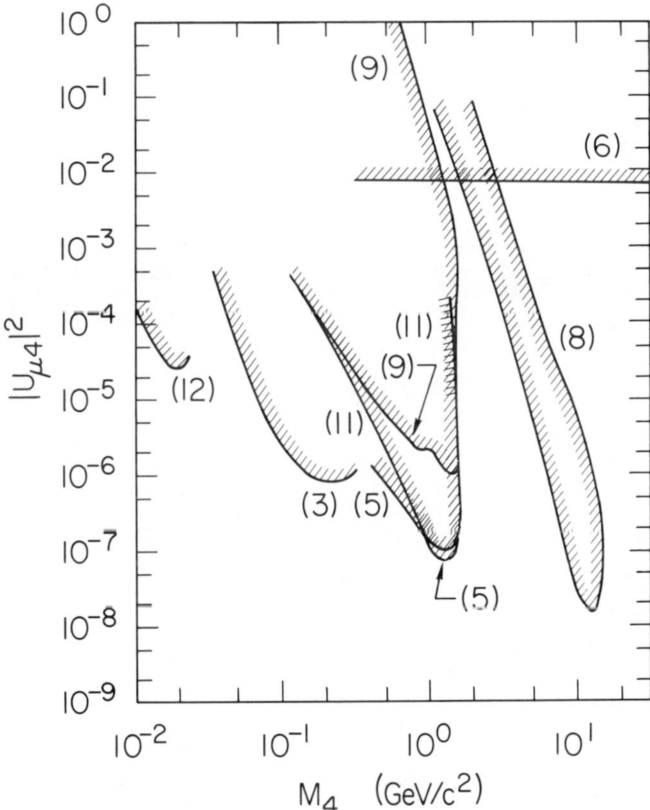

FIGURE 3. Limits on $|U_{\mu4}|^2$ as a function of the mass M_4 of a sequential, fourth generation neutrino as obtained from (3) KEK $K \rightarrow \mu\nu$,[48] (5) CHARM experiment at CERN using a beam dump,[53] (6) Universality,[63] (8) Mark II secondary vertex search at PEP,[64] (9) Beam dump experiment at Fermilab,[54] (11) BEBC experiment at CERN using a beam dump,[55] and (12) SIN $\pi \rightarrow \mu\nu$.[49]

in association with ordinary charged leptons. Hence, the relevant mixing matrix elements and amplitudes at the production vertex are independent of whether a doublet, a singlet, or a mirror neutral lepton is under scrutiny.

To extend the search to the domain of yet higher masses, we turn to electron-positron colliders, where we are limited kinematically only by the center-of-mass energy for $e^+e^- \rightarrow \bar{\nu}_4\nu_e$ (and half of it for pair production of heavy neutrinos). If the

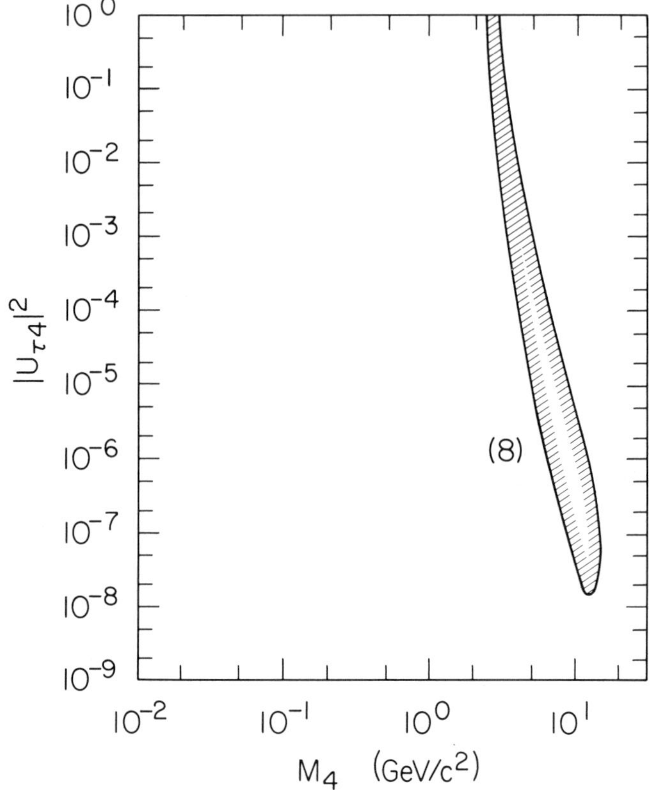

FIGURE 4. Limits on $|U_{\tau 4}|^2$ as a function of the mass M_4 of a sequential, fourth generation neutrino as obtained from (8) Mark II secondary vertex search at PEP.[64]

neutrino is in a left-handed doublet, as for a fourth generation neutrino, then its production proceeds in lowest order through W exchange with a cross section[57] for $s \ll M_W^2$:

$$\sigma(e^+e^- \rightarrow \bar{\nu}_4\nu_e) = |U_{e4}|^2 \frac{G_F^2 s}{6\pi} (1 - m^2/s)^2 (1 + m^2/2s), \qquad (15)$$

where m is the neutrino mass and s is the square of the center-of-mass energy.

Such a process will be observed as an electron-positron collision that results in an event with missing energy and momentum (due to the ν_e) on one side and, if m is not too large, a jet of decay products of ν_4 on the other side, that is, a monojet event. Such events have been searched for recently at PEP and PETRA in another context.[58–61] By combining the PEP data, one obtains[62] the upper limit shown as curve (7) in FIGURE 2. This is already a considerable improvement over the limit following from universality[63] shown as curve (6), and can be improved still further, as we will discuss below.

In contrast with equation 15, the cross section for production of a pair of "sequential" Dirac neutrinos through a "virtual" Z,

$$\sigma(e^+e^- \to \bar{\nu}_4\nu_4) = \frac{G_F^2 s}{24\pi} \frac{(1 - 4m^2/s)^{1/2}\,(1 - m^2/s)}{1 - s/M_Z^2}\,(1 - 4\sin^2\theta_W + 8\sin^4\theta_W), \quad (16)$$

does not contain a factor of $|U_{e4}|^2$. With the presently accumulated luminosity at PEP, dozens of such heavy neutrino pairs should have been produced in each experimental region as long as their mass was at least slightly below the beam energy. Because the decay width is still proportional to the square of a mixing matrix element, a search for secondary decay vertices sweeps out a region in the $|U|^2$ versus m plane. The results of such a search between 0.2 and 10 cm from the interaction point are shown[64] as the diagonal region going up to masses of almost 14 GeV/c^2, which is bounded by curve (8) in FIGURES 2 and 3. The analogous result for a "sequential" neutrino mixed with the tau neutrino is shown in FIGURE 4.

What happens to all these bounds if we had been considering a singlet-neutrino rather than a "sequential" neutrino in a doublet? All the bounds that come from hadron leptonic or semileptonic decay remain the same (aside from slight shifts in limits for those experiments that rely on detecting particular decays whose predicted branching ratios change somewhat due to additional neutral current interactions).[65] The cross section for $e^+e^- \to Z_{\mathrm{virtual}} \to \overline{N}N$ picks up a factor of $|U|^4$ over equation 16, making it negligibly small given present experimental sensitivities and interesting values of $|U|^2$, and we, unfortunately, lose the restrictions obtained from the search for secondary vertices in electron-positron annihilation. In the case of $e^+e^- \to N\bar{\nu}_e$, we now have both W exchange and direct-channel Z contributions, and the expression for the cross section becomes

$$\sigma(e^+e^- \to N\bar{\nu}_e) = |U_{eN}|^2 \frac{G_F^2 s}{24\pi}\left(1 - \frac{m^2}{s}\right)^2\left(1 + \frac{m^2}{2s}\right)(1 + 4\sin^2\theta_W + 8\sin^4\theta_W). \quad (17)$$

If $\sin^2\theta_W = 0.22$, then the cross section in equation 17 is 0.57 times that in equation 16, and the monojet searches have somewhat reduced sensitivity.

In contrast to the reduced sensitivity to mixing of N with the electron neutrino in monojet searches, we gain, from the neutrino flavor-changing coupling of the Z, a direct-channel Z contribution to $e^+e^- \to N\bar{\nu}_\mu$, which is proportional to $|U_{\mu N}|^2$:

$$\sigma(e^+e^- \to N\bar{\nu}_\mu) = |U_{\mu N}|^2 \frac{G_F^2 s}{24\pi}\left(1 - \frac{m^2}{s}\right)^2\left(1 + \frac{m^2}{2s}\right)(1 - 4\sin^2\theta_W + 8\sin^2\theta_W). \quad (18)$$

An identical expression with τ in place of μ follows from potential mixing with the tau neutrino. The limit that is obtained for $|U_{\mu N}|^2$ from equation 18 plus existing experiments is not better than that from universality.

However, the presence of neutrino flavor-changing couplings of the Z does lead to a new limit on $|U_{\mu N}|^2$ from the lack of observation of the process ν_μ + nucleus $\rightarrow N + \ldots$, followed by decay of the N. An additional region has been excluded by the CHARM collaboration[53] in this way.

The experimental situation for mirror neutrinos is somewhat of a cross between those for sequential and right-handed singlet heavy neutral leptons. Because mirror neutrinos lie in right-handed doublets, they have full-strength couplings to the Z (like sequential neutrinos) and the limits from electron-positron annihilation experiments pertain to them [curve (8) in FIGURES 2, 3, and 4]. On the other hand, being left-handed singlets, they also have flavor-changing couplings to the Z, and limits following from production through such couplings are also applicable.

In the near future, it should be possible to considerably extend many of the limits by further analysis of existing data. Combining all the experiments, it should be possible, for example, to exclude values of $|U_{e4}|^2$ and $|U_{\mu 4}|^2$ above 10^{-6} up to masses of about 20 GeV/c^2. Upper limits on $|U_{e4}|^2$ of 10^{-2} to 10^{-3} should be obtainable from PETRA data for masses up to 30 GeV/c^2 from monojet searches.

A dramatic increase in sensitivity will occur when the SLC and LEP come into full operation. With 6% of Z decays going into neutrino-antineutrino for each light neutrino in a weak doublet, additional sequential or mirror neutrinos will be very amenable to detection. The presence of such particles with low masses can be ascertained by "neutrino counting", that is, by tagging Z decays into unseen neutrals and seeing if the resulting decay width is accounted for by the known neutrinos. Those with higher masses will generally decay through mixing, but here the technique of using vertex detectors will enable us to exclude masses up to about 40 GeV/c^2 and mixing matrix elements squared down to around 10^{-10}. Monojet searches looking for $Z \rightarrow N\bar{\nu}$ could be sensitive to weak singlet heavy neutral leptons with masses up to a large fraction of the Z mass and with squares of mixing matrix elements to the known neutrinos down to about 10^{-5}.

EXTRA Z'S

Over the years, there have been numerous suggestions in the direction of extending the electroweak gauge group beyond $SU(2) \times U(1)$. In particular, many grand unified theories have extra Z's, as do left-right symmetrical theories such as $SU(2)_L \times SU(2)_R \times U(1)$.

The investigation of this area has received a big boost from the advent of superstrings, where the combined low energy gauge group is generally larger than $SU(3)_c \times SU(2)_L \times U(1)_Y$.[66-69] An early favorite in this regard is to have $SU(3)_c \times SU(2)_L \times U(1)_Y \times U(1)$ at "low" energies, but this is by no means the only possibility.[67-69] The patterns of bosons and fermions following under various assumptions has been analyzed by many people.[70] A number of these possibilities can already be ruled out, depending in part on what other theoretical assumptions are made (such as the number of scales of symmetry breaking). Still, a range of possibilities with at least an extra $U(1)$ and an associated neutral gauge boson remains, and it is interesting to examine how they are constrained by existing experiments.

The constraints on an extra Z that we have at our disposal come from neutral current experiments, the measured W and Z masses, and the lack of observation of a second A in collider experiments up to now. One needs to apply these constraints to each proposed low energy model in turn to see what limits can be placed on the mass and mixing of the extra Z' with the Z of the standard model. The results[71] are somewhat surprising in that a Z' with a mass as low as about 130 GeV is allowed, just as long as the mixing with the standard model Z is fairly small (so as not to displace it too far from its nominal mass). This comes about in part because the couplings of the Z' given by such models tend to be small to ordinary fermions. This also makes them harder to produce in ordinary lepton or hadron collisions than the usual Z, and it makes their decay branching ratios to ordinary leptons smaller if the "exotic" channels for decay are open. In spite of this, up to masses of several hundred GeV, they have production cross sections in $\bar{p}p$ collisions that are adequate enough that we will be watching the Tevatron collider to see if it will turn up direct evidence of their existence in the next few years.

CONCLUSIONS

As we noted at the beginning, the agreement of the standard model and experiment is extremely impressive. The pickings are pretty slim if one is looking for areas of potential discrepancy.

Nevertheless, it is worth keeping an eye on the situation with respect to CP violation and, to a lesser extent, the decays of the tau. While there is not yet any smoking gun, another round of experiments is in progress. We are guaranteed continued improvement in the accuracy of our knowledge of various quantities, and the possibility for surprises along the way certainly exists.

We have also seen that there are very sensitive limits on certain possibilities that go beyond the standard model. In particular, the limits on heavy neutral leptons are very impressive and promise to improve by several orders of magnitude in sensitivity to small mixing angles, as well as to extend considerably further in mass range during the next few years.

However, the limits on other possibilities are rather poor. In particular, I have in mind the constraints that present experiments put on extra Z's. Thus, new physics may be just around the corner.

NOTES AND REFERENCES

1. See, for example, the excellent recent review by: LANGACKER, P. 1986. *In* Proceedings of the 1985 Int. Symp. on Lepton and Photon Interactions at High Energies. Kyoto University, Kyoto, Japan. M. Konuma & K. Takahashi, Eds.: 186.
2. ARNISON, G. *et al.* 1983. Phys. Lett. **122B**: 103; **129B**: 273; BANNER, M. *et al.* 1983. Phys. Lett. **122B**: 476.
3. ARNISON, G. *et al.* 1983. Phys. Lett. **126B**: 398; BAGNAIA, P. *et al.* 1983. Phys. Lett. **129B**: 130.
4. See the compilation in table 2 of reference 1.
5. The data are reviewed by: MARSHALL, R. 1985. Invited talk at the Int. Symp. on Multiparticle Dynamics, Kiryat Anavim, Israel, June 9–14, 1985, and Rutherford Laboratory report no. RAL-85-078. Unpublished.
6. EICHTEN, E. J., K. D. LANE & M. E. PESKIN. 1983. Phys. Rev. Lett. **50**: 811.
7. The data and references to the recent PETRA measurements are given in reference 5.
8. KOBAYASHI, M. & T. MASKAWA. 1973. Prog. Theor. Phys. **49**: 652.

9. In particular, such a connection was conjectured by: FRITZSCH, H. 1977. Phys. Lett. **70B:** 436; this topic has since developed a substantial literature of its own.

10. The following is a brief version of the section of the 1986 Review of Particle Properties prepared by: GILMAN, F. J. & K. KLEINKNECHT. 1986. Phys. Lett. **170B:** 74.

11. MARCIANO, W. J. & A. SIRLIN. 1986. Phys. Rev. Lett. **56:** 22.

12. LEUTWYLER, H. & M. ROOS. 1984. Z. Phys. **C25:** 91.

13. ABRAMOWICZ, H. *et al.* 1982. Z. Phys. **C15:** 19.

14. THORNDIKE, E. 1986. *In* Proceedings of the 1985 Int. Symp. on Lepton and Photon Interactions at High Energies. Kyoto University, Kyoto, Japan. M. Konuma & K. Takahashi, Eds.: 406.

15. MAIANI, L. 1977. *In* Proceedings of the 1977 Int. Symposium on Lepton and Photon Interactions at High Energies. DESY, Hamburg, p. 867.

16. DONOGHUE, J. F. & B. R. HOLSTEIN. 1984. Phys. Rev. **D29:** 2088; FRERE, J. M. *et al.* 1985. Phys. Lett. **151B:** 161.

17. GILMAN, F. J. & M. B . WISE. 1980. Phys. Lett. **93B:** 129; 1983. Phys. Rev. **D27:** 1128.

18. THORNDIKE, E. 1986. See reference 14; BERKELMAN. K. 1987. These proceedings.

19. DONOGHUE, J. F., E. GOLOWICH & B. R. HOLSTEIN. 1982. Phys. Lett. **119B:** 412; APPELQUIST, T., J. D. BJORKEN & M. S. CHANOWITZ. 1973. Phys. Rev. **D7:** 2225.

20. BIJNENS, J., H. SONODA & M. B. WISE. 1984. Phys. Rev. Lett. **53:** 2367.

21. See table 6 in reference 1 for a sample of these calculations.

22. GILMAN, F. J. & J. S. HAGELIN. 1983. Phys. Lett. **133B:** 443.

23. GILMAN, F. J. & M. B. WISE. 1979. Phys. Lett. **83B:** 83; Phys. Rev. **D20:** 2392.

24. BERNSTEIN, R. H. *et al.* 1985. Phys. Rev. Lett. **54:** 1631.

25. BLACK, J. K. *et al.* 1985. Phys. Rev. Lett. **54:** 1628.

26. PHAM, T. N. 1984. Phys. Lett. **145B:** 113; PHAM, T. N. & Y. DUPONT. 1984. Phys. Rev. **D29:** 1368; DONOGHUE, J. F. 1984. Phys. Rev. **D30:** 1499.

27. For other recent reviews, see: LANGACKER, P. 1986. Reference 1; WOLFENSTEIN, L. 1985. Comments Nucl. Part. Phys. **14:** 135.

28. See, for example: BERNARD, C. *et al.* Phys. Rev. Lett. **55:** 2770; SHARPE, S. R. 1985. Phys. Lett. **165B:** 371.

29. Recent reviews of the properties of the tau and references to specific experiments are found in: THORNDIKE, E. 1986. Reference 14; GAN, K. K. 1985. Invited talk at the 1985 Annual Mtg. of the Div. of Particles and Fields of the APS, Eugene, Oregon, August 12–15, 1985, and Purdue University report no. PU-85-539. Unpublished.

30. GILMAN, F. J. & S. H. RHIE. 1985. Phys. Rev. **D31:** 1066, and references to other previous work therein.

31. Since the conference, this possibility has received some support from an analysis of Mark II data using a different approach of "tagging" the tau decays; however, within errors, there is no disagreement with the earlier results of other experiments. See: BURCHAT, P. 1986. Stanford University Ph.D. thesis and SLAC report no. 292.

32. GLASHOW, S. L., J. ILIOPOULOS & L. MAIANI. 1970. Phys. Rev. **D2:** 1285; GLASHOW, S. L. & S. WEINBERG. 1977. Phys. Rev. **D15:** 1958; PASCHOS, E. A. 1977. Phys. Rev. **D15:** 1966.

33. More precisely, it is the left-handed antilepton that is in the same 16-dimensional representation as the left-handed neutrino. For the question of neutrino masses in $O(10)$, see: WITTEN, E. 1980. Phys. Lett. **91B:** 81.

34. MOHAPATRA, R. N. & G. SENJANOVIC. 1980. Phys. Rev. Lett. **44:** 912; 1981. Phys. Rev. **D23:** 165; RIAZUDDIN, R., E. MARSHAK & R. N. MOHAPATRA. 1981. Phys. Rev. **D24:** 1310.

35. GELL-MANN, M., P. RAMOND & R. SLANSKY. 1979. *In* Supergravity. D. Z. Freedman & P. Van Nieuwenhuizen, Eds.: 317. North-Holland. Amsterdam; YANAGIDA, T. 1979. Proceedings of the Workshop on "The Unified Theory and the Baryon Number in the Universe". KEK, Ibaraki. O. Sawada & A. Sugamoto, Eds.: 95.

36. WYLER, D. & L. WOLFENSTEIN. 1983. Nucl. Phys. **B218:** 205.

37. Recent phenomenological work in this case has been done by: LEUNG, C. N. & J. L. ROSNER. 1983. Phys. Rev. **D28:** 2205; FISHBANE, P. M. *et al.* 1985. Phys. Rev. **D32:** 1186.

38. PATI, J. C. & A. SALAM. 1975. Phys. Lett. **58B**: 333; MAALAMPI, J. & K. ENQVIST. 1980. Phys. Lett. **97B**: 217; ENQVIST, K., K. MURSULA & M. ROOS. 1983. Nucl. Phys. **B226**: 121; ROOS, M. 1984. Phys. Lett. **135B**: 487; WILCZEK, F. & A. ZEE. 1982. Phys. Rev. **D25**: 553; SENJANOVIC, G., F. WILCZEK & A. ZEE. 1984. Phys. Lett. **141B**: 389.

39. BAGGER, J. & S. DIMOPOULOS. 1984. Nucl. Phys. **B244**: 247; BAGGER, J. *et al.* 1985. Phys. Rev. Lett. **54**: 2199; Nucl. Phys. **B258**: 565.

40. BJORKEN, J. D. & C. H. LLEWELLYN SMITH. 1973. Phys. Rev. **D7**: 887.

41. GRONAU, M., C. N. LEUNG & J. L. ROSNER. 1984. Phys. Rev. **D29**: 2539. These authors consider especially heavy neutral leptons corresponding to "right-handed neutrinos". Fourth generation leptons are discussed by: BARGER, V., W. Y. KEUNG & R. J. N. PHILLIPS. 1984. Phys. Lett. **141B**: 126; BARGER, V., H. BAER, K. HAGIWARA & R. J. N. PHILLIPS. 1984. Phys. Rev. **D30**: 947.

42. GRONAU, M. 1985. *In* Neutrino Mass and Low Energy Weak Interactions, Telemark, 1984. V. Barger & D. Cline, Eds.: 276. World Scientific. Singapore; WOLFENSTEIN, L. 1984. *In* Flavor Mixing in Weak Interactions. L. L. Chau, Ed.: 187. Plenum. New York; KAYSER, B. 1984. *In* Massive Neutrinos in Astrophysics and in Particle Physics. J. Tran Thanh Van, Ed.: 11. Editions Frontiers. Gif-sur-Yvette, France.

43. PERL, M. L. 1984. *In* Proceedings of the Santa Fe Meeting. T. Goldman & M. M. Nieto, Eds.: 159. World Scientific. Singapore.

44. See, for example: JAROS, J. A. 1985. *In* Proceedings of the Twelfth SLAC Summer Institute on Particle Physics. SLAC, Stanford, California. P. McDonough, Ed.: 427.

45. SHROCK, R. E. Phys. Rev. **D24**: 1232.

46. BRYMAN, D. A. *et al.* Phys. Rev. Lett. **50**: 1546.

47. Data from SIN reported in: SARKAR-COOPER, A. M. 1984. *In* Proceedings of the XXII International Conference on High Energy Physics. Vol. I. A. Meyer & E. Wieczorek, Eds.: 263. Akademie der Wissenschaften. Zeuthen, East Germany.

48. YAMAZAKI, T. *et al.* 1984. Proceedings of the XIth International Conference on Neutrino Physics and Astrophysics. K. Kleinknecht & E. A. Paschos, Eds.: 183. World Scientific. Singapore.

49. ABELA, R. *et al.* 1981. Phys. Lett. **105B**: 263.

50. GRONAU, M. 1983. *In* Electroweak Interactions at High Energies. R. Kogerler & D. Schildknecht, Eds.: 143. World Scientific. Singapore; Phys. Rev. **D28**: 2762.

51. BERGSMA, F. *et al.* 1983. Phys. Lett. **128B**: 361.

52. BERNARDI, G. *et al.* 1986. Phys. Lett. **166B**: 479.

53. DORENBOSCH, J. *et al.* 1986. Phys. Lett **166B**: 473.

54. BALL, R. C. *et al.* 1985. University of Michigan preprint no. UM HE 85-09. Unpublished.

55. COOPER-SARKAR, A. M. *et al.* 1985. **160B**: 207.

56. ROSNER, J. L. 1985. Phys. Rev. **D31**: 2372.

57. BJORKEN, J. D. & C. H. LLEWELLYN SMITH. 1973. See reference 40; HUMPERT, B. & P. SCHARBACH. 1975. Phys. Lett. **68B**: 71.

58. FELDMAN, G. J. *et al.* 1985. Phys. Rev. Lett. **54**: 2289.

59. AKERLOF, C. *et al.* 1985. Phys. Lett. **156B**: 271.

60. ASH, W. W. *et al.* 1985. Phys. Rev. Lett. **54**: 2477.

61. BARTEL, W. *et al.* 1983. Phys. Lett. **133B**: 353; 1985. Phys. Lett. **155B**: 288.

62. GILMAN, F. J. & S. H. RHIE. 1985. Phys. Rev. **D32**: 324.

63. We use the 2σ limits of Gronau, Leung, and Rosner; see reference 41.

64. FELDMAN, G. J. 1985. SLAC preprints nos. SLAC-PUB-3684 and SLAC-PUB-3822. Unpublished; WENDT, C. *et al.* To be published.

65. Care must also be taken on whether Dirac or Majorana neutrinos are under consideration because, for example, limits on mixing matrix elements derived from those experiments that detect heavy lepton decays in a given fiducial volume depend on their decay rates, which generally differ in the two cases. See: SHROCK, R. E. 1981. Phys. Rev. **D24**: 1275; 1982. Phys. Lett. **112B**: 382.

66. CANDELAS, P., G. HOROWITZ, A. STROMINGER & E. WITTEN. 1985. Nucl. Phys. **B258**: 46.

67. DINE, M., V. KAPLUNOVSKI, M. MANGANO, C. NAPPI & N. SEIBERG. 1985. Nucl. Phys. **B259**: 549.

68. SCHWARZ, J. H. 1987. These proceedings and references therein.

69. NANOPOULOS, D. V. 1987. These proceedings and references therein.
70. See, for example: DINE, M. *et al.* 1985. See reference 67; ROSNER, J. L. 1985. University of Chicago report no. EFI-85-34-CHICAGO. Unpublished; BINETRUY P., S. DAWSON, I. HINCHLIFFE & M. SHER. 1985. LBL report no. LBL-20317. Unpublished; COHEN E., J. ELLIS, K. ENQVIST & D. V. NANOPOULOS. 1985. Phys. Lett. **165B:** 76; ELLIS, J., K. ENQVIST, D. V. NANOPOULOS & F. ZWIRNER. 1985. CERN report no. CERN-TH.4323/85. Unpublished; 1986. CERN-TH.4350/86. Unpublished.
71. BARR, S. M. 1985. Phys. Rev. Lett. **55:** 2778; ELLIS, J. *et al.* See reference 70; DURKIN, L. S. & P. LANGACKER. 1986. Phys. Lett. **166B:** 436; BARGER, V., N. G. DESHPANDE & K. WHISNANT. 1985. Phys. Rev. Lett. **56:** 30.

e^+e^-: Survey of Experimental Results

KARL BERKELMAN

Laboratory of Nuclear Studies
Cornell University
Ithaca, New York 14853

ELECTROWEAK: MAINLY HEAVY QUARK DECAYS

Quick Survey of Some Other Topics

As the PETRA and PEP rings continue to run at high energies, the data on weak-electromagnetic inferences improve in accuracy. In particular, the forward-backward asymmetry results[1] for the lepton pair reactions, $e^+e^- \rightarrow e^+e^-$, $\mu^+\mu^-$, and $\tau^+\tau^-$, now provide measurements of the Weinberg angle that are comparable in accuracy to values obtained from neutrino experiments. The agreement with the accepted $\sin^2\theta_W = 0.22$ is excellent.

Our experimental picture of tau decays is becoming rather detailed. There are many measurements of the τ lifetime, all from e^+e^- experiments. The world average is now 3.0 ± 0.2 ps,[2] to be compared with the predicted 2.8 ± 0.2 ps. A new more accurate measurement of the leptonic branching ratio, Br $(\tau \rightarrow \ell\nu\bar{\nu}) = 0.178 \pm 0.005$ (average of e and μ), has been reported by the MAC collaboration,[3] which is in excellent agreement with theory. CLEO has also made a new measurement[4] of the Michel parameter describing the shape of the electron and muon spectra in the leptonic decays of the tau: $\rho = 0.70 \pm 0.10 \pm 0.03$, to be compared with the values of 0.75, 0, and 0.375 expected for $V - A$, $V + A$, and V or A, respectively. Branching fractions have been accurately measured for presumably all of the significant decay modes of the tau. Where theoretical predictions exist, the agreement is excellent. The only trouble is that the measured branching fractions sum to only 93%.[2] No one has a plausible explanation for the missing 7%. Except for this, one could say that the τ lepton is perfectly understood.

Motivations for the Study of Heavy Quark Decays

There are two reasons for studying heavy quark decays. (1) We want to check the standard electroweak model of six quarks and six leptons. So far, no discrepancies have been found. (2) We want to measure the unknown parameters of the model: the three angles and one phase of the Kobayashi-Maskawa (K-M) matrix generalization[5] of the Cabibbo theory. The decays of states containing only u, d, s, and c are sensitive mainly to the well-known θ_1. The decays involving the transitions $b \rightarrow c$ and $b \rightarrow u$ are sensitive to the angles θ_2 and θ_3. In order to extract information on b quark decays, we must understand the mechanisms of decay of heavy quarks bound in mesons. To do this, we study charmed meson decays, where the relevant K-M angle is known.

21

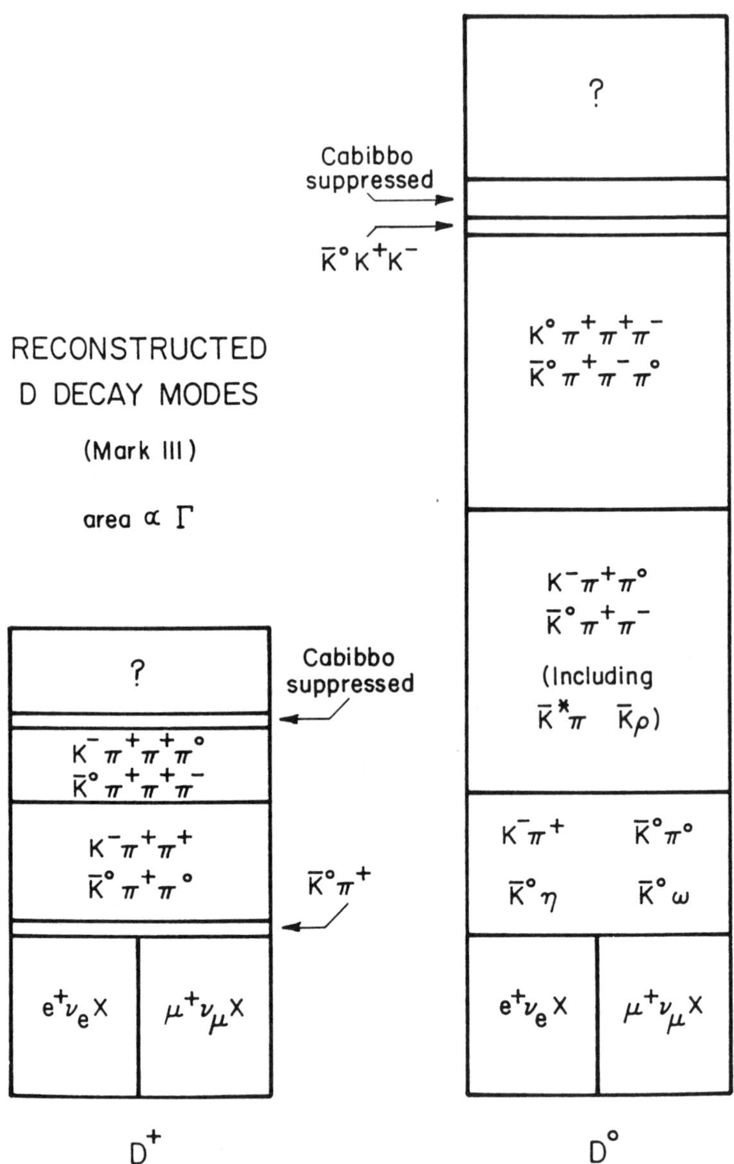

FIGURE 1. D^+ and D^0 decay modes reconstructed by Mark III.[6] Relative areas are proportional to inverse lifetimes or partial widths.

D *Meson Decays*

The new data on D decays come mainly from Mark III at SPEAR,[6] running on the $\psi''(3770)$ D-factory. Besides having smaller statistical errors than earlier data, the new measurements use a much more reliable normalization. Instead of dividing observed rates by the total D event rate inferred from the shape of the ψ'' peak, Mark III uses the events in which one D decay is reconstructed as a pure unbiased event sample for studying the other D. This only has to be done for a few higher-rate modes to serve as a normalization for all modes. The new branching ratios are summarized in FIGURE 1. They are about a factor of two higher than previous numbers; that is, the ψ'' cross section implied by the Mark III double-tag data is about half of that assumed in normalizing the old data. This will have the effect of halving the D production cross sections reported in other experiments.

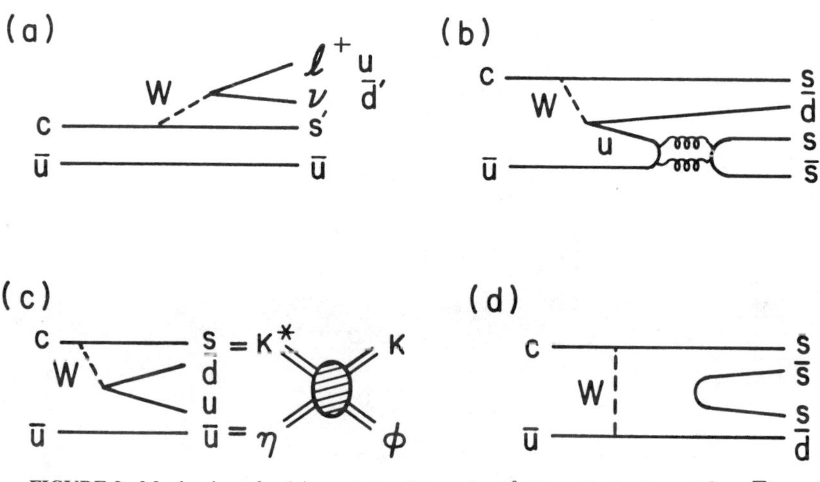

FIGURE 2. Mechanisms for (a) spectator decay of a D^0, (b, c, d) the decay $D^0 \rightarrow \bar{K}^0\phi$.

Thanks to the factor of two from the renormalization, as well as the increase in the number of modes measured, Mark III has accounted for 87% of all D decays. Summed over all observed modes, the ratio of Cabibbo suppressed to favored hadronic branching fractions is 0.08. This rich mine of information will keep theorists busy for years, with the result (I hope) that we will really understand the mechanisms of decay of mesons containing heavy quarks.

A *Nonspectator Decay*

One particular mode deserves more comment. The decay $D^0 \rightarrow \bar{K}^0\phi$ cannot occur through the simple spectator graph (FIGURE 2a) because the final state does not contain the spectator $\bar{u}$ quark. You either have to (FIGURE 2b) annihilate a u from the W with the spectator $\bar{u}$ to produce an $s\bar{s}$ pair, which is color suppressed and

Zweig-violating, or (FIGURE 2c) produce $\bar{K}^0\phi$ by rescattering from the $K^*\eta$ state, or (FIGURE 2d) exchange a W with the spectator and then pop an $s\bar{s}$ pair. We know that the simple spectator graph cannot explain all D decays (because it implies equal lifetimes for the charged and neutral D), but it would be nice to see a decay that cannot be a spectator. The ARGUS group[7] reports a signal for $e^+e^- \rightarrow D^0X$, $D^0 \rightarrow \bar{K}^0\phi$, with

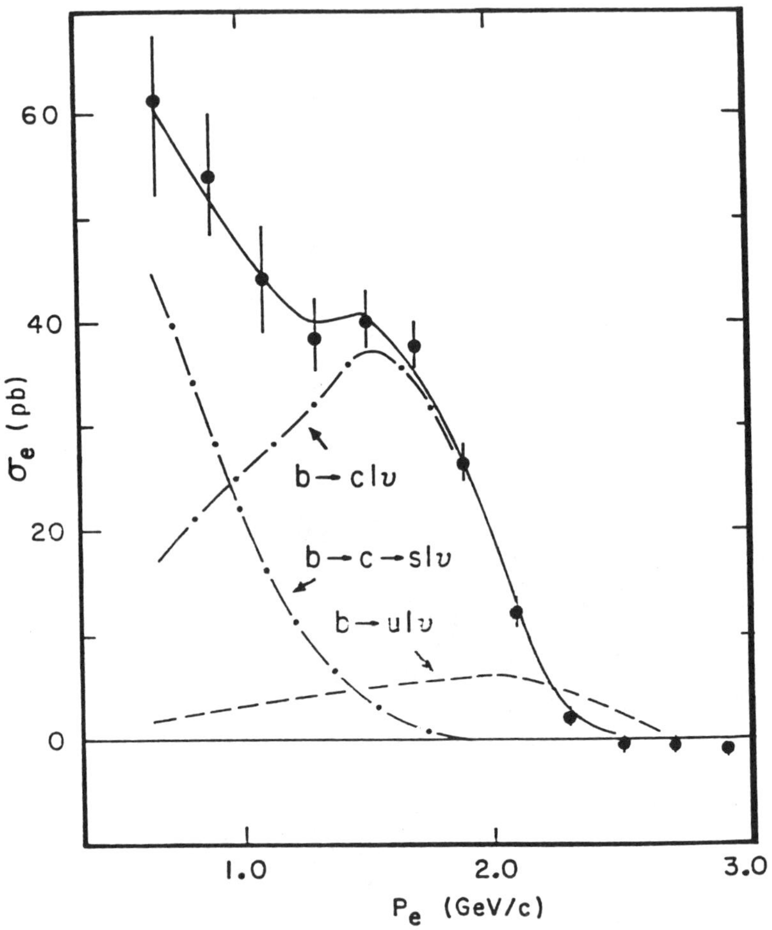

FIGURE 3. Electron momentum spectrum in semileptonic B decay (CLEO[12]).

the $\bar{K}^0$ detected through its decay to two charged pions and the ϕ detected in K^+K^-. They calculate a branching ratio for $D^0 \rightarrow \bar{K}\phi$ of $1.4 \pm 0.5\%$. CLEO[8] and Mark III[9] confirm it with branching ratios of $0.9 \pm 0.3 \pm 0.3\%$ and $1.1^{+0.7}_{-0.5}{}^{+0.4}_{-0.2}\%$, respectively. Although the results appear consistent, there is controversy over how to treat the background from $D^0 \to \bar{K}^0 K^+ K^-$, where the $K^+ K^-$ do not form a ϕ. I think the $\bar{K}^0 \phi$ signal is real, but it is certainly much smaller than the 14% rate for the spectator-allowed $D^0 \to K^- \rho^+$ mode.

B *Meson Decay: The Noncharm/Charmed Ratio*

Two measurements are made in B decays to extract information about the K-M angles θ_2 and θ_3: the ratio of decay rates to noncharm and charmed decays, $[b \to u]/[b \to c]$, is measured at CESR and DORIS; and the b lifetime is measured at PEP and PETRA. Actually, all we have is an upper limit on $[b \to u]/[b \to c]$. It is obtained by looking at the momentum spectrum (FIGURE 3) of electrons or muons in the semileptonic decays of the B and noting that if the B decays to a noncharm hadron state plus $\ell\nu$, then the lepton momentum can be higher than if the B decays to a charmed hadron state (typically more massive) plus $\ell\nu$. Until last summer, the (90% confidence) upper limit was 4% from CLEO[10] and 5.5% from CUSB,[11] which perhaps could be combined to give a 3% limit. The good news is that the data are getting better: there are new results from CLEO[12] and ARGUS,[13] which if analyzed as before, would give 1% and 4% limits. The bad news is that it is now clear that the extraction of a limit from the measured spectrum by fitting to a theoretical model is flawed by the uncertainties in the model. This was made obvious when the more accurate CLEO data did not fit well the popular Altarelli model[14] with the usually assumed value of the parameter describing the Fermi momentum of the quarks in the meson. Letting this parameter float in the fit (there is no reason to take it as known) weakens the $[b \to u]/[b \to c]$ upper limit to 4%. A similar limit is obtained by fitting only the spectrum above the maximum momentum available in charmed final states. Significant progress in pushing down this limit (or actually measuring $[b \to u]/[b \to c]$) by this method depends on better understanding of the theory. If this ratio is actually as low as 1%, it will imply that θ_3 is too small to allow the CP violation in K^0 decays to be explained by the phase in the K-M matrix.

There is another way to get evidence for the $b \to u$ coupling. CLEO[16] has searched for exclusive noncharm decays. The upper limits (90% confidence) on several low-multiplicity modes are given below; combinatoric background makes the higher multiplicities less useful:

$$
\begin{array}{lll}
B^- \to \rho^0 \pi^- & <2 \times 10^{-4} & \\
\quad\quad \rho^0 A_1^- & 8 & \\
\quad\quad \rho^0 A_2^- & 4 & \\
\quad\quad \rho^0 X & 2 & (1.0 < M_x < 1.6 \text{ GeV}, \Gamma_x = 100 \text{ MeV}), \\
\bar{B}^0 \to \pi^+ \pi^- & 2 & \\
\quad\quad \pi^+ A_1^- & 24 & \\
\quad\quad \pi^+ A_2^- & 20 & \\
\quad\quad \pi^+ X^- & 7 & (1.0 < M_x < 1.6 \text{ GeV}, \Gamma_x = 100 \text{ MeV}).
\end{array}
$$

Although these limits are as much as two orders of magnitude smaller than for the corresponding low multiplicity charmed modes, one does not know how the relative probability of various multiplicities depends on the masses.

B *Meson Lifetime*

Over the past few years, there has been a steady accumulation of b lifetime measurements from most of the PEP and PETRA detectors. To get an enriched sample of b jets, some use leptons with high p_T relative to the jet axis as a signature, while others use an event shape criterion to pick up the broader b jets. The lifetime is measured either from the lepton impact parameter or by using vertices. All the data are consistent. At the 1984 Leipzig conference, the measurements averaged about 1.5 ± 0.3 psec; measurements since then[17] bring the mean down to 1.06 ± 0.17 psec.

Two B meson decays have been observed in an event recorded by the CERN WA75 experiment[18] using emulsion, silicon microstrips, and a magnetic spectrometer: $B^- \rightarrow D^0 \mu \nu$ after about 0.08 psec, and $B^0 \rightarrow D^- X^+$ after about 0.5 psec. Although the first decay time is rather short compared to the mean life measured in the e^+e^- experiments, one such event is not enough to challenge those measurements.

With only an upper limit for $[b \rightarrow u]/[b \rightarrow c]$ and no measurement of the K-M phase parameter δ, one cannot derive values for the angles θ_2 and θ_3; one can derive only limits of an allowed region in θ_2 versus θ_3 that are different for each assumed value of δ. Since useful b decay lifetimes first became available three years ago,[19] the measured lifetime value has decreased and become more accurate, and the $[b \rightarrow u]/[b \rightarrow c]$ limit has perhaps decreased. These trends cause the allowed values for θ_2 to increase (0.05 ± 0.02) and for θ_3 to decrease (<0.04).

QCD: MAINLY FRAGMENTATION

Quick Survey of QCD in e$^+$e$^-$ *Physics*

The short-distance behavior of QCD, presumably predictable from low-order perturbation calculations, is checked in a number of measurements made in e^+e^- experiments: the R ratio of $e^+e^- \rightarrow$ hadrons and $e^+e^- \rightarrow \mu^+\mu^-$ cross sections, the gluon bremsstrahlung in $e^+e^- \rightarrow q\bar{q}g$, and the three-gluon and radiative two-gluon decay widths of the upsilon. These measurements have given results for α_s consistent with data from other experiments. Except in the case of the R measurement where the experimental uncertainties dominate, the comparisons between experiment and QCD are now limited by theoretical uncertainties. There is nothing new to report from recent experiments.

In the intermediate distance range from 0.1 to 1 Fermi, our best experimental information comes from the study of the spectroscopy of hadrons containing heavy quarks. The most recent results from e^+e^- include the measurement of the χ_b ($1^3P_{0,1,2}$) states of the $b\bar{b}$ system at CESR[20] and DORIS,[21] the discovery of the F^* (the ^3S of $c\bar{s}$) and D^{**} (presumably, the P state of $c\bar{u}$ and $c\bar{d}$) by ARGUS,[22,23] and the discovery of the B^* (the ^3S of $b\bar{u}$ and $b\bar{d}$) by CUSB.[24] These results have been in the published literature for some time now, so I will not dwell on them here.

The process by which a quark or gluon produced in a high energy collision evolves into a jet of hadrons involves the action of QCD at long distances in a stochastic chain of successive multiplications of the number of particles. For quark jets of energies up to 20 GeV, e^+e^- experiments are our best source of information on fragmentation topologies, inclusive particle momentum spectra and abundances, two-particle correlations, etc. The accuracy and detail of the available experimental data are steadily improving as the various experiments continue to accumulate statistics. Obviously, I do not have time to discuss the implications of all of these measurements. Let me just concentrate on some recent results on charm fragmentation and on Bose-Einstein correlations.

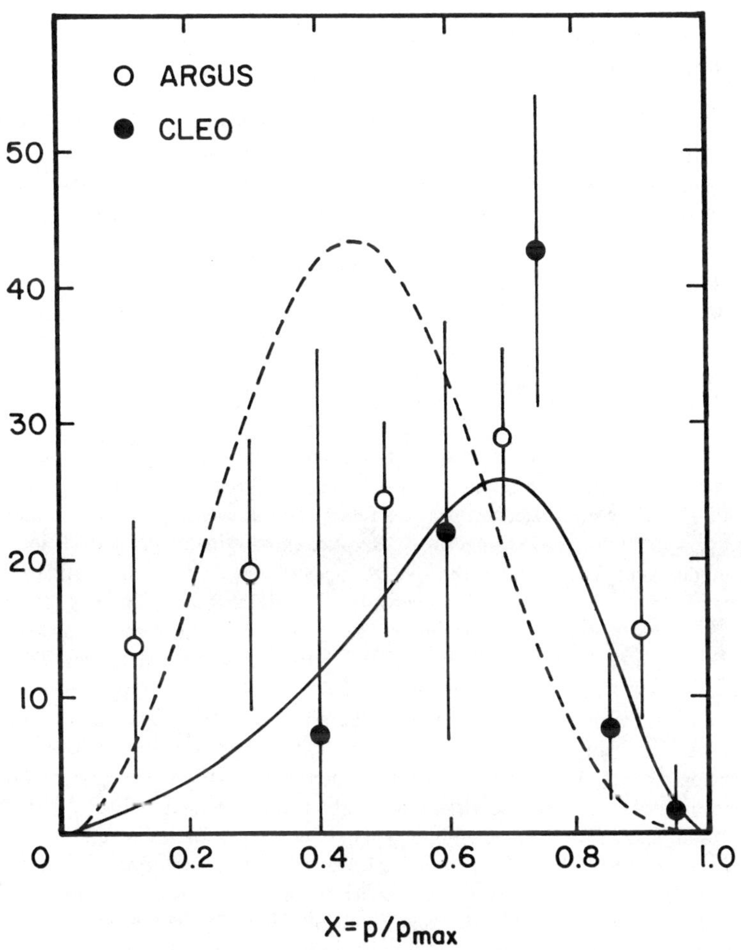

FIGURE 4. Fragmentation function for Λ_c in e^+e^- annihilation.[25,26]

Charm Fragmentation

CLEO and ARGUS have now produced the first fragmentation function for the charmed baryon Λ_c. CLEO[25] detects the Λ_c in the $\Lambda\pi\pi\pi$ mode; ARGUS[26] uses the $pK^-\pi^+$ mode. Because the branching ratios are not very well known, I plot the two fragmentation function results (FIGURE 4) with arbitrary scales chosen to make the magnitudes compatible. The spectrum peaks at an x ($= p/p_{max}$) of about 0.7, which is somewhat higher than for the D, and is inconsistent with the prediction of deGrand.[27]

The same groups also now have rather accurate fragmentation function data for the $D*$,[26,28] which is the most cleanly detected charmed particle because of the low Q in the $D*^+ \rightarrow D^0\pi^+$ decay. Several PEP and PETRA groups also have $D*$ data at higher c.m. energies, but of lower accuracy.[29] ARGUS has recently reported fragmentation functions for D^0, D^+, $D*^0$, $D**^0$, F, and $F*$.[26,30] What then can we learn from a comparison of these momentum spectra?

At the present levels of experimental accuracy, the measured charm fragmentation functions can be fit to the Peterson formula.[31] This parameterization is probably not the best, but most of the published results use it. It is based on a simple picture of heavy quark hadronization due to Bjorken[32] and Suzuki.[33] One assumes that the charmed hadron contains the leading charmed quark from the original $e^+e^- \rightarrow c\bar{c}$. An antiquark $\bar{u}$, $\bar{d}$, or $\bar{s}$ (or diquark ud in the case of a charmed baryon) popped from the QCD flux tube binds to the c quark, thus forming a decelerated hadron. Simple kinematics then leads to the Peterson formula for the charmed hadron momentum spectrum or, more properly, the spectrum in the light-cone variable $x = (E_h + p_{L,h})/(E_c + p_c)$:

$$f(x) \propto x^{-1}\left[1 - (1/x) - \frac{\epsilon}{(1-x)}\right]^{-2}.$$

The single parameter ϵ should be the square of the ratio of the mass of the partner antiquark and the charmed quark, but it is generally fitted to the experimental data. The value at which the spectrum peaks is related to ϵ: for example, $\epsilon = 0.15$ and 0.45 imply $x_{peak} = 0.67$ and 0.52, respectively. Because the original c quark direction is not very well defined by experiment, the spectra measured at PETRA and PEP are given in terms of $x = E_h/E_{beam}$, while the spectra from DORIS and CESR are expressed as functions of $x = p_h/p_{h,max}$. At the higher energy machines, all three x variables are essentially the same, but at the lower energies, where the c mass is not negligible, the data seem to scale [i.e., $f(x)$ is independent of E_{beam}] best if $x = p_h/p_{h,max}$ is used.

FIGURE 5 shows a comparison of ϵ values obtained for the various charmed hadrons at c.m. energies from 10 GeV to 34 GeV. The first observation, and perhaps the only one we should make, is that the present experimental errors are too large to permit any further statements. One should also be wary about the effects of radiative corrections and different choices for the meaning of x. However, let us forge ahead fearlessly and see what we might be able to learn in the future when the uncertainties are less. First, in comparing the ϵ values for the F and $F*$ with those for the various D states (ARGUS and CLEO data), there seems to be some confirmation of the expectation that ϵ should increase with the mass of the partner antiquark; the heavier strange antiquark decelerates the charmed hadron more than do the light antiquarks. Whether the same kind of deceleration effect is occurring in Λ_c production is less clear. In comparing the DORIS-CESR $D*$ data below $b\bar{b}$ threshold with the higher energy $D*$ data from PETRA-PEP, we may be seeing the effect on the $D*$ spectrum of the charmed mesons from B decays; these cascade $D*$ add mainly to the lower x region, decrease x_{peak}, and increase the fitted ϵ value relative to the expectation assuming only $e^+e^- \rightarrow c\bar{c}$. Similarly, the effect of the feed-down of D^0 from $D*$ decays may be visible in the higher fitted ϵ values for D^0 compared with $D*$ and D^+ (ARGUS and CLEO data). It will be interesting to see how these conjectures hold up as the data get better.

Bose-Einstein Correlations

It has been 27 years since Bose-Einstein correlations were first observed in pion production.[34] One sees an enhancement in the rate of production of pairs of like-sign pions when the difference in the momenta is small. Results from e^+e^- production in the psi region were first reported by Mark II[35] some years ago; there are new data now at higher energies.[36-38] It is customary to display the ratio R of the rate for $\pi^+\pi^+$ and $\pi^-\pi^-$ divided by a "reference spectrum," plotted against either Q (the absolute value of the momentum difference in the dipion rest frame) or Q^2, or q_T (the component of Q transverse to the dipion net momentum). Q^2 is actually an invariant, which can be

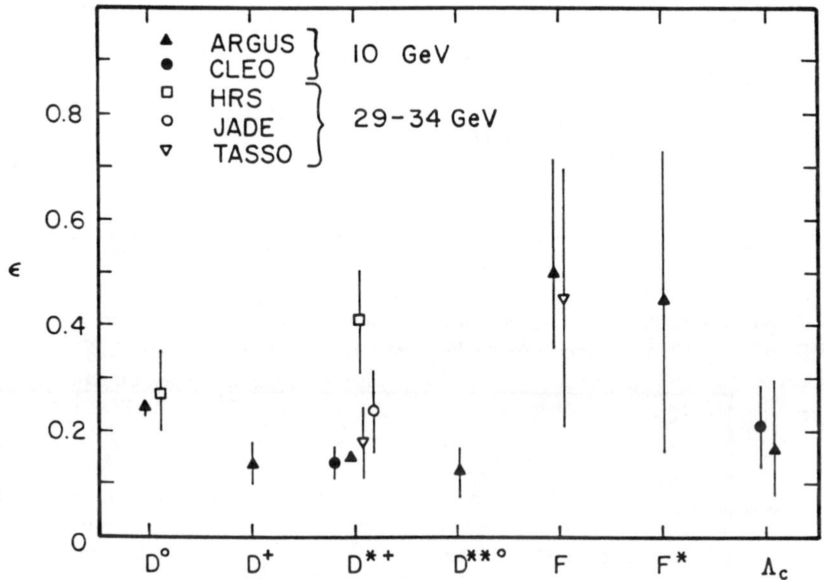

FIGURE 5. Fitted values of the ϵ parameter in the Peterson function[31] for charm fragmentation in e^+e^- annihilation.[25-30]

written as $M_{\pi\pi}^2 - 4m_\pi^2$. Several such plots are shown in FIGURE 6. The various experimenters use different choices for the reference spectrum: the measured $\pi^+\pi^-$ spectrum (with K_s and ρ decays excluded), or pion pairs made by combining different events, or a Monte Carlo spectrum. The experiments are mainly sensitive to the directions of the dipion momentum difference perpendicular to the net dipion momentum.

The dependence of the enhancement on Q tells us something about the size and shape of the emitting space-time region. In the simplest parametrization, $R(Q)$ is fit to the form,

$$R(Q) = 1 + \alpha e^{-r^2 Q^2}.$$

With this assumption, r is the root mean square transverse radius of the emitting

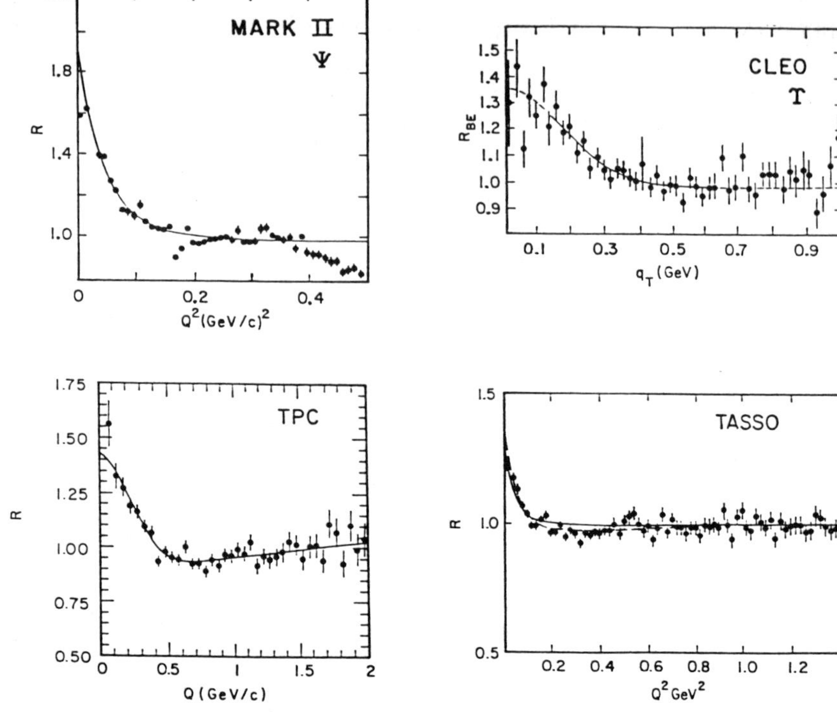

FIGURE 6. Measurements of the ratio of the like-sign pion pair rate to the reference spectrum (see text) as a function of the relative momentum of the pair.[35–38]

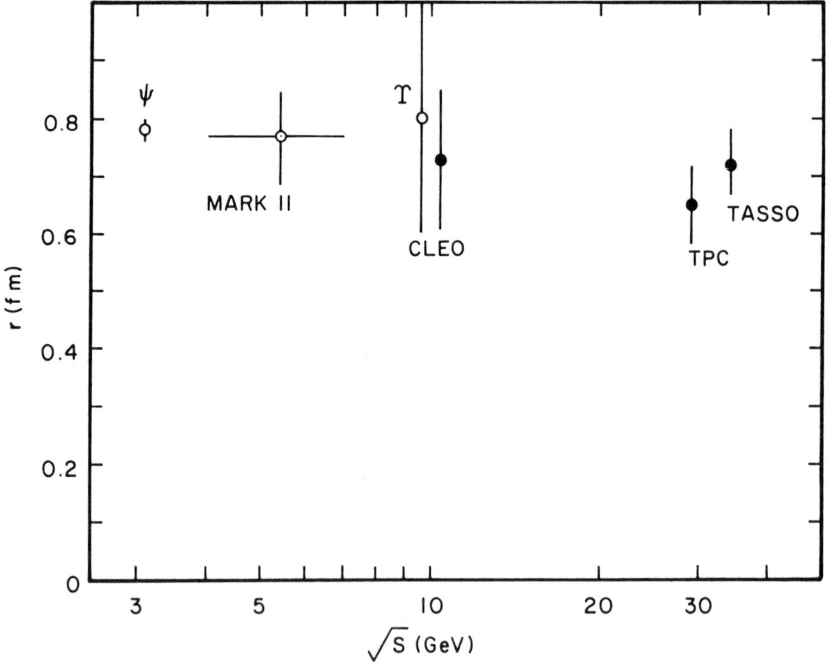

FIGURE 7. Measured source radius from Bose-Einstein correlations as a function of total c.m. e^+e^- collision energy.[35–38]

region. FIGURE 7 shows the fitted values for r from the various experiments, including results from ψ and Υ decays, as well as from the continuum over the energy range from SPEAR to PETRA. There seems to be little or no dependence of the source size on either the c.m. energy or on whether the pions come from quark or gluon jets; all experiments are consistent with $r = 0.7$ to 0.8 fm. The r measurements in FIGURE 8 are taken by binning the data in the angle θ between the Q vector ($\mathbf{p}_1 - \mathbf{p}_2$) and the jet axis;

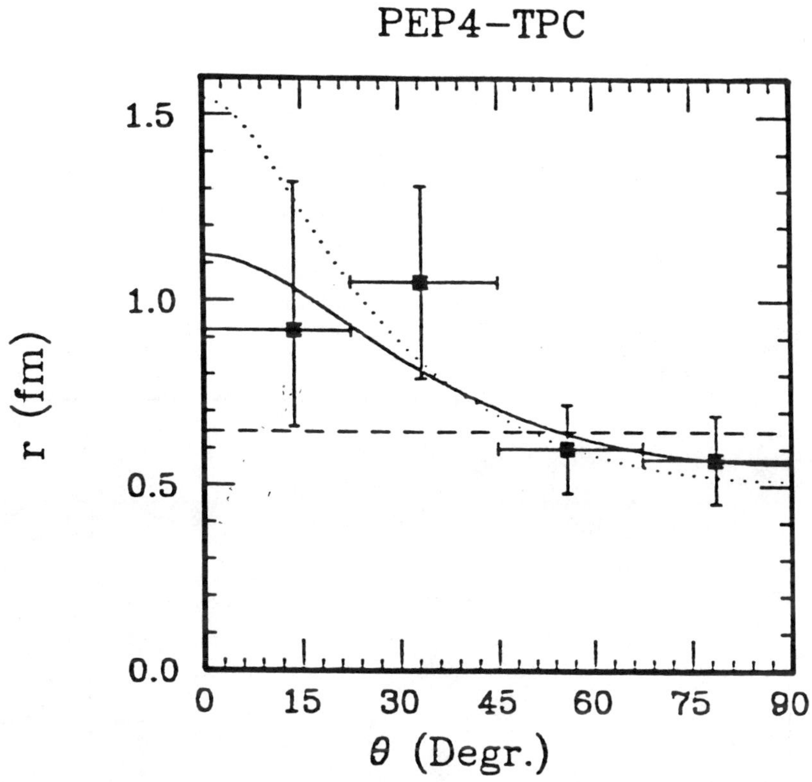

FIGURE 8. Dependence of the measured source radius on the angle between the pair momentum difference and the jet axis (TPC[36]).

the results are consistent with an almost spherical emitting region. Apparently, although pions may be emitted from an elongated string, pairs spanning widely separated points along the string do not overlap enough in momentum space to participate in the Bose-Einstein effect.

The parameter α is a measure of the Bose-Einstein correlation strength; ideally, $\alpha = 1$. Experimentally, the value from the fitted $R(0) = 1 + \alpha$ is generally consistent with $\alpha = 0.5$. The measured α is reduced by the effect of long-lived resonances, ω, η, K, etc., which contribute an unobservable narrow peak at $Q = 0$, and by the effect of dynamical correlations in the reference sample. Moreover, what one gets for α depends crucially on what one takes for the functional form of $R(Q)$. If one assumes that $R(Q)$ is more sharply peaked at $Q = 0$ than a Gaussian, the same data can give fits for α near unity. An example is a parametrization of the TPC $R(Q)$ data[36] by Andersson and Hoffmann[39] suggested by a variation of the Lund string model.

As the data get more accurate, we can expect that Bose-Einstein correlations will become a rather revealing probe of the space-time behavior of multihadron production.

MISSING PIECES OF THE STANDARD MODEL

The Third Generation

Until Tristan, SLC, and LEP begin operating, the t quark will not be seen in electron-positron experiments. The lower limit on its mass, from PETRA experiments,[40] is still 23.3 GeV.

Although it would be a surprise if the tau-neutrino did not exist, there is no experimental evidence for it yet; that is, no detected reaction initiated by ν_τ. There are, however, new upper limits on the tau-neutrino mass from ARGUS[41] (70 MeV, using

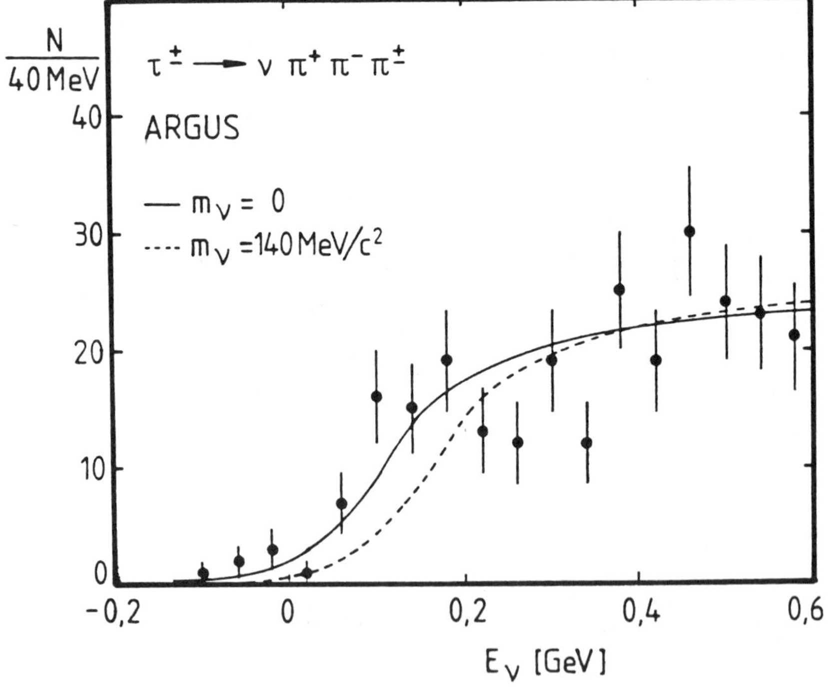

FIGURE 9. Spectrum of missing tau-neutrino energies in $\tau \to \nu\pi\pi\pi$ decays (ARGUS[31]).

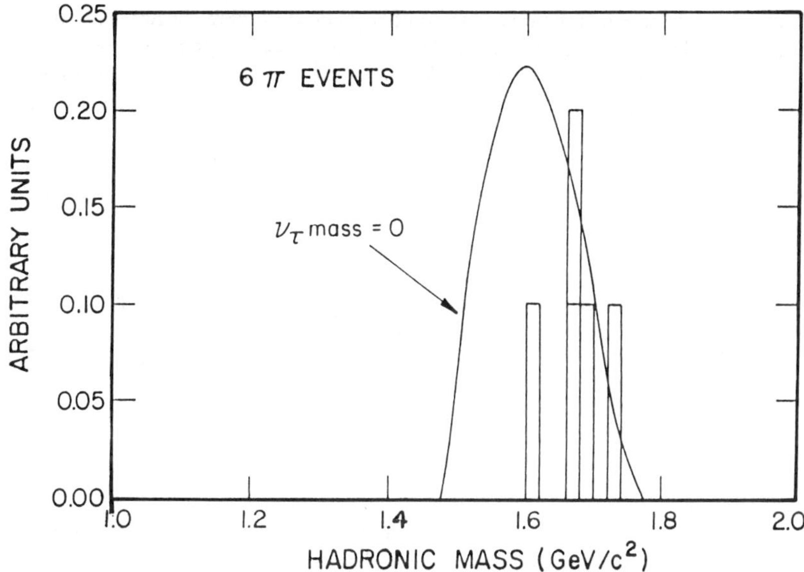

FIGURE 10. Spectrum of hadronic mass observed in $\tau \rightarrow \pi\pi\pi\pi\pi\pi\nu$ decays (HRS[32]).

the missing energy spectrum in the decay $\tau \rightarrow A_1\nu$; see FIGURE 9) and from HRS[42] [85 MeV, using the distribution of hadronic mass in $\tau \rightarrow 5\pi^\pm(\pi^0)\nu$; see FIGURE 10].

The Higgs Boson(s)

The "zeta" peak at 1.07 GeV in the single photon inclusive spectrum from upsilon decays (reported at the 1984 Leipzig conference by the Crystal Ball[43]) is apparently not really there.[44,45] Before it disappeared, it occasioned some excitement through the possibility that it might be a Higgs boson, even though the 0.5% branching ratio for $\Upsilon \rightarrow \gamma H$ implied by the Crystal Ball data was many times larger than expected for a standard Higgs. In the course of trying to confirm the zeta, the CUSB collaboration[44] was able to collect enough statistics at higher photon energies to rule out the upsilon radiative decay to a standard neutral Higgs of mass below about 5 GeV (FIGURE 11), at least if you ignore QCD corrections in the cross-section calculation.

The Axion

The axion is a scalar particle invented by Wilczek[46] and Weinberg[47] to solve the "strong CP" problem through the Peccei-Quinn mechanism.[48] Its assumed properties depend on the ratio X of vacuum expectation values of the two Higgs doublets. The mass should be

$$m_a = N_g[X + (1/X)] \cdot 25 \text{ keV} \qquad (N_g = \text{number of generations} = 3).$$

Unless X is very large or very small, we would expect m_a to be of the order of 100 keV, and only the decay $a \to \gamma\gamma$ would be kinematically allowed. The coupling to "up" quarks of mass m should be proportional to mX; the coupling to "down" quarks or leptons should go as m/X. Thus, the expected ratio of the rate of radiative decay for the

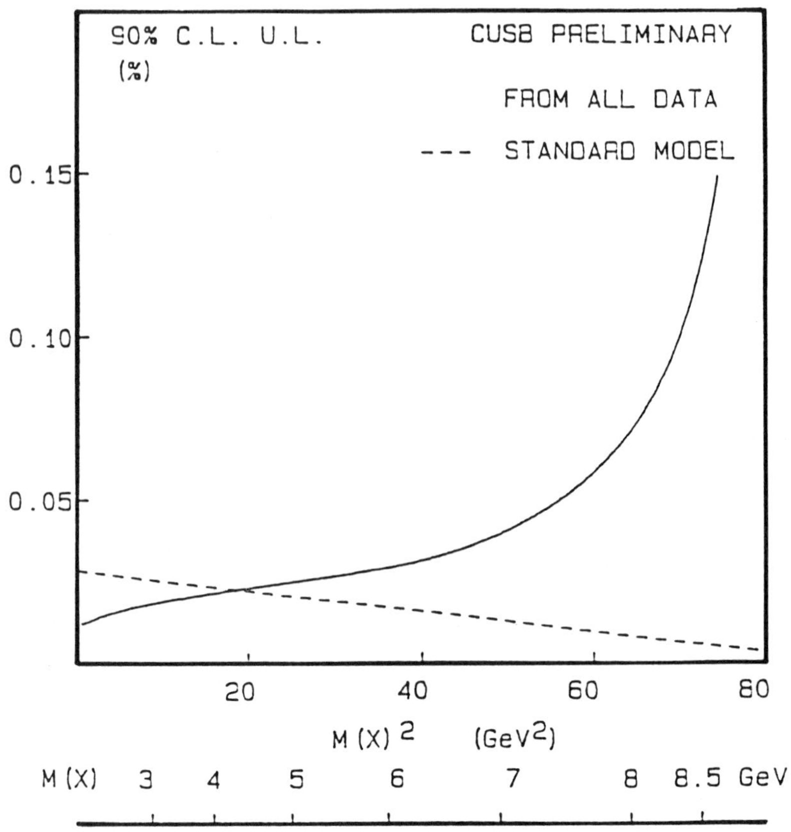

FIGURE 11. Experimental branching ratio upper limit for the decay $\Upsilon \to \gamma X$ (CUSB[44]) compared with the standard Higgs expectation, as a function of the mass of X.

upsilon into the axion to the dilepton decay rate would be

$$\frac{\Gamma(\Upsilon \to \gamma a)}{\Gamma(\Upsilon \to \mu\mu)} = \frac{G_F m_q^2}{2\pi\alpha M} \cdot \frac{1}{X^2},$$

and the corresponding formula for ψ decays is the same with X replacing $1/X$. The product of the ψ and Υ ratios would be independent of X and would have the value of $(1.6 \pm 0.3) \times 10^{-8}$.

Several years ago, the Crystal Ball group at SPEAR[49] searched for $\psi \to \gamma a$, and the CUSB and CLEO groups at CESR[50] looked for $\Upsilon \to \gamma a$. The ψ or Υ was tagged by the

$\pi^+\pi^-$ transition from the ψ' or Υ', and the signature for the decay was a high energy photon opposite the missing momentum of the presumed unseen long-lived axion. The measured upper limit on the product of ψ and Υ decay ratios was 0.4×10^{-8}, thus ruling out this long-lived axion.

Recently, results from heavy ion collision experiments at the GSI linac in Darmstadt[51] have been interpreted as the production of a particle of mass about 1.7 MeV decaying into electron and positron. The suggestion has been made[52] that this may be the axion. If it is the standard Wilczek-Weinberg axion, its mass implies a very small X value of 0.044. The coupling to leptons then implies a short lifetime for decay into e^+e^- of about 2 ps and, therefore, a mean laboratory decay length of about 2 m for axions produced in upsilon radiative decay. Because a substantial fraction of such axions produced in the CESR experiments would not have decayed within the experimental apparatus, the previous limits still apply, thus again ruling out this axion. However, if somehow the lifetime were much shorter than 2 psec, the axion would always decay into e^+e^- inside the apparatus, and its existence would not have been ruled out by the published upsilon decay searches.

CUSB and CLEO have now reexamined their data to look for a short-lived axion. The small X value associated with rapid decay into electron pairs implies a large branching ratio for $\Upsilon \rightarrow \gamma a$, say 20% if $X = 0.03$. Although the sequence, $e^+e^- \rightarrow \Upsilon$, $\Upsilon \rightarrow \gamma a$, $a \rightarrow e^+e^-$, is indistinguishable from the ordinary QED process, $e^+e^- \rightarrow \gamma\gamma$, followed by the pair conversion of one of the photons, the large branching ratio expected was easily ruled out by CUSB; their branching ratio upper limit is between 0.36% and 0.78%, depending on the assumed axion lifetime.[53] In the CLEO experiment, the QED $\gamma\gamma$ background is eliminated by tagging the Υ with the $\pi^+\pi^-$ transition from the Υ'. CLEO sees no candidates for $\Upsilon \rightarrow \gamma a$ and sets a branching ratio upper limit of less than 0.2%, depending on axion mass and lifetime.[54] Whatever the effect seen in the heavy ion experiments is, it is not an axion coupling similarly to leptons and b quarks.

REFERENCES

1. CELLO COLLABORATION. 1985. Contributed paper to International Symposium on Lepton and Photon Interactions at High Energies, Kyoto, Japan; BARTEL, W. *et al.* (JADE). 1985. Phys. Lett. **161B:** 188; ASH, W. W. *et al.* (MAC). 1985. Phys. Rev. Lett. **55:** 1831.
2. GAN, K. K. 1985. Annual Meeting of the Division of Particles and Fields (Eugene, Oregon), and Purdue Univ. preprint no. PU-85-539.
3. ASH, W. W. *et al.* (MAC). 1985. Phys. Rev. Lett. **55:** 2118.
4. BEHRENDS, S. *et al.* (CLEO). 1985. Phys. Rev. **D32:** 2468.
5. KOBAYASHI, M. & T. MASKAWA. 1973. Prog. Theor. Phys. **49:** 652.
6. COWARD, D. (MARK III). 1985. SLAC preprint no. SLAC-PUB-3818; BALTRUSAITIS, R. M. *et al.* 1985. Phys. Rev. Lett. **55:** 150.
7. ALBRECHT, H. *et al.* (ARGUS). 1985. Phys. Lett. **158B:** 525.
8. BEBEK, C. *et al.* (CLEO). 1986. Preprint no. CLNS-86/715.
9. BALTRUSAITIS, R. M. 1986. Preprint no. SLAC-PUB-3858.
10. CHEN, A. *et al.* (CLEO). 1984. Phys. Rev. Lett. **52:** 1084.
11. KLOPFENSTEIN, C. *et al.* (CUSB). 1983. Phys. Lett. **130B:** 444.
12. THORNDIKE, E. H. (CLEO). 1985. International Symposium on Lepton and Photon Interactions at High Energies, Kyoto, Japan, p. 405.
13. ARGUS COLLABORATION. 1985. Preprint DESY.

14. ALTARELLI, G. *et al.* 1982. Nucl. Phys. **B208:** 365.

15. GRINSTEIN, B., M. B. WISE & N. ISGUR. 1986. Phys. Rev. Lett. **56:** 298.

16. CLEO COLLABORATION. 1985. Preprint DESY.

17. BARANKO, G. 1985. Review of HRS, JADE, MAC, Mark II, and TASSO results presented at the Annual Meeting of the Division of Particles and Fields (Eugene, Oregon).

18. ALBANESE, J. P. *et al.* (WA75). 1985. Phys. Lett. **158B:** 186.

19. STONE, S. 1983. Proceedings of the 1983 International Symposium on Lepton and Photon Interactions at High Energies (Cornell). F.R. Newman Laboratory of Nuclear Physics. Cornell University.

20. KLOPFENSTEIN, C. *et al.* (CUSB). 1983. Phys. Rev. Lett. **51:** 160; HAAS, P. *et al.* (CLEO). 1984. Phys. Rev. Lett. **52:** 799.

21. NERNST, R. *et al.* (Crystal Ball). 1985. Phys. Rev. Lett. **54:** 2195; ALBRECHT, H. *et al.* (ARGUS). 1985. Phys. Lett. **160B:** 331.

22. ALBRECHT, H. *et al.* (ARGUS). 1984. Preprint no. DESY 84-052.

23. ALBRECHT, H. *et al.* (ARGUS). 1986. Phys. Rev. Lett. **56:** 549.

24. HAN, K. *et al.* (CUSB). 1985. Phys. Rev. Lett. **55:** 36.

25. BOWCOCK, T. *et al.* (CLEO). 1985. Phys. Rev. Lett. **55:** 923.

26. ORR, R. S. (ARGUS). 1985. International Europhysics Conference on High Energy Physics (Bari, Italy).

27. DEGRAND, T. A. 1982. Phys. Rev. **D26:** 3298.

28. AVERY, P. *et al.* (CLEO). 1983. Phys. Rev. Lett. **51:** 1139.

29. ALTHOFF, M. *et al.* (TASSO). 1983. Phys. Lett. **126B:** 493; 1984. Phys. Lett. **136B:** 130; BARTEL, W. *et al.* (JADE). 1984. Phys. Lett. **146B:** 121; DERRICK, M. *et al.* (HRS). 1984. Phys. Rev. Lett. **53:** 1971; ABACHI, S. *et al.* (HRS). 1985. Purdue Univ. preprint no. PU-85-536.

30. ALBRECHT, H. *et al.* (ARGUS). 1985. Phys. Lett. **153B:** 343.

31. PETERSON, C. *et al.* 1983. Phys. Rev. **D27:** 105.

32. BJORKEN, J. D. 1978. Phys. Rev. **D17:** 171.

33. SUZUKI, M. 1977. Phys. Lett. **71B:** 139.

34. GOLDHABER, G. *et al.* 1959. Phys. Rev. Lett. **3:** 181; 1960. Phys. Rev. **120:** 300.

35. GOLDHABER, G. (MARK II). 1981. International Conference on High Energy Physics (Lisbon), p. 767.

36. AIHARA, H. *et al.* (TPC). 1985. Phys. Rev. **D31:** 966.

37. AVERY, P. *et al.* (CLEO). 1985. Phys. Rev. **D32:** 2294.

38. ALTHOFF, M. *et al.* (TASSO). 1985. Preprint no. DESY 85-126.

39. ANDERSSON, B. & W. HOFFMANN. 1985. Preprint no. LBL-20272.

40. SILVERMAN, A. 1984. XXIIth International Conference on High Energy Physics (Leipzig), vol. II, p. 91. Akademie der Wissenschaften. Zeuthen, East Germany.

41. ALBRECHT, H. *et al.* (ARGUS). 1985. Phys. Lett. **163B:** 404.

42. ABACHI, S. *et al.* (HRS). 1986. Phys. Rev. Lett. **56:** 1039.

43. TROST, H. J. 1984. (Crystal Ball). 1984. XXIIth International Conference on High Energy Physics (Leipzig).

44. YOUSSEF, S. *et al.* (CUSB). 1984. Phys. Lett. **139B:** 332.

45. BLOOM, E. D. (Crystal Ball). 1985. Preprint no. SLAC-PUB-3686; ALBRECHT, H. *et al.* (ARGUS). 1985. Phys. Lett. **154B:** 452; Preprint no. DESY 85-083; BESSON, D. *et al.* (CLEO). 1986. Phys. Rev. **D33:** 300.

46. WILCZEK, F. 1978. Phys. Rev. Lett. **40:** 279.

47. WEINBERG, S. 1978. Phys. Rev. Lett. **40:** 223.

48. PECCEI, R. D. & H. R. QUINN. 1977. Phys. Rev. Lett. **38:** 1440.

49. EDWARDS, C. *et al.* (Crystal Ball). 1982. Phys. Rev. Lett. **48:** 903.

50. SIVERTZ, M. *et al.* (CUSB). 1982. Phys. Rev. **D26:** 717; ALAM, M. S. *et al.* (CLEO). 1983. Phys. Rev. **D27:** 1665.

51. SCHWEPPE, J. *et al.* 1983. Phys. Rev. Lett. **51:** 2261; CLEMENTE, M. *et al.*, 1984. Phys. Lett. **137B:** 41; COWAN, T. *et al.* 1985. Phys. Rev. Lett. **54:** 1761; 1986. Phys. Rev. Lett. **56:** 444.

52. BALANTEKIN, A. B. *et al.* 1985. Phys. Rev. Lett. **55:** 461.

53. MAGERAS, G. (CUSB). Private communication.

54. BOWCOCK, T. *et al.* (CLEO). 1986. Preprint no. CLNS-86/719.

Classical Pictures of Confinement[a]

F. ZACHARIASEN

Department of Physics
California Institute of Technology
Pasadena, California 91125

INTRODUCTION

The fact that confinement of color is a consequence of QCD is, by now, doubted by practically nobody. However, the demonstration of confinement is still, after all these years, incomplete, and therefore, there are still conjectures as to what the physical mechanism is that makes it happen. Two possible mechanisms, both based on analogy with known phenomena in classical physics, dominate present-day thinking. One of these is that confinement works like a classical ferroelectric; the other is that it works like a dual superconductor.

The ferroelectric picture is based on the idea that the QCD vacuum is a medium with a field-dependent dielectric constant $\epsilon(E)$ that vanishes at some nonzero electric field E_0, as in a classical ferroelectric. The color electric displacement of a point charge q is $\vec{D} = q\vec{r}/4\pi r^3$: As $r \rightarrow \infty$, $\vec{D}$ vanishes and $\vec{E} \rightarrow \vec{E}_0 \neq 0$. Consequently, for large r, the electrostatic energy density becomes $\frac{1}{2}\vec{E} \cdot \vec{D} = q\vec{r} \cdot \vec{E}_0/8\pi r^3 \sim 1/r^2$, and the total energy is thus infinite:[1] color is therefore confined.

The distinguishing feature of this picture is that the field configurations are purely electric; no magnetic structure exists.[2] They are also Abelian, so all fields effectively point in a single color direction. Finally, flux is not quantized in these models. The QCD vacuum is purely electric: the electric field is E_0; the magnetic field is zero.

The second of the two major pictures of confinement is that of a dual superconductor; that is, a superconductor with electricity and magnetism interchanged.[3-6] The dual Meissner effect forces color electric flux into vortices, or flux tubes. In particular, if two opposite color electric charges are placed far apart in a dual superconductor, the color electric flux joining them is compressed into a thin flux tube, giving rise to a linear potential between them. As in the superconductor, the color electric flux in the tubes is quantized; however, in QCD, because of the non-Abelian structure of the theory, only a finite number of quanta of flux can exist [up to $N - 1$ units for the color group $SU(N)$; this is known as $Z(N)$ flux tubes]. The field configurations are that the electric field and the electric displacement, as well as the magnetic field $\vec{H}$, are confined within the flux tube, but the magnetic induction $\vec{B}$ exists outside the flux tube, vanishing on the axis.

The vacuum is characterized by a magnetic field, but no electric field; however, the detailed physics of what gives rise to this field—that is, what is the analog of the Cooper pairs in the superconductor—is still uncertain. One suggestion is that the QCD

[a]This work was supported in part by the U.S. Department of Energy under Contract No. DEAC-03-81-ER40050.

vacuum is full of magnetic monopoles.[4] Another is that it is a writhing spaghetti of tubes of magnetic flux.[7] On the other hand, it may be made up of instantons.[8] Whatever the details, the result is a nonzero magnetic field. In order to respect Lorentz invariance, this field may occur in a domain structure,[7] which can then be lined up by disturbances in the vacuum having spatial and color orientations.

Which of the two pictures that we have described obtains in QCD can be tested by QCD lattice calculations. Flower and Otto[9] have calculated the gauge invariant field configurations $\langle \mathrm{tr}\, E_\perp^2 \rangle$, $\langle \mathrm{tr}\, E_\parallel^2 \rangle$, $\langle \mathrm{tr}\, B_\perp^2 \rangle$, and $\langle \mathrm{tr}\, B_\parallel^2 \rangle$ in and around the flux tube joining a static quark and antiquark in a $10^2 \times 12^2$ lattice. ($\parallel$ and $\perp$ refer to components parallel and perpendicular to the flux tube axis.) The results are shown in FIGURE 1 ($\langle \mathrm{tr}\, B_\parallel^2 \rangle$ is statistically not different from zero, and is not shown; $\langle \mathrm{tr}\, B_\perp^2 \rangle$ is shown relative to its vacuum value). Evidently, $\langle \mathrm{tr}\, E_\perp^2 \rangle$ near the center of the tube is very small (note the scale), and apart from end effects near the quarks, it probably vanishes. The dominant components are clearly the parallel E field and the transverse B field. The existence of a large magnetic field argues strongly against the ferroelectric picture and, as the pictures show, the lattice calculation strongly supports the dual superconductor version of confinement.

Both pictures give rise to baglike models of hadrons.[10] In the ferroelectric case, the bag boundary is the surface on which the electric field from the quarks inside takes on the critical value E_0 at which the dielectric constant vanishes. All electric flux is confined within the bag, and the field is everywhere tangent to the surface. The dielectric constant is $\epsilon(E^2)$ inside and zero outside.

The bag in the dual superconductor is not exact, but only represents an approximate solution to the analog of the Landau-Ginsberg equations. In this approximation, the fields are given their asymptotic (or QCD vacuum) values outside of some surface, while inside the surface, the dual Higgs potential is made to vanish. Therefore, no symmetry breaking occurs and the perturbative vacuum is obtained. The surface itself is defined by the requirement that the vacuum magnetic pressure outside is balanced by the electrostatic energy density of the quarks inside.

It is worth noting that neither of these pictures is exactly the same as the original bag model, which took $\epsilon = 1$ inside the bag and $\epsilon = 0$ outside. For the ferroelectric, $\epsilon = \epsilon(E^2)$ inside and zero outside; for the dual superconductor, the bag is characterized by the vanishing of the Higgs potential rather than by a change in ϵ. Nevertheless, most of the bag phenomenology is predicted by both pictures.

THE FERROELECTRIC

Because of asymptotic freedom, the one-loop Yang-Mills effective Lagrangian in a constant field can be calculated. It is known to be[11]

$$L_{\mathrm{eff}} \propto F^2 \log F^2/F_0^2, \tag{2.1}$$

where $F^2 \equiv F_{\mu\nu}^a F_{\mu\nu}^a$ and F_0^2 is a constant. The dielectric constant, ϵ, identified from the formula,

$$L_{\mathrm{eff}} = -\tfrac{1}{4}\, \mathrm{tr}\, (F_{\mu\nu}\epsilon F_{\mu\nu}), \tag{2.2}$$

is therefore proportional to $\log F^2/F_0^2$ and vanishes at $F^2 = F_0^2 \neq 0$. The ferroelectric

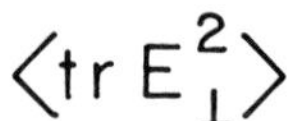

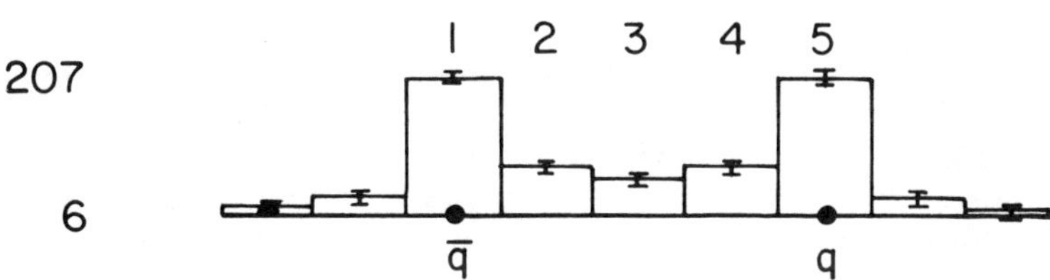

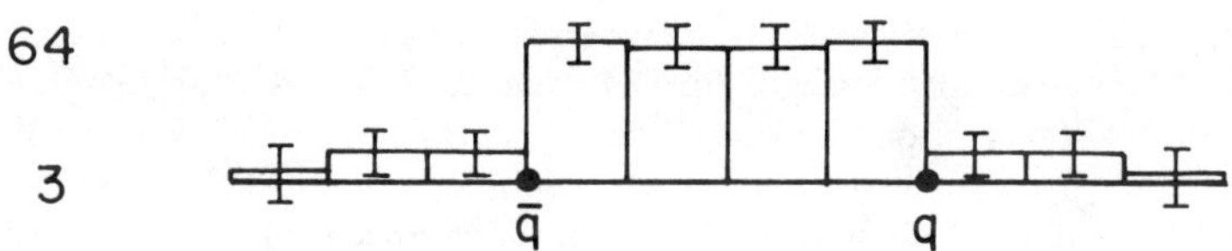

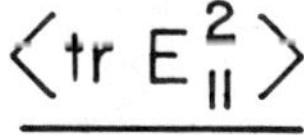

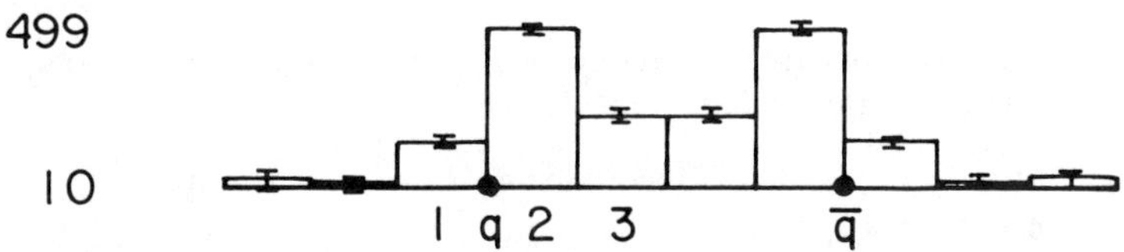

FIGURE 1. This figure shows $\langle \operatorname{tr} E_\perp^2 \rangle$, $\langle \operatorname{tr} B_\perp \rangle^2$ (relative to its vacuum value), and $\langle \operatorname{tr} E_\parallel^2 \rangle$ as calculated in $SU(3)$ lattice theory with a quark and antiquark source. The values on the $q\bar{q}$ axis are plotted. The fields fall off very rapidly away from the axis. (From reference 9.)

model of confinement advocated by Adler[2,12] consists of assuming that equation 2.1 is valid.

One looks for static Abelian solutions to the field equations resulting from equation 2.1. All fields are therefore taken to be in a single color direction, and the field equations are simply Maxwell's equations,

$$\vec{\nabla} \cdot \vec{D} = \rho \qquad \vec{\nabla} \times \vec{H} = 0$$

$$\vec{\nabla} \cdot \vec{B} = 0 \qquad \vec{\nabla} \times \vec{E} = 0, \tag{2.3}$$

together with $\vec{D} = \epsilon \vec{E}$, $\vec{B} = \mu \vec{H}$, and $\epsilon\mu = 1$. Here, ρ is a color electric charge density consisting of two oppositely charged heavy quarks: $\rho = q\delta(x)\delta(y)[\delta(z - R/2) - \delta(z + R/2)]$.

The desired solution of equation 2.3 is purely electric, with $\vec{B} = \vec{H} = 0$. Therefore, the dielectric constant is

$$\epsilon = \log E^2/E_0^2 \tag{2.4}$$

and vanishes when $E^2 = E_0^2 \neq 0$. Maxwell's equations thus become nonlinear equations for E. The solution has the following properties: All of the electric flux is confined inside of an azimuthally symmetric surface. Outside this surface, or bag, the electric field squared is E_0^2 and $\epsilon = 0$. Inside the boundary surface, $E^2 \geq E_0^2$ and approaches E_0^2 on the surface, where it is tangential, thus confining the flux. Therefore, the electrostatic energy [which is just the static quark potential $V(R)$] grows linearly with quark separation R as $R \rightarrow \infty$.

While the flux is confined, it is, however, not quantized; that is, solutions of this type to Maxwell's equations exist for any value of the quark charge. There are no magnetic fields; there is only an electric field E_0^2 outside the bag.

The static potential between heavy quark sources can be calculated in this model[12] by solving equation 2.3 numerically for $\vec{D}$ and $\vec{E}$ and by computing the potential $V(R)$ from the formula,

$$V(R) = \int d^3\vec{x} \int_0^D E(D')dD'. \tag{2.5}$$

For large R, the potential is bounded below by

$$V(R) \geq qE_0R. \tag{2.6}$$

The potential obtained from equation 2.5 is shown in FIGURE 2.

THE DUAL SUPERCONDUCTOR

The effective Lagrangian describing the dual superconductor can be obtained as follows. Suppose the weak field Yang-Mills dielectric constant in some gauge behaves at long range, or small momentum, like this: $\epsilon(q) \rightarrow q^2/M^2$ as $q^2 \rightarrow 0$. (This is a guess frequently made[13] because it is equivalent to assuming that the gluon propagator, in the same gauge, behaves like $1/q^4$, which is in turn equivalent to assuming a linear potential at the one gluon exchange level. Such a behavior cannot quite yet be derived from Yang-Mills theory, though it can be made plausible.[14]) At this level, then, the effective Lagrangian would be

$$L_{\text{eff}}^{(0)} = -\frac{1}{4} (\partial_\mu A_\nu - \partial_\nu A_\mu) \frac{\Box}{M^2} (\partial_\mu A_\nu - \partial_\nu A_\mu) \tag{3.1}$$

because in coordinate space, $\epsilon = \Box/M^2$. Unfortunately, the vector potential is very singular, as can be seen from the fact that the gluon propagator is $\Delta_A \sim \langle A^2 \rangle \sim 1/q^4$ as $q^2 \to 0$.[15]

When the vector potential is singular, one should describe physics with the dual potential C_μ,[6] defined (in the weak field limit) by $G_{\mu\nu} = \partial_\mu C_\nu - \partial_\nu C_\mu$, where $G_{oi} = H_i = \epsilon B_i$ and $G_{ij} = G_{ijk} D_k = \epsilon_{ijk} \cdot \epsilon E_k$. Because A is singular, C is (exponentially) convergent at long range. In terms of C,

$$L_{\text{eff}}^{(0)} = \frac{1}{4} (\partial_\mu C_\nu - \partial_\nu C_\mu) \frac{M^2}{\Box} (\partial_\mu C_\nu - \partial_\nu C_\mu) \tag{3.2}$$

because the permeability, $\mu = 1/\epsilon$, is $\mu = M^2/\Box$.

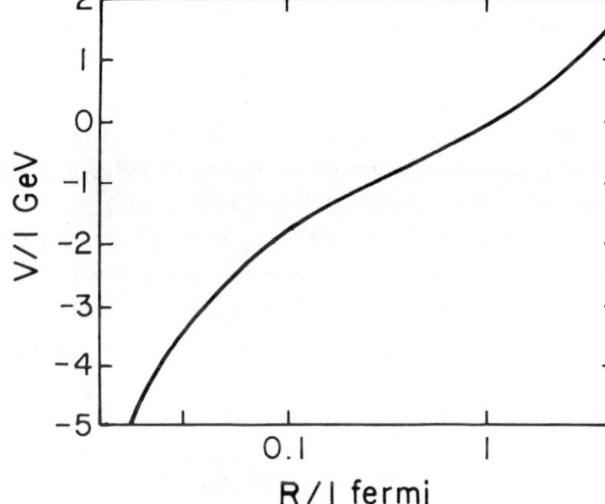

FIGURE 2. The static potential between heavy quarks in the fer-roelectric model. (From reference 12.)

The Lagrangian is now obtained as the minimal gauge invariant extension of equation 3.2: we have

$$L_{\text{eff}} = \frac{1}{4} \text{ tr } G_{\mu\nu} \frac{M^2}{D^2(C)} G_{\mu\nu}, \tag{3.3}$$

where

$$G_{\mu\nu} = \partial_\mu C_\nu - \partial_\nu C_\mu - i[C_\mu, C_\nu] \tag{3.4}$$

and

$$D_\mu(C) = \partial_\mu - i[C_\mu,]. \tag{3.5}$$

To deal with the operator $1/D^2(C)$ is hard, so it is convenient to define an auxiliary field, $F_{\mu\nu}$, treated as an independent dynamical variable, and rewrite equation 3.3 as

$$L_{\text{eff}} = \frac{1}{2} G_{\mu\nu} \tilde{F}_{\mu\nu} - \frac{1}{4} \tilde{F}_{\mu\nu} \frac{D^2(C)}{M^2} \tilde{F}_{\mu\nu}. \tag{3.6}$$

The last move is to add two phenomenological terms to equation 3.6. One of these represents the next term in the long-range expansion of the permeability: $\mu \rightarrow M^2/q^2 + 1/f^2$ as $q^2 \rightarrow 0$. The second is a phenomenological function of the auxiliary field $\tilde{F}_{\mu\nu}$. (C_μ vanishes at long range; therefore, any additional C dependence in L_{eff} beyond what we have already included disappears at long range. $\tilde{F}_{\mu\nu}$, on the other hand, does not vanish, so an additional function of $\tilde{F}$ can occur at long range.) Finally, the effective Lagrangian has thus become

$$L_{\text{eff}} = \frac{1}{2} G\tilde{F} - \frac{1}{4} \tilde{F} \frac{D^2(C)}{M^2} \tilde{F} - \frac{1}{4f^2} G_{\mu\nu} G_{\mu\nu} - W(\tilde{F}). \tag{3.7}$$

It is this Lagrangian that describes the dual superconductor. The field equations derived from it are the analogs of the Landau-Ginsberg equations. As we have indicated, it cannot yet be derived from the underlying QCD quantum field theory (the analog of the BCS theory), but such was also the case for many years in superconductivity.

The function $W(\tilde{F})$ that appears in equation 3.7 is analogous to the Higgs potential in the relativistic superconductor. Spontaneous symmetry breaking takes place if $W(\tilde{F})$ has a minimum at a nonzero value of $\tilde{F}_{\mu\nu}$—call it $\tilde{F}_{\mu\nu}^{(0)}$. Let us suppose that this occurs; further details of W are unimportant.

In the perturbative vacuum, all fields are zero; if spontaneous symmetry breaking occurs, there is a nonperturbative vacuum with $\tilde{F}_{\mu\nu} = \tilde{F}_{\mu\nu}^{(0)}$. However, $\tilde{F}_{\mu\nu}^{(0)}$ is a Lorentz tensor, which must vanish in any true vacuum. Thus, the nonperturbative vacuum can exist only in the presence of some excitation, which provides an orientation in space. Far from the excitation, the (oriented) nonperturbative vacuum can exist. It will be related to the unoriented nonperturbative vacuum through $\operatorname{tr}[\tilde{F}_{\mu\nu}^{(0)} \tilde{F}_{\mu\nu}^{(0)}] = \langle \tilde{F}_{\mu\nu} \tilde{F}_{\mu\nu} \rangle_{\text{true vacuum}}$. If the domain picture[7] of the QCD vacuum is correct, then in the true vacuum, $\langle \tilde{F}_{\mu\nu} \rangle = 0$ because the domain orientations are averaged over. The perturbation, with its preferred direction, rotates the domains so that they line up; hence, in the oriented vacuum, $\langle \tilde{F}_{\mu\nu} \rangle$ does not have to vanish.

A specific illustration of this is the existence of quantized color electric flux tube solutions to equation 3.7. (These are analogous to magnetic flux tubes in a superconductor.) The oriented vacuum associated with such flux tubes has a nonzero value of the "magnetic field" $B^i \equiv \tilde{F}_{oi}$ and of the vector potential: $\vec{C} = 1/\rho \, \hat{e}_\phi$ (in an arbitrary color direction) in cylindrical coordinates. Far from the flux tube, the fields take on these values. Inside the flux tube, $\vec{C}$ and $\vec{B}$ drop smoothly to zero at the origin. The other independent fields, $E_i \equiv \frac{1}{2} G_{ijk} \tilde{F}_{jk}$ and C_0, vanish exponentially asymptotically and approach constants as $\rho \rightarrow 0$. The physical fields, $D_i = \frac{1}{2} \epsilon_{ijk} G_{jk}$ and $H_i = G^{oi}$, as well as the energy density, vanish exponentially as $\rho \rightarrow \infty$. (See FIGURE 3.)

The color structure is crucial to the existence of quantized flux tubes; in particular, the flux tubes are not Abelian in color and this is a major difference between equation 3.7 and a relativistic superconductor. As a result of the non-Abelian structure, there are $Z(N)$ flux tubes for $SU(N)$ of color; there are only N topologically distinct quantized flux tubes, having 0 (the perturbative vacuum), 1, $1/N$, $2/N$, up to $(N-1)/N$ units of quantized flux. In a relativistic superconductor, in contrast, any number of units of magnetic flux is possible.

The specific field and color configurations predicted by equation 3.7 are [in

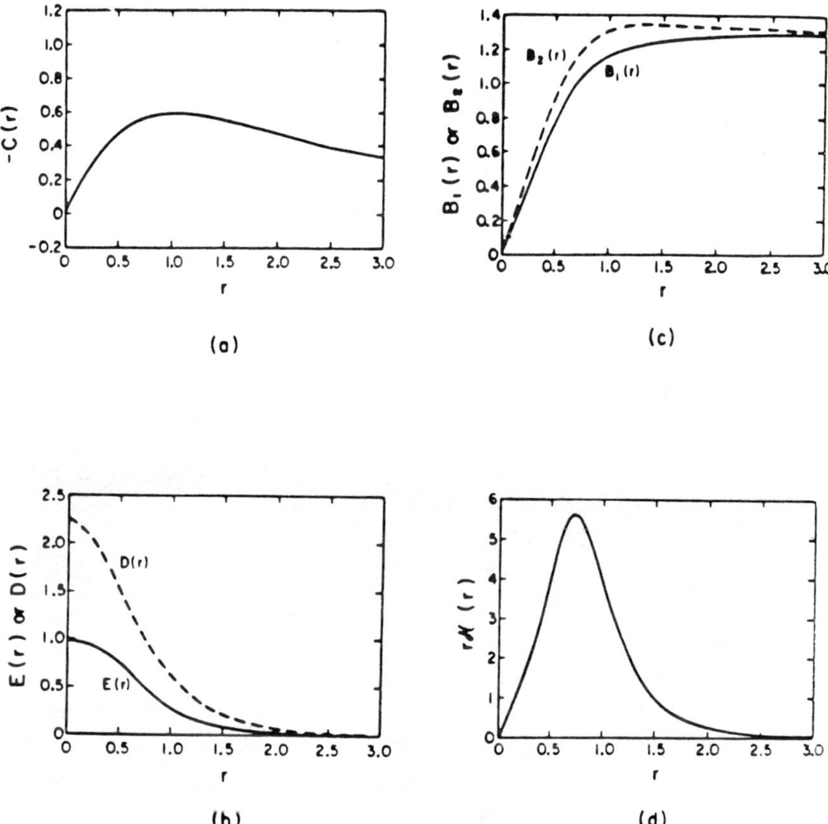

FIGURE 3. (a) $C(r)$, (b) $E(r)$ or $D(r)$, (c) $B_1(r)$ or $B_2(r)$, and (d) the energy density $rH(r)$ are shown from numerical solutions to the $n = 1$ flux tube equations. (From reference 14.)

cylindrical coordinates and for $SU(2)$ of color]:

$$C_0 = C_0(\rho)T^1,$$

$$\vec{C} = C(\rho)\hat{e}_\phi T^3,$$

$$\vec{E} = E(\rho)\hat{e}_z T^3,$$

$$\vec{B} = B_1(\rho)\hat{e}_r T^1 + B_2(\rho)\hat{e}_\phi T^2.$$

Note that the gauge invariant quantities tr $E_{\parallel}^2$, etc., agree with the field configurations suggested by the lattice calculations[9] (see FIGURE 1). The scalar functions C_0, C, etc., are shown in FIGURE 3.

The interaction energy between different flux tubes is also different from the superconductor case. Here, for example, for $SU(2)$ of color, the interaction energy of two flux tubes (shown in FIGURE 4) and of a flux tube and an antiflux tube are the same. Note that the interaction energy falls exponentially with flux tube separation; there are no long-range van der Waals–type forces present. In a superconductor, flux tubes either just repel (in type II superconductors) or attract (in type I). Here, in contrast, there is an attractive short-range force, outside of which is a repulsive barrier. Consequently, one can expect toroidal flux rings to exist in QCD; in a superconductor, they cannot.

Other excitations, besides quantized color electric flux tubes, are predicted by equation 3.7. Some are spherically symmetric and are probably to be interpreted as glueballs. They are not topologically stable as are the flux tubes. Again, the excitations have an orientation (in this case, a correlation between space and color) so that an oriented nonperturbative vacuum exists. [It has the same value of tr $(\tilde{F}_{\mu\nu}^{(0)}\tilde{F}_{\mu\nu}^{(0)})$ as the previous case.]

A number of quantities of physical interest can be computed for all of these solutions. These include: the vacuum energy density ϵ_{vac}; the string tension (i.e., the energy per unit length in the quantized flux tube) κ; the glueball mass M_G; the value of the gluon condensate $G_2 = \alpha/\pi\,(\tilde{F}_{\mu\nu}^{(0)}\tilde{F}_{\mu\nu}^{(0)})$; the radius of a flux tube r_{cyl}; and the radius of a glueball r_{sph}. They depend on four parameters: the original mass scale M; the phenomenological parameter f^2 introduced in equation 3.7; and two parameters giving the behavior of $W(\tilde{F})$ near its minimum (W can just be taken to be a polynomial; the parameters are the coefficients in the polynomial).

If any four of the physical quantities are taken from "experiment" (in this case, lattice gauge theory calculations) and are used to fix the parameters, the remaining quantities can be predicted. A number of sets of consistent values for the physical quantities are shown in TABLE 1 [for $SU(2)$ as the color group].

Input values in TABLE 1 are the string tension $\kappa = 0.2\,\text{GeV}^2$ and the vacuum energy density $\epsilon_{\text{vac}} = -0.096\kappa^2$. The table then lists consistent sets of values for the mass scale M, the glueball mass, the magnetic condensate, the flux tube radius, and the glueball

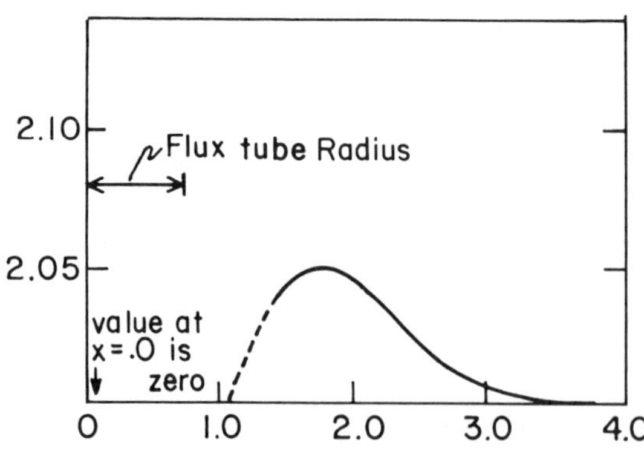

FIGURE 4. The interaction energy between two $n = 1$ flux tubes as a function of separation. (From reference 16.)

TABLE 1. Quantities of Physical Interest[a]

Solution	M (MeV)	M_{Glueball} (GeV)	G_2 (GeV)^4	r_{cylinder} fm	r_{sph} fm
a	128	2.44	0.0212	0.48	0.25
b	141	4.60	0.0308	0.53	0.27
c	153	4.35	0.0178	0.77	0.22
d	114	2.00	0.0048	0.28	0.09
e	29	2.03	0.0004	0.32	0.30
f	50	2.07	0.0012	0.32	0.29
g	221	1.15	0.0085	0.20	0.19

[a] $\kappa = 0.2 \text{ GeV}^2$; $\epsilon_{\text{vac}} = -0.096 \, \kappa^2$.

radius. The numbers are evidently reasonable. Solution (d) is perhaps the best and has a glueball mass of around 2 GeV. (This is of course the 0^+ ground state glueball.)

A "bag model" can be derived as an approximate solution to equation 3.7. In the exact solutions to equation 3.7 (as we have mentioned), the magnetic field $\vec{B}$ and therefore the potential W have a nonzero constant value far from excitations. Inside the excitations, they vanish, as FIGURE 3 shows. An approximate version of equation 3.7, then, is to write for static problems:

$$L_{\text{eff}} = \int_V [\vec{D} \cdot \vec{E} + \frac{1}{2} \vec{E} \cdot \frac{\nabla^2}{M^2} \vec{E} + \frac{1}{2f^2} \vec{D}^2 + \vec{B} \cdot \vec{H} + \frac{1}{2} \vec{B}$$

$$\cdot \frac{\nabla^2}{M^2} \vec{B} + \frac{1}{2f^2} \vec{H}^2 \,] \, d^3\vec{x} - BV, \quad (3.8)$$

where $B = -\epsilon_{\text{vac}}$. This neglects the non-Abelian structure, and replaces the physical fields $\vec{D}, \vec{H}$ etc., by zero outside the volume V of the excitation (the bag), while it replaces $W(\tilde{F})$ by zero inside the bag.

Note that the interior of the bag is not characterized by a normal dielectric constant that is zero outside the bag, as in the MIT bag model.[10] Here, the bag is associated with the vanishing of the potential W inside, while it has a finite value outside; the dielectric constant is D^2/M^2 (or $-\nabla^2/M^2$ in the static Abelian limit) everywhere, both inside and outside.

The bag approximation to equation 3.7 is simple enough so that static quark sources can be introduced into the problem; however, a comment about how this is to be done when the dynamics is described by the dual potential needs to be made first.

If one simply writes $\vec{D} = -\vec{\nabla} \times \vec{C}$, then one cannot have electric sources because $\vec{\nabla} \cdot \vec{D} = 0$ is automatic. One instead writes $\vec{D} = -\vec{\nabla} \times \vec{C} + \vec{D}_s$, where $\vec{\nabla} \cdot \vec{D}_s = \rho$, and ρ is the charge density that we wish to introduce. For a single point charge, this decomposition just amounts to writing (in spherical coordinates)

$$\frac{1}{4\pi r^2} \hat{e}_r = \vec{\nabla} \times \left[\frac{1 + \cos\theta}{4\pi r \sin\theta} \hat{e}_\phi \right] + \hat{e}_z \delta(x)\delta(y)\theta(z). \quad (3.9)$$

Thus, D_s is just present to cancel out the string associated with the attempt to represent a point charge as a curl of a dual potential. Using this representation of $\vec{D}$ and equation 3.8 as an approximate Lagrangian, one can insert static charges. In particular, by minimizing equation 3.8 with respect to the bag boundary shape, one can calculate the static potential between a heavy quark and an antiquark in an overall color singlet state. The result is in good agreement with phenomenologically determined potentials.

REFERENCES

1. Wu, Y. S. & A. Zee. 1983. Univ. of Washington preprint no. 40048-13 P3.
2. Adler, S. 1982. Phys. Lett. **110B:** 302.
3. Mandelstam, S. 1979. Phys. Rev. **D19:** 2391.
4. 't Hooft, G. 1979. Nucl. Phys. **B153:** 141.
5. Englert, F. 1977. Cargese Lectures, p. 503.
6. Nair, V. P. & C. Rosenzweig. 1984. Phys. Lett. **B135:** 450.
7. Nielsen, H. B. & P. Olesen. 1979. Nucl. Phys. **B160:** 57; Ambjørn, J. & P. Olesen. 1980. Nucl. Phys. **B170:** 60, 265; Cornwall, J. M. 1982. Phys. Rev. **D26:** 1453.
8. Callan, C. G., R. Dashen & D. Gross. 1977. Phys. Lett. **66B:** 375.
9. Flower, J. & S. Otto. 1985. Caltech preprint no. CALT-68-1271.
10. See, for example: DeTar, C. & J. Donoghue. 1983. Ann. Rev. Nucl. Part. Sci. **33:** 235; Hasenfratz, P. & J. Kuti. 1978. Phys. Rep. **40C:** 75.
11. Matinyan, S. G. & G. K. Savidy. 1978. Nucl. Phys. **B134:** 539; Savidy, G. K. 1977. Phys. Lett. **71B:** 133.
12. Adler, S. & T. Piran. 1982. Phys. Lett. **113B:** 405; **117B:** 91; 1984. Rev. Mod. Phys. **56:** 1.
13. See, for example: Richardson, J. L. 1979. Phys. Lett. **82B:** 272.
14. Baker, M., J. S. Ball & F. Zachariasen. 1985. Phys. Rev. **D31:** 2575.
15. This singularity (indeed, the nonexistence) of Δ_A in axial gauge is supported by lattice calculations. See: DeTar, C., J. King, S. P. Li & L. McLerran. 1984 (August). Fermilab preprint no. PUB-84/79-T.
16. Baker M., J. S. Ball & F. Zachariasen. 1986 (January). Univ. of Utah preprint

Glueball Searches in e^+e^- Annihilation[a]

DAVID HITLIN

Department of Physics
California Institute of Technology
Pasadena, California 91125

INTRODUCTION

There has been widespread interest in finding evidence for hadronic states composed either entirely of gluons or of gluons and quarks (hybrids). Estimates of glueball and hybrid masses are becoming more refined, and there is broad agreement on the expected J^{PC} sequence of these states as calculated using several different models. I will not attempt to review these calculations here because they are treated in the accompanying paper by Meshkov.[1] One comment is, however, in order, as it touches on the central experimental problem in glueball or hybrid searches.

We will discuss below the experimental signatures of glueballs. The most striking such signature would be the isolation of a meson whose quantum numbers are not allowed by the quark model (an "oddball"), such as $J^{PC} = 1^{-+}$. No oddball candidates have been found to date. In fact, all mesons that have been advanced as glueball candidates have a J^{PC} of either 0^{++}, 0^{-+}, or 2^{++}, that is, quantum numbers allowed for quark states.

This greatly complicates the experimental problem because the phenomenological quark-gluon couplings that are responsible for the decay of these states will *ipso facto* produce mixing of quark, glueball, and hybrid states with the same quantum numbers and similar masses. Thus, despite certain estimates of very narrow decay widths for pure glueball states,[2] it seems clear that physical 0^{-+} and 2^{++} candidates are likely to have typical hadronic widths, if only due to mixing with quark states.

Mixing also, of course, tends to blur other unique characteristics of pure glueball or hybrid states. One criterion that remains valid is that such states should be preferentially produced in gluon-rich reactions such as radiative J/ψ or Υ decay or in OZI-forbidden hadronic reactions (although the definition of the latter has been the subject of some controversy). A pure glueball must be an isoscalar flavor singlet, but mixing can certainly undermine the criterion of flavor independent couplings to final state mesons. Similarly, pure glueball states should have small radiative widths and small production in two-photon processes, but mixed states may have larger electromagnetic couplings.

An improved understanding of the ordinary $q\bar{q}$ spectrum, including the properties of radial excitations, is vital to an evaluation of glueball candidates, as are improved mixing models in the pseudoscalar and tensor sector. This review, however, will be restricted to a discussion of the current glueball candidates, emphasizing the experimental facts and pointing out those areas in which improved calculations or experiments are likely to lead to improved understanding.

[a]This work was supported in part by Department of Energy Contract No. DE-AC03-81-ER40050.

48 ANNALS NEW YORK ACADEMY OF SCIENCES

$$\iota(1460)$$

There is quite a lot of information on this state, from radiative J/ψ decay, direct J/ψ decay, two-photon production, and hadronic reactions. Experimental data on the iota are summarized in TABLE 1. There remain a large number of experimental uncertainties, however. It is still not entirely clear whether there are one or more states in the iota region.

The iota is seen most directly in radiative J/ψ decay in the $K\overline{K}\pi$ final state. New measurements from Mark III and DM2, summarized in TABLE 1, clearly establish the mass as higher than originally claimed. The three $K\overline{K}\pi$ modes seen by Mark III have branching fractions in a ratio that is consistent with an isoscalar meson. Analysis of the angular distribution[5,6] clearly indicates that the $\iota(1460)$ has $J^{PC} = 0^{-+}$. Analysis of the Dalitz plot in the $K_s^0 K^{\mp}\pi^{\pm}$ decay mode by the Mark III group also indicates that the decay proceeds through $\delta^{\pm}\pi^{\mp}$; a limit on the K^*K decay mode,

$$\frac{B(\iota \rightarrow K^*K)}{K(\iota \rightarrow KK\pi)} < 0.35 \text{ at } 90\% \text{ C.L.},$$

has been set.

This limit leads to one of the central mysteries of iota decay. Because the decay $\delta \rightarrow \eta\pi$ is well established, it should be possible to observe the decay sequence $J/\psi \rightarrow \gamma\iota$,

TABLE 1. Iota Experimental Summary

Exp.	Mark II[3]	Crystal Ball[4]	Mark III[5]	DM2[6]
Produced $J/\psi(\times 10^6)$	1.3	2.2	2.7 + (3.2)	8.6
$K\overline{K}\pi$ Mode				
$K_s K^{\mp}\pi^{\pm}$	$\surd$		$\surd$	$\surd$
$K^{\pm}K^{\mp}\pi^0$		$\surd$	$\surd \} \Longrightarrow I = 0$	$\surd$
$K_s K_s \pi^0$			$\surd$	
Mass (MeV)	1440^{+10}_{-15}	1440^{+20}_{-15}	$1456 \pm 5 \pm 6$	$1460 \pm 3 \pm 8$
Γ (MeV)	50^{+30}_{-20}	55^{+20}_{-30}	$95 \pm 10 \pm 15$	$100 \pm 12 \pm 15$
J^{PC}		0^{-+}	0^{-+}	Consistent with 0^{-+}
$B(J/\psi \rightarrow \gamma\iota) \times B(\iota \rightarrow K\overline{K}\pi) \times 10^3$	4.3 ± 1.7	$4.0 \pm 0.7 \pm 1.0$	$5.0 \pm 0.3 \pm 0.8$	$3.9 \pm 0.6 \pm 0.9$
$M(\gamma\rho^0)$(MeV)		1390 ± 25	$1420 \pm 15 \pm 20$	1401 ± 18
$\Gamma(\gamma\rho^0)$(MeV)		185^{+110}_{-88}	$133 \pm 55 \pm 30$	174 ± 44
$B(J/\psi \rightarrow \gamma\text{``}\iota\text{''}) \times B(\text{``}\iota\text{''} \rightarrow \gamma\rho^0) \times 10^4$		$1.9 \pm 0.5 \pm 0.4$	$1.0 \pm 0.2 \pm 0.2$	$0.9 \pm 0.2 \pm 0.14$
$\Gamma(\text{``}\iota\text{''} \rightarrow \gamma\rho^0)$(MeV)		3.8 ± 1.7	2.0 ± 0.7	1.8 ± 0.7

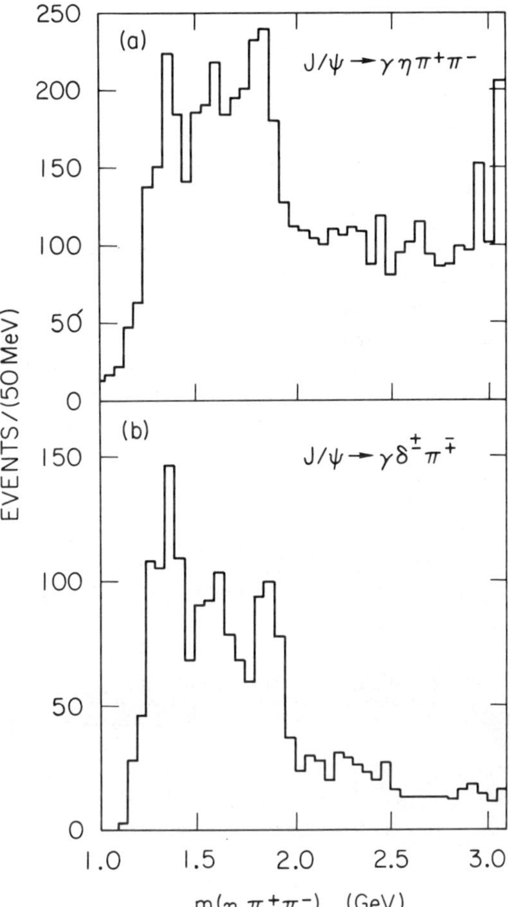

FIGURE 1. (a) Invariant mass of $\eta\pi^+\pi^-$ observed in radiative J/ψ decay by Mark III in a sample of 5.8×10^6 produced J/ψ's; (b) The same sample with a $\delta^{\pm} \rightarrow \eta\pi^{\pm}$ required [$0.93 \leq m(\eta\pi^{\pm}) \leq 1.02$ GeV$/c^2$]. No peak is seen at the mass of $\iota(1460)$.

$\iota \rightarrow \delta\pi$, $\delta \rightarrow \eta\pi$. While the Mark III indeed observes structure in this decay mode (see FIGURE 1), there is no evidence of a peak at the mass of $\iota(1460)$. The limit placed is

$$B(J/\psi \rightarrow \gamma\iota)\,B(\iota \rightarrow \delta^{\pm}\pi^{\mp})\,B(\delta^{\pm} \rightarrow \eta\pi^{\pm}) < 3.9 \times 10^{-4} \text{ at } 90\% \text{ C.L.}$$

This discrepancy has fostered at least three explanations:

(a) Palmer and Pinsky[7] propose an interfering pole model.
(b) Cahn and Landshoff[8] point out that the standard factorization hypothesis for production and decay of a resonance need not apply under certain conditions. More data on $\delta(980)$ production in different processes is clearly indicated.
(c) Frank *et al.*[9] propose that the δ (and the S^*) are 4-quark $K\bar{K}$ molecules and that $K\bar{K}$ final state interactions distort the $\iota(1460)$ Dalitz plot.

The object seen in this region in hadronic reactions in the $K\bar{K}\pi$ final state has had a checkered history.[10] There has been disagreement as to the J^{PC} of this "E/ι", but the mass has consistently been found to be 1420 MeV/c^2. It is not clear, due to the different mass and width observed, that the 0^{-+} object seen in hadronic reactions is identical with that observed in radiative J/ψ decay.

The situation is further complicated by the observation of the $\gamma\rho$ final state in radiative J/ψ decay. Three experiments have now measured this decay mode (see TABLE 1), and all agree that the mass and width of the $\gamma\rho$ state are below 1460 MeV and have a width substantially greater than that seen in the $K\overline{K}\pi$ channel. This could be due to the $\gamma\rho$ decay of the $\eta(1275)$ or other background below 1400 MeV, but it is nonetheless difficult to unambiguously identify the $\gamma\rho$ and $K\overline{K}\pi$ final states with the same parent object. If we do so for the moment, however, and further assume that $K\overline{K}\pi$ is the only other ι decay mode, then we obtain the "ι" $\rightarrow \gamma\rho$ decay widths shown in TABLE 1, using the $K\overline{K}\pi$ width of 100 MeV. These are large values for an unmixed pure glueball and could be taken to indicate the presence of $q\overline{q}$ in the iota wave function.

Another indication of charged constituents in a meson could be the observation of production in two-photon reactions. TABLE 2 summarizes recent limits on iota production in $\gamma\gamma$ reactions in both $K\overline{K}\pi$ and $\gamma\rho^0$ final states. On the assumption that the η' and ι are $SU(3)$ singlets, a simple vector dominance prediction for $\Gamma_{\iota\gamma\gamma}$ is approximately 10 keV. The experimental limits are substantially below this value.

Still another way to investigate the structure of the iota is to search for direct production in J/ψ decays. The Mark III has searched for iota production in association with ω and ϕ mesons. In the final state

$$\omega K^{\pm}K^0_s\pi^{\mp} \\ \quad\longrightarrow \pi^+\pi^-\pi^0,$$

there is indeed evidence for structure in the iota region, with a cut of ± 30 MeV around the ω mass for $\pi^+\pi^-\pi^0$. The recoiling $K\overline{K}\pi$ spectrum is shown in FIGURE 2a. A fit to a Breit-Wigner resonance yields the result,

$$M = 1449 \pm 8 \text{ MeV}, \qquad \Gamma = 44\,^{+29}_{-16} \text{ MeV},$$

with a product branching ratio of

$$B[J/\psi \rightarrow \omega X(1450)]\,B[X(1450) \rightarrow K\overline{K}\pi] = (7.5 \pm 2.5 \pm 2.0) \times 10^{-4},$$

assuming $I_X = 0$. As yet, there has been no J^P determination for this structure. It is tempting to identify this structure with $\iota(1460)$, but note that the width is smaller than that observed in the $K\overline{K}\pi$ final state.

TABLE 2. Limits on $\gamma\gamma$ Production of $\iota(1460)$

Final State	Limit	Experiment
$\Gamma_{\iota\gamma\gamma} \cdot B(\iota \rightarrow K\overline{K}\pi)$	<2.2 keV at 95% C.L.	TASSO[11]
	<2.0 keV at 90% C.L.	Mark II[12]
	<1.6 keV at 95% C.L.	PEP-9[13]
$\Gamma_{\iota\gamma\gamma} \cdot B(\iota \rightarrow \gamma\rho^0)$	<1.5 keV at 95% C.L.	TASSO[14]
	<0.2 keV at 90% C.L.	Mark II[12]

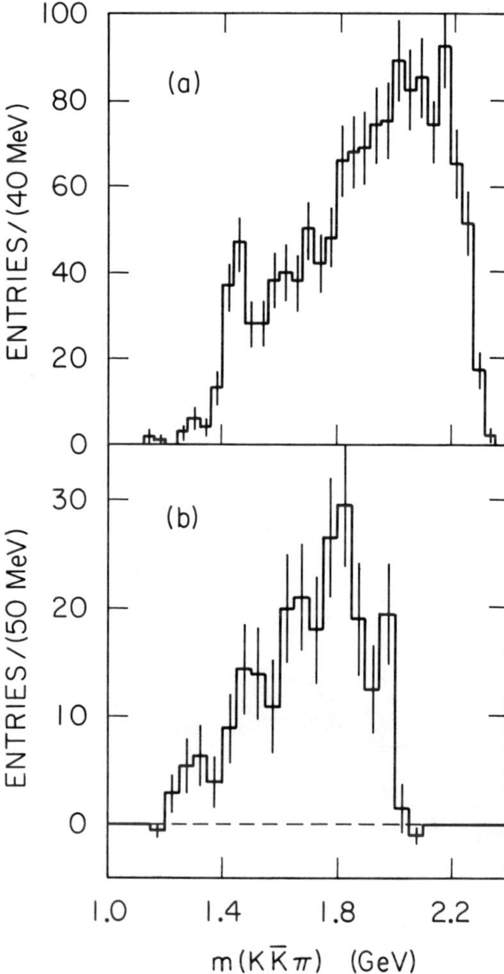

FIGURE 2. (a) Invariant mass of $K^\pm K^0_s \pi^\mp$ seen in $J/\psi \to \omega K^\pm K^0_s \pi^\mp$; (b) Invariant mass of $K\overline{K}\pi$ seen in $J/\psi \to \varphi K\overline{K}\pi$.

The final $K\overline{K}\pi$ state recoiling against the ϕ (into the K^+K^- channel) as observed by Mark III is also shown in FIGURE 2b. No structure is observed in this channel, leading to the limit,

$$B(J/\psi \to \phi\iota)B(\iota \to K\overline{K}\pi) < 2.3 \times 10^{-4} \text{ at } 90\% \text{ C.L.}$$

This investigation ties in nicely to another Mark III study of direct J/ψ decays to pseudoscalar-vector (PV) final states.[15] The ensemble of such decays may be described in terms of two strong interaction amplitudes, g and h, with relative phase, ϕ, and an electromagnetic amplitude, e. Following Rosner,[16] we may define the wave functions of the η and η' mesons in terms of their light and strange quark components and a possible third component, which we will call $|G\rangle$, suggesting gluon, although it could represent

TABLE 3. $J/\psi \rightarrow$ PV Amplitudes

Decay Mode	Amplitude
$\rho^+\pi^-$; $\rho^0\pi^0$; $\rho^-\pi^+$	$g + e$
$K^{*+}K^-$; $K^{*-}K^+$	$g - h + e \cdot (2 - x)$
$K^{*0}\bar{K}^0$; $\bar{K}^{*0}K^0$	$g - h - 2e \cdot \left[\dfrac{1 + x}{2}\right]$
$\omega\eta$	$(g + e)X_\eta$
$\omega\eta'$	$(g + e)X_{\eta'}$
$\phi\eta$	$(g - 2h - 2e \cdot x)Y_\eta$
$\phi\eta'$	$(g - 2h - 2e \cdot x)Y_{\eta'}$
$\rho^0\eta$	$3eX_\eta$
$\rho^0\eta'$	$3eX_{\eta'}$
$\omega\pi^0$	$3e$
$\phi\pi^0$	0

an admixture of non–ground state $q\bar{q}$:

$$\eta = X_\eta\left|\frac{u\bar{u} + d\bar{d}}{\sqrt{2}}\right\rangle + Y_\eta|s\bar{s}\rangle + Z_\eta|G\rangle,$$

$$\eta = X_{\eta'}\left|\frac{u\bar{u} + d\bar{d}}{\sqrt{2}}\right\rangle + Y_{\eta'}|s\bar{s}\rangle + Z_{\eta'}|G\rangle,$$

with normalization $X_\eta^2 + Y_\eta^2 + Z_\eta^2 = X_{\eta'}^2 + Y_{\eta'}^2 + Z_{\eta'}^2 = 1$.

TABLE 4. $J/\psi \rightarrow$ PV Branching Ratios[a]

Decay Mode	Final State	Branching Ratio (in units of 10^{-3})
$\rho\pi$	$\pi^0\pi^+\pi^-$	$13.3 \pm 0.3 \pm 1.5$
$K^{*+}\bar{K}^-$ + c.c.	$K^\pm\pi^\mp K_s$	$5.0 \pm 0.25 \pm 0.6$
	$K^+K^-\pi^0$	$6.0 \pm 0.3 \pm 0.7$
$K^{*0}\bar{K}^0$ + c.c.	$K^\pm\pi^\mp K_s$	$3.9 \pm 0.2 \pm 0.6$
$\omega\eta$	$\pi^0\pi^+\pi^-\eta$	$1.9 \pm 0.2 \pm 0.3$
$\omega\eta'$	$\pi^0\pi^+\pi^-\gamma\rho^0$	$0.39 \pm 0.11 \pm 0.06$
	$\pi^0\pi^+\pi^-\eta\pi^+\pi^-$	$0.43^{+0.19}_{-0.22} \pm 0.07$
$\phi\eta$	K^+K^- neutrals	$0.69 \pm 0.07 \pm 0.08$
	$K^+K^-\eta$	$0.64 \pm 0.15 \pm 0.08$
	$K^+K^-\pi^0\pi^+\pi^-$	$0.61 \pm 0.14 \pm 0.08$
$\phi\eta'$	$K^+K^-\gamma\rho^0$	$0.39 \pm 0.10 \pm 0.06$
	$K^+K^-\eta\pi^+\pi^-$	$0.365 \pm 0.04 \pm 0.05$
$\omega\pi^0$	$\pi^0\pi^+\pi^-\pi^0$	$0.67 \pm 0.06 \pm 0.11$
$\rho^0\eta$	$\pi^+\pi^-\eta$	$0.18 \pm 0.02 \pm 0.04$
$\pi^0\eta'$	$\pi^+\pi^-\eta\pi^+\pi^-$	<0.1
$\phi\pi^0$	$K^+K^-\pi^0$	<0.013

[a]Unless specified, the π^0 and η in the final state are identified by their $\gamma\gamma$ decay, and the K_s by its $\pi^+\pi^-$ decay mode. The first error is statistical and the second one is systematic. The upper limits are given at the 90% confidence level.

TABLE 5. Results of the Global Fit

$\|g\| = 1.17 \pm 0.06$	$\|X_\eta\| = 0.63 \pm 0.06$
$\|h\| = 0.24 \pm 0.07$	$\|Y_\eta\| = 0.83 \pm 0.13$
$\|e\| = 0.15 \pm 0.01$	$\|X_{\eta'}\| = 0.36 \pm 0.05$
$\phi = 1.19 \pm 0.20$	$\|Y_{\eta'}\| = 0.72 \pm 0.12$

The various PV decay amplitudes are then shown in TABLE 3. The Mark III has measured a large number of PV final states, which are shown in TABLE 4. It is thus possible to solve for $\|g\|, \|h\|, \phi, \|e\|, \|X_\eta\|, \|Y_\eta\|, \|X_{\eta'}\|$, and $\|Y_{\eta'}\|$, as shown in TABLE 5.

The result is interesting in that $\|X_\eta\|^2 + \|Y_\eta\|^2 = 1.1 \pm 0.2$, indicating that the η meson is a mixture of light and strange quark (as is conventionally assumed), while $\|X_{\eta'}\|^2 + \|Y_{\eta'}\|^2 = 0.65 \pm 0.18$, indicating that there may be a mixture of nonquark amplitude in the η'. Such an admixture would likely arise from mixing with nearby pseudoscalars, such as the $\iota(1460)$, which would therefore have a major gluonic component. The $\omega"\iota"$ and $\phi\iota$ measurements may be interpreted in a similar way, yielding

$$|X_\iota| = \frac{(0.65 \pm 0.20)}{\sqrt{B(\iota \to K\overline{K}\pi)}}, \qquad |Y_\iota| < \frac{0.6}{\sqrt{B(\iota \to K\overline{K}\pi)}}.$$

Thus, it appears that the pseudoscalar meson sector observed in radiative J/ψ decay may be strongly mixed, with some states being composed predominantly of quarks and others predominantly of glue.

$\theta(1700)$

The experimental information on another glueball candidate, $\theta(1700)$, is summarized in TABLE 6. While Mark III makes an unambiguous J^{PC} measurement of 2^{++} for the θ, DM2 claims that both spin 0 and 2 are possible. I will point out only two interesting points relative to the θ. The first is that the helicity amplitudes determined

TABLE 6. $\theta(1700)$ Experimental Summary

Mode	Mass (MeV/c^2)	Γ (MeV/c^2)	J^{PC}	$B(J/\psi \to \gamma\theta)B(\theta \to \text{Mode})$ ($\times 10^{-4}$)	Exp.
$\eta\eta$	1640 ± 50	220^{+100}_{-70}	2^{++}	$4.9 \pm 1.4 \pm 1.0$	CB[17]
	$[f', \theta$ not resolved]				
K^+K^-	1700 ± 30	156 ± 20	2^{++}	$6.0 \pm 0.9 \pm 2.5$	Mark II[18]
	1720 ± 10	130 ± 20	2^{++}	$4.8 \pm 0.6 \pm 0.9$	Mark III[10]
	1707 ± 10	166 ± 33	$[0^{++}, 2^{++}]$	$4.6 \pm 0.7 \pm 0.7$	DM2[6]
$K^0_s K^0_s$	1720 (Fixed)	130 (Fixed)		$4.5 \pm 1.2 \pm 1.1$	Mark III
	1711 ± 9	173 ± 22		$3.1 \pm 0.7 \pm 0.6$	DM2
$\pi^+\pi^-$	1713 ± 15	130 (Fixed)		$2.4 \pm 0.5 \pm 0.4$	Mark III
	1688 ± 9	225 ± 13		$1.8 \pm 0.2 \pm 0.3$	DM2

by Mark III (see TABLE 7) indicate that the production process in J/ψ radiative decay is quite different for the θ than it is for the well-established 2^+ f and f' mesons. The second is that Mark III and DM2 have now observed the decay $\theta \rightarrow \pi^+\pi^-$. This observation would naively tend to restore flavor symmetry in θ decay, strengthening a glueball interpretation, but when phase-space differences are considered, the result is still rather far from flavor symmetry. Mark III and DM2 have also observed a higher mass $\pi^+\pi^-$ resonance, with

$$M_X = 2086 \pm 15 \text{ MeV}, \Gamma_X = 210 \pm 63 \text{ MeV},$$
$$\text{and } B(J/\psi \rightarrow \gamma X)B(X \rightarrow \pi^+\pi^-) = (4.5 \pm 0.8 \pm 1.0) \times 10^{-4} \quad \text{(Mark III)}$$

and

$$M_X = 2025 \pm 21 \text{ MeV}, \Gamma_x = 312 \pm 30 \text{ MeV},$$
$$\text{and } B(J/\psi \rightarrow \gamma X)B(X \rightarrow \pi^+\pi^-) = (2.1 \pm 0.3 \pm 0.4) \times 10^{-4} \quad \text{(DM2)},$$

that may be consistent with the $H(2030)$ isoscalar, $J^{PC} = 4^{++}$ meson.

TABLE 7. Production Characteristics of f, f', θ in Radiative J/ψ Decay

State	$x \equiv \dfrac{A_1}{A_0}$	ϕ_x	$y \equiv \dfrac{A_2}{A_0}$	ϕ_y	Exp.
$f(1270)$	0.96 ± 0.12	-0.5 ± 0.7	0.06 ± 0.13	-0.4 ± 1.9	Mark III
$f'(1515)$	0.63 ± 0.10	~ 0	$0.17 + - + 0.20$	~ 0	Mark III
	1.08 ± 0.10		0.19 ± 0.11		DM2
$\theta(1700)$	-1.07 ± 0.20	0.6 ± 0.8	-1.09 ± 0.25	-0.1 ± 0.5	Mark III
	-1.3 ± 0.14	0 ± 1.4	$-1.1 + - 60.18$	1.3 ± 0.3	DM2
	0.87 ± 0.20	$\equiv 0$	-0.64 ± 0.39	$\equiv 0$	CB
					$[f', \theta$ not resolved]

DM2 has also observed the θ in the $J/\psi \rightarrow \phi K_s^0 K_s^0$ reaction, with

$$B(J/\psi \rightarrow \gamma\theta)B(\theta \rightarrow K\overline{K}) = (3.6 \pm 0.7 \pm 0.7) \times 10^{-4}.$$

The θ has not been seen in $\gamma\gamma$ reactions, with $\Gamma_{\theta\gamma\gamma} \cdot B(\theta \rightarrow K\overline{K}) < 0.28$ keV at 95% confidence level.[11]

$\xi(2230)$

In addition to the f' and θ structures seen in K^+K^- and $K_s^0K_s^0$ final states in radiative J/ψ decay by Mark III and DM2, a few years ago,[19-21] Mark III reported a surprisingly narrow structure, called ξ, at a mass of 2.2 GeV. DM2, with a larger sample of J/ψ's, has not observed the ξ, thereby placing limits on ξ production (assuming spin 0) of

$$B(J/\psi \rightarrow \gamma\xi)B(\xi \rightarrow K^+K^-) < 1.2 \times 10^{-5} \text{ at 95% C.L.}$$

FIGURE 3. Invariant mass distribution of K^+K^- in the Mark III sample of $J/\psi \rightarrow \gamma K^+K^-$ decays: (a) First sample of 2.7×10^6 produced J/ψ's; (b) Recent sample of 3.1×10^6 additionally produced J/ψ's.

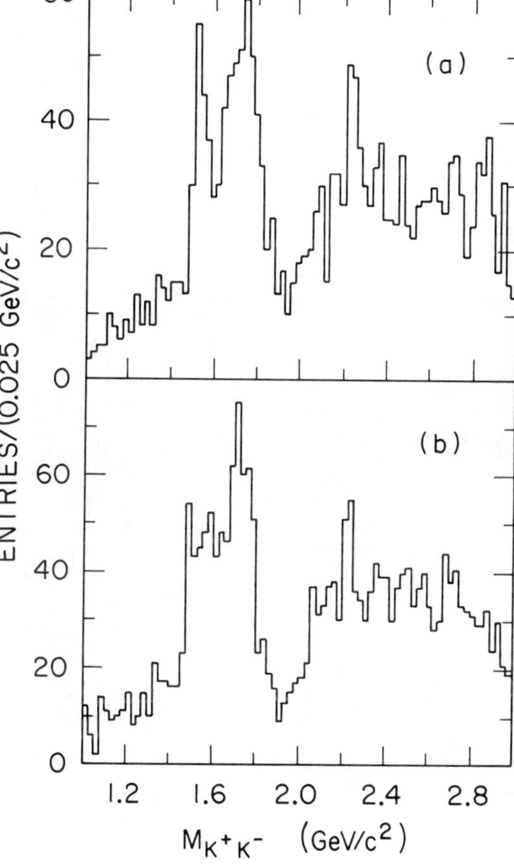

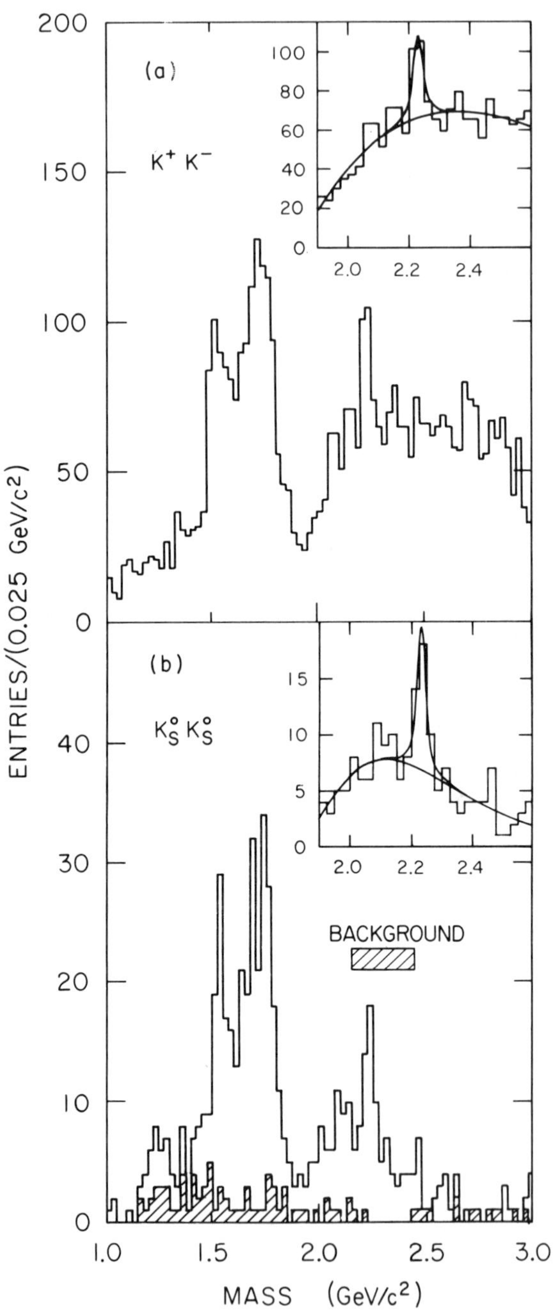

FIGURE 4. Invariant mass distributions of $\bar{K}K$ in the total Mark III $\gamma\bar{K}K$ data sample: (a) K^+K^- final state; (b) $K^0_s K^0_s$ final state. The insets show fits to the 1.9 to 2.6 GeV/c^2 mass region of a polynomial background plus a Breit-Wigner folded with a Gaussian resolution function.

and

$$B(J/\psi \rightarrow \gamma\xi) B(\xi \rightarrow K_s^0 K_s^0) < 2.0 \times 10^{-5} \text{ at } 95\% \text{ C.L.}$$

Mark III has now more than doubled its J/ψ data sample and once again sees the ξ signal.[22] FIGURE 3 shows the K^+K^- mass spectrum in old and new data, while FIGURE 4 shows the K^+K^- and $K_s^0 K_s^0$ mass spectra derived from the total data sample. The parameters derived from a fit to the spectrum with a 10-MeV Gaussian resolution

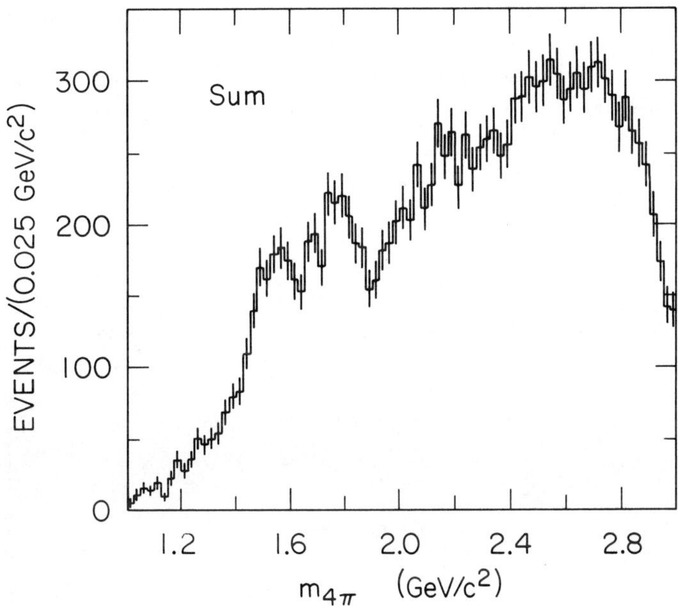

FIGURE 5. Invariant mass of 4π in $J/\psi \rightarrow \gamma 4\pi$ decay. The histogram is the sum of distributions for $J/\psi \rightarrow \gamma\pi^+\pi^-\pi^+\pi^-$ and $J/\psi \rightarrow \gamma\pi^+\pi^0\pi^-\pi^0$.

folded with a Breit-Wigner and a polynomial background are:

K^+K^-: $M(\xi) = 2230 \pm 6 \pm 14$ MeV

         $\Gamma(\xi) = 26^{+20}_{-16} \pm 17$ MeV

         $B(J/\psi \rightarrow \gamma\xi) B(\xi \rightarrow K^+K^-) = (4.2^{+1.7}_{-1.4} \pm 0.8) \times 10^{-5},$

$K_s^0 K_s^0$: $M(\xi) = 2232 \pm 7 \pm 7$ MeV

         $\Gamma(\xi) = 18^{+23}_{-15} \pm 10$ MeV

         $B(J/\psi \rightarrow \gamma\xi) B(\xi \rightarrow K_s^0 K_s^0) = (3.2^{+1.6}_{-1.3} \pm 0.7) \times 10^{-5}.$

From a maximum likelihood analysis, the significance of the signals is found to be

4.5 SD in K^+K^- and 3.6 SD in $K^0_s K^0_s$. As yet, no spin-parity determination has been made. The discrepancy between Mark III and DM2 results remains unresolved.

$\rho\rho$ AND $\omega\omega$ FINAL STATES IN RADIATIVE J/ψ DECAY

The 4π final state in radiative J/ψ decay was observed some years ago by Mark II[23] to have structure in the $\rho\rho$ channel. As this structure is broad and is centered above 1600 MeV, it is interesting to inquire whether it is related to the $\theta(1700)$. To answer this question, Mark III has subjected both $\gamma\pi^+\pi^-\pi^+\pi^-$ and $\gamma\pi^+\pi^0\pi^-\pi^0$ samples to a full 10-channel spin-parity analysis after subtraction of background due to the five-pion direct decay.[24]

FIGURE 5 shows the 4π invariant mass distribution, with structure evident in the 1.6 to 1.8 GeV/c^2 region. The total branching fraction to $\gamma 4\pi$ is quite large. For $m_{4\pi} < 2$ GeV/c^2, the branching fractions are

$$B(J/\psi \rightarrow \gamma\pi^+\pi^-\pi^+\pi^-) = (3.05 \pm 0.8 \pm 0.45) \times 10^{-3}$$

and

$$B(J/\psi \rightarrow \gamma\pi^+\pi^0\pi^-\pi^0) = (8.3 \pm 0.2 \pm 3.1) \times 10^{-3};$$

these ratios are consistent with an isoscalar 4π state. When these spectra are subjected

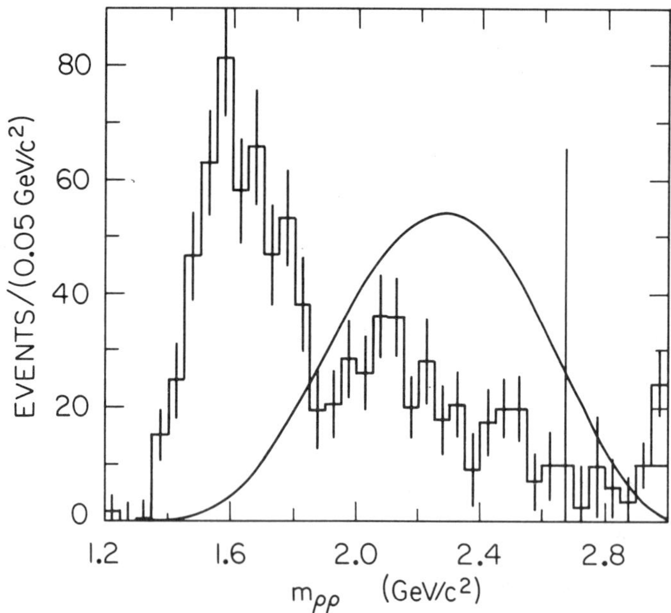

FIGURE 6. Invariant mass distribution for the 0^- $\rho\rho$ component in radiative $J/\psi \rightarrow \gamma 4\pi$ extracted from the 10-channel spin-parity analysis. The solid line is the expected p wave phase-space contribution.

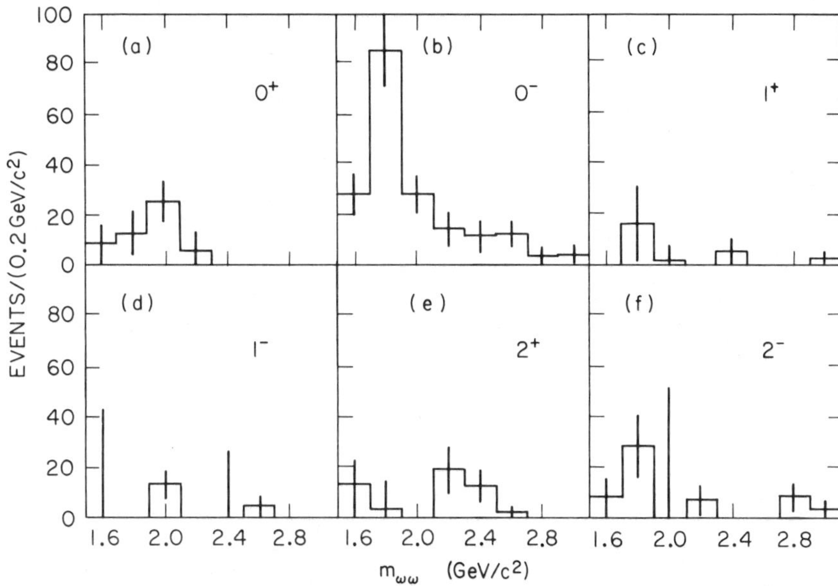

FIGURE 7. Results of the spin-parity analysis of $J/\psi \rightarrow \gamma\omega\omega$. Six spin-parity channels are shown; $(85 \pm 15)\%$ of the $\omega\omega$ signal below 2 GeV/c^2 is 0^-.

to the spin-parity analysis, the region below 2 GeV/c^2 is found to be dominated by $\rho\rho$ with $J^{PC} = 0^{-+}$. FIGURE 6 shows the 0^- $\rho\rho$ spectrum. The structure at 1.6 GeV/c^2 is thus predominantly pseudoscalar and is not, therefore, to be identified with the θ. The higher mass region also shows no evidence of a 2^+ channel, allowing a limit to be placed on radiative decay of the J/ψ to the g_T states, on the assumption that the g_T's, seen in $\phi\phi$, would also decay to $\rho\rho$. The branching ratios and limits discussed above are

$$B(J/\psi \rightarrow \gamma X_{0^-})B(X^0 \rightarrow \rho\rho) = (4.7 \pm 0.3 \pm 0.9) \times 10^{-3},$$

$$B(J/\psi \rightarrow \gamma\theta)B(\theta \rightarrow \rho\rho) < 5.5 \times 10^{-4} \text{ at } 90\% \text{ C.L.,}$$

and

$$B(J/\psi \rightarrow \gamma g_T)B(g_T \rightarrow \rho\rho) < 6.0 \times 10^{-4} \text{ at } 90\% \text{ C.L.,}$$

where the g_T mass region is defined to be between 2.1 and 2.4 GeV/c^2.

The $\gamma\omega\omega$ channel has also been subjected to a spin-parity analysis, and is found to be predominantly pseudoscalar as well (see FIGURE 7).

CONCLUSION

Recent experimental results on structures seen in radiative J/ψ decays have been reviewed. There are a large number of prominent structures observed in the 0^{-+} and 2^{++} channels. Evidence is mounting to support the contention that mixing between quark states and glueballs or hybrids is responsible for the properties of these states.

REFERENCES

1. MESHKOV, S. 1987. These proceedings.
2. ROBSON, D. 1977. Nucl. Phys. **B130:** 328.
3. SCHARRE, D. L. *et al.* 1980. Phys. Lett. **B97:** 329.
4. EDWARDS, C. *et al.* 1982. Phys. Rev. Lett. **49:** 259.
5. RICHMAN, J. D. 1985. Ph.D. thesis, California Institute of Technology. Unpublished.
6. AUGUSTIN, J. E. *et al.* 1985. LAL/85-27.
7. PALMER, W. & S. PINSKY. 1983. Phys. Rev. **D27:** 2219.
8. CAHN, R. & P. V. LANDSHOFF. 1985. CERN-TH.419/85.
9. FRANK, M., N. ISGUR, P. J. O'DONNELL & J. WEINSTEIN. 1985. UTPT-85-06.
10. For a recent review, see: CHUNG, S. U. 1987. These proceedings.
11. ALTHOFF, M. *et al.* 1985. DESY no. 85-093.
12. GIDAL, G. 1985. Proceedings of the 16th Symposium on Multiparticle Dynamics, Kiryat-Anavim, Israel.
13. AIHARA, H. *et al.* 1986. UCSB-HEP-86-2.
14. BINNIE, D. M. *et al.* 1979. Phys. Lett. **83B:** 141.
15. BALTRUSAITIS, R. M. *et al.* 1985. Phys. Rev. **D32:** 2883.
16. ROSNER, J. L. 1983. Phys. Rev. **D27:** 1101.
17. EDWARDS, C. *et al.* 1982. Phys. Rev. Lett. **49:** 458.
18. FRANKLIN, M. E. B. 1982. SLAC-PUB-254.
19. EINSWEILER, K. Ph.D. thesis, Stanford University. Unpublished.
20. TOKI, W. 1983. SLAC-PUB-3262.
21. HITLIN, D. 1984. Proceedings of the 1983 International Symposium on Lepton and Photon Interactions, Ithaca, New York.
22. BALTRUSAITIS, R. M. *et al.* 1985. Phys. Rev. Lett. **56:** 107.
23. BURKE, D. L. *et al.* 1982. Phys. Rev. Lett. **49:** 632.
24. BALTRUSAITIS, R. M. *et al.* 1985. SLAC-PUB-3682. Submitted to Phys. Rev. D.
25. BALTRUSAITIS, R. M. *et al.* 1985. Phys. Rev. Lett. **55:** 1723.

Hadronic Production
of $J^{PC} = 2^{++}$ Glueballs[a]

S. J. LINDENBAUM

Physics Department
Brookhaven National Laboratory
Upton, New York 11973
and
Physics Department
City College of New York
New York, New York 10031

INTRODUCTION

Color confinement and the running coupling constant make the existence of multigluon resonances or glueballs[1a,2] inescapable in QCD.[1a-c] Lattice gauge calculations[3,4] have quantitatively demonstrated this. Thus, in spite of the fact that perturbative QCD has had many successes (which include quantitative ones to ~10–20%), the glueball missing link must be found if the theory is to survive.

BNL/CCNY use an OZI[5-8] suppressed channel with variable mass, namely, the reaction $\pi^- p \rightarrow \phi\phi n$ as a filter that allows resonating gluons or glueballs to pass, while strongly rejecting conventional quark-built hadronic states.[b] The breakdown of the OZI suppression signals a glueball.

THE OZI RULE

In the u, d, s quark system, $q\bar{q}$ meson nonets are well established for $J^{PC} = 0^{-+}, 1^{--}$, 2^{++}, and 3^{--}. Except for the 0^{-+} nonet, all those with $J \geq 1$ are nearly ideally mixed[9] and representable by Zweig's Quark Line Diagrams.

In disconnected diagrams, the $s\bar{s}$ pair in the ϕ or f' has to be annihilated or created by at least two or three hard gluons, respectively, to conserve all quantum numbers including color. Asymptotic freedom strongly decouples hard glue from quarks at relatively moderate gluon energies;[c] this is observed to occur at energies as low as those involved in the three-gluon decay of the ϕ, and thus is referred to as "precocious". The resultant relatively weak coupling constants of the hard gluons naturally explain the observed OZI suppression factors of ~100 for both ϕ and f' decay and production, and the even larger OZI suppression (~1,000) in the decay of the J/ψ and Υ.

[a]This research was supported by the U.S. Department of Energy under Contract Nos. DE-AC02-76CH00016 (BNL) and DE-AC02-83ER40107 (CCNY).

[b]Provided $\phi\phi$ system $J \geq 1$ so that vacuum mixing is negligible; otherwise, this vacuum mixing could possibly lead to apparent violations of Zweig suppression because it could lead to large departures from ideal mixing.

[c]$\Lambda_{\text{eff}} \leq 100$–200 MeV.

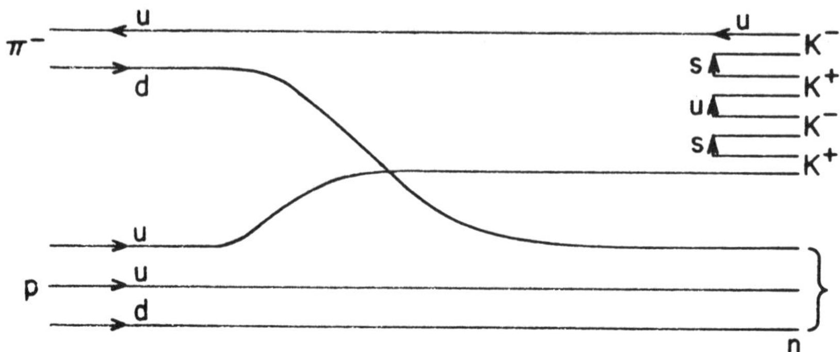

FIGURE 1. The Zweig Quark Line Diagram for the reaction $\pi^- p \to K^+ K^- K^+ K^- n$, which is connected and OZI allowed.

In $q\bar{q}$ meson states, departures from ideal mixing can only be expected to occur when flavor-changing diagrams (which convert $s\bar{s}$ quark pairs into $u\bar{u}$ or $d\bar{d}$ quark pairs, or vice versa) have gluons relatively strongly coupled.

One such flavor-mixing mechanism is vacuum effects,[10] which are expected to be important for $J = 0$ (e.g., the $J^{PC} = 0^{-+}$ nonet), but which are unimportant for $J \geq 1$; this is certainly consistent with the experimental results. In QCD, the only other known basic flavor-mixing mechanism is the presence of glueballs with the same quantum numbers near enough to the nonet singlet masses and with the appropriate width to effectively mix with the singlets.

It is a well-known experimental fact that in all Zweig disconnected diagrams in the ϕ, f', J/ψ, and Υ systems, the OZI rule[11] appears universal.[8,12] As previously discussed, the OZI rule appears on paper to be defeatable by two-step processes, of which each is OZI allowed. However, experimentally, there is no evidence for this occurring, and reasons why it should probably not occur have been given.[13,14]

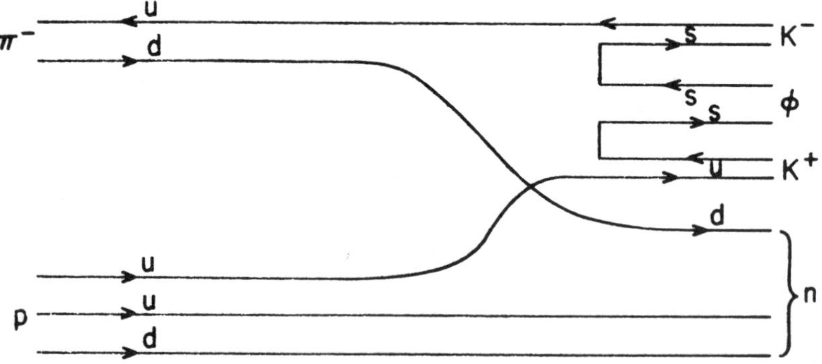

FIGURE 2. The Zweig Quark Line Diagram for the reaction $\pi^- p \to \phi K^+ K^- n$, which is connected and OZI allowed.

FIGURES 1–3 are the Zweig Quark Line Diagrams for the three reactions studied by the BNL/CCNY group in three generations of experiments since 1978.[5,6,15] In FIGURE 4, which shows a scatterplot of K^+K^- masses from the BNL/CCNY experiment, we see the general approximately uniform background from the reaction (a) $\pi^-p \rightarrow K^+K^-K^+K^-n$, which is OZI allowed, and the two ϕ bands representing (b) $\pi^-p \rightarrow \phi K^+K^-n$, which is also OZI allowed. Where the two ϕ bands cross, we have the Zweig forbidden reaction $\pi^-p \rightarrow \phi\phi n$. The black spot shows an obviously more or less complete breakdown of the Zweig suppression.

The black spot is approximately 1,000 times the density of reaction (a) and approximately 50 times the density of reaction (b). If one projects out the ϕ bands (even with rather wide cuts $\pm$ 14 MeV), there is a huge $\phi\phi$ signal compared to the $\approx 13\%$ background from reaction (b). The recoil neutron signal is also very clean ($\approx 97\%$ neutron).

FIGURE 5 shows the acceptance corrected $\phi\phi$ mass spectrum in the ten mass bins that were used for the final partial wave analysis of the 3,652 events.[15]

All waves (i.e., 52) with $J = 0$–4, $L = 0$–3, $S = 0$–2, $P = \pm$, and η (exchange naturality) = $\pm$ were allowed in the partial wave analysis (PWA). The Gottfried-Jackson frame angles [β(polar) and γ(azimuthal)], the polar angles (θ_1, θ_2) of the K^+ decay in the $\phi_{1,2}$ rest systems relative to the ϕ direction, and the corresponding azimuthal angles (α_1 and α_2) were used to specify an event.

In the PWA, we find a unique solution (FIGURES 6a and 6b) that consists of only three $J^{PC} = 2^{++}$ waves, an S-wave with $S = 2$, a D-wave with $S = 2$, and a D-wave with $S = 0$, which is a good fit. All three waves have $J_Z = 0$ in the Gottfried-Jackson frame and the exchange naturality = (-1), which are the characteristics of pion exchange. The observed $(d\sigma/dt')_{\phi\phi} = e^{(9.4\pm0.7)t'}$ for $t' < 0.3$, which is the low t'-region that contains most of the data. The only charged particle exchange that will give this is pion exchange. The best two-wave fit is $\approx 30\ \sigma$ away. Our selected three-wave fit is $\approx 15\ \sigma$ better than the next best three-wave fit.

The few percent background was estimated to be entirely (within errors) composed of the reaction $\pi^-p \rightarrow \phi K^+K^-n$. A partial wave analysis of this background in the region where the K^+K^- mass was slightly larger than that of the ϕ revealed that $\approx 65\%$ of it was structureless and incoherent, which is what one would expect from the addition of many possible partial waves. Only approximately 7% of the background was 2^{++}, which had an ≈ 0 amplitude in the threshold region and which peaked at ≈ 2.4 GeV (see FIGURE 7). Approximately 28% of the background was 1^{--}, which are the expected quantum numbers for a ϕK^+K^- system where all particles are in an S-wave with respect to each other and the production is via π-exchange (i.e., a kinematic effect). Thus, the ϕK^+K^- background reaction was totally different than the 2^{++} states observed in the $\phi\phi$ system.

The partial wave amplitudes and phase behavior of the $\phi\phi$ system (shown in FIGURES 6a and 6b) clearly suggest that these three waves are produced by resonances. A two-pole K-matrix fit that allows all three observed waves to mix in each pole was rejected by $\approx 15\ \sigma$ fit.

A three-pole K-matrix fit gave a good fit. The Argand diagram for this fit is shown in FIGURE 8 and exhibits the phase motion characteristics typical of resonance pole behavior. A four-pole K-matrix fit did not lead to any further improvement. The

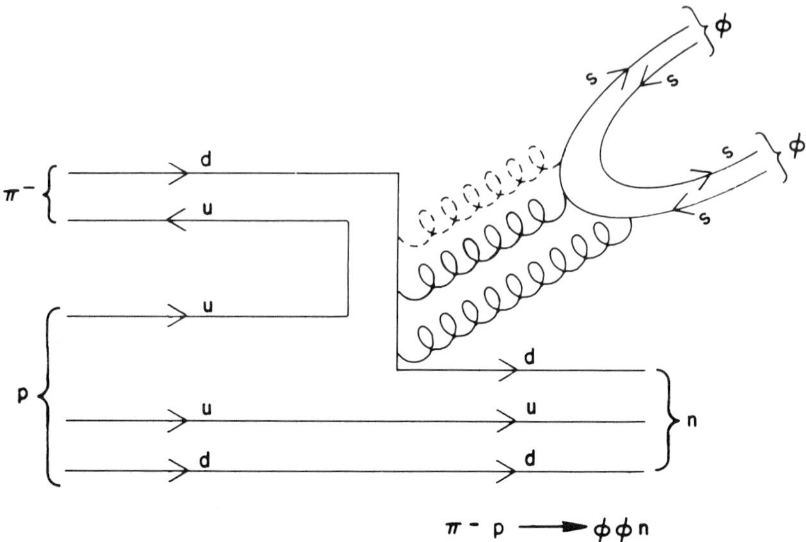

FIGURE 3a. The Zweig Quark Line Diagram for the reaction $\pi^- p \to \phi\phi n$, which is disconnected (i.e., a double hairpin diagram) and is OZI forbidden. Two or three gluons are shown connecting the disconnected parts of the diagram depending upon the quantum numbers of the $\phi\phi$ system. For the g_T's, $J^{PC} = 2^{++}$ and only two gluons are required. From the data analysis, they come from the annihilation of the incident π^- and a π^+ exchanged between the lower and the upper parts of the diagram.

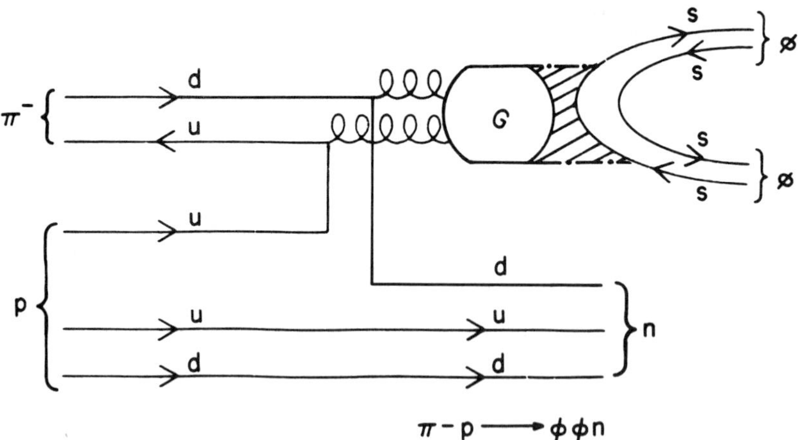

FIGURE 3b. The $J^{PC} = 2^{++}$ glueball intermediate state in $\pi^- p \to \phi\phi n$. The region of crosshatched lines within the dash-dot lines indicates that we do not know the details of the glueball hadronization into $\phi\phi$.

three-pole K-matrix fit was used to fit the data contained in 90 angular variable bins for each of the ten mass bins used, thus yielding a total of 900 data bins. The fit was good to $\lesssim 1\ \sigma$. The resonance parameters of this fit are given in TABLE 1.

Due to the small mostly incoherent background, the S-wave that dominates the $g_T(2050)$ must be used as a phase reference. The phase difference of this and the other

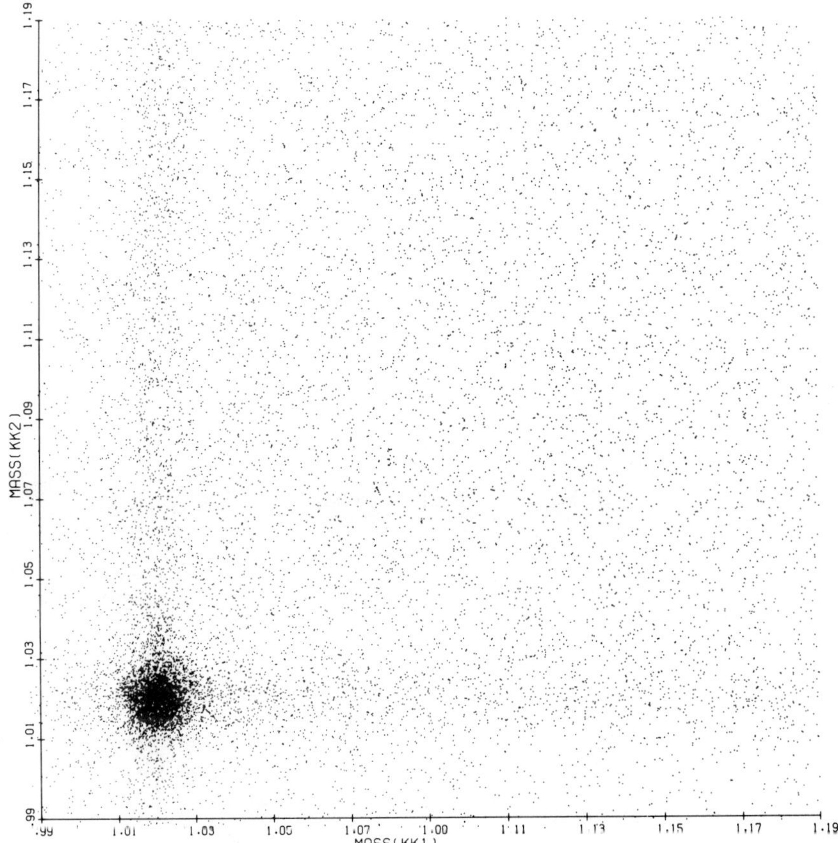

FIGURE 4. Scatterplot of K^+K^- effective mass for each pair of K^+K^- masses. Clear bands of $\phi(1020)$ are seen with an enormous enhancement (black spot) where they overlap (i.e., $\phi\phi$). This overlap essentially represents a complete breakdown of OZI suppression.

two D-waves precisely matches the three-pole K-matrix fit, thus clearly demonstrating that all three states have the pole behavior that is the best and only critical definition of a resonance.

In addition, attributing the production of these states to 1–3 primary $J^{PC} = 2^{++}$ glueballs has explained all their features in a clear-cut and simple manner.[7,8,13,14]

GLUEBALL MASSES

The constituent (i.e., the gluon has effective mass) gluon models[16] would predict three low-lying $J^{PC} = 2^{++}$ glueballs. The mass estimates from the lattice gauge groups cover the range ≈ 1.7–2.5 GeV for $J^{PC} = 2^{++}$ glueballs,[3,4,17] with which we are clearly consistent.

GLUEBALL WIDTH

The hadronization process (i.e., the creation of one or more $q\bar{q}$ pairs) must occur near the outer region of confinement involving strongly interacting soft glue (including collective interactions) if we are to have resonances decay with typical hadronic widths

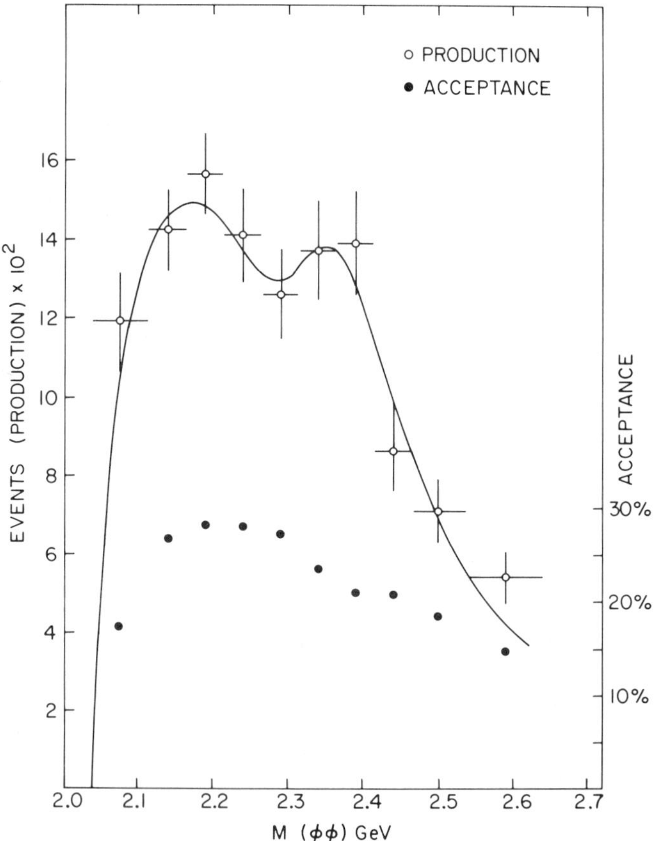

FIGURE 5. The $\phi\phi$ mass spectrum corrected for acceptance. The solid line is the fit to the data with the three resonant states to be described later. The points at the bottom of the diagram are the acceptance for each mass bin to be read with the scale at the right.

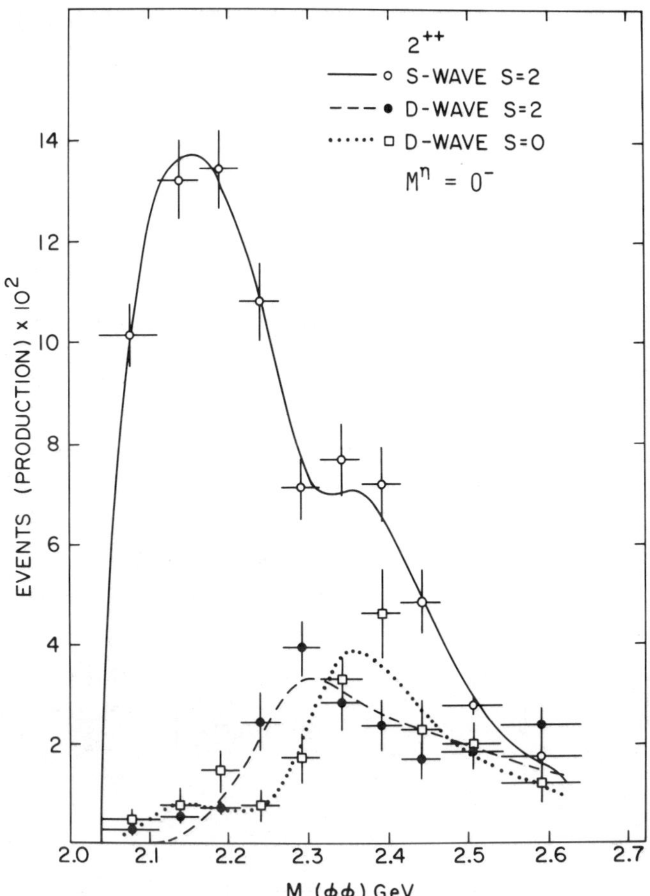

FIGURE 6a. The three $J^{PC} = 2^{++}$ partial waves at production in 50-MeV mass bins (except at ends). $J_Z = 0$ in the Gottfried-Jackson frame and the exchange naturality is $(-)$, corresponding to pion exchange for all three waves. The smooth curves are derived from a three-pole K-matrix fit.

(Γ hadronic ~ 100 to several hundred MeV) that are more or less independent of the number of particles in the dominant decay mode as observed.

The gluon-gluon coupling is stronger than the quark-gluon coupling, and thus it would be expected (via gluon splittings before the final hadronization) to have a similar hadronization process to a $q\bar{q}$ hadron and to have typical hadronic widths for nonexotic J^{PC} states. For exotic J^{PC} states, this argument may not hold because no one yet knows what suppresses the unobserved exotic sector. Therefore, Meshkov's oddballs[16] may be narrow.

One can ask why we have not seen a Meshkov oddball (i.e., exotic J^{PC}). Approximately 90% of our $\phi\phi$ data is characteristic of being produced by pion exchange. Exotic J^{PC} cannot be produced by pion exchange. Our data exhibits great selectivity for $M = 0$. However, for A_1 exchange (for example), one can create an exotic J^{PC}. Estimating the ratio of A_1 exchange to π exchange would lead to a guesstimate of a few percent or less. Thus, we need much more statistics and we may eventually find evidence for a 1^{-+} or 3^{-+} oddball.

SUMMARY OF CHARACTERISTICS OF $\pi^- p \rightarrow \phi\phi n$

The observed characteristics of the reaction $\pi^- p \rightarrow \phi\phi n$ are very unusual and striking in the following respects:

(1) The expected OZI suppression is completely broken in a very unusual manner because virtually all parts of the $\pi^- p \rightarrow \phi\phi n$ cross section are composed of three resonant $\phi\phi$ states, the $g_T(2050)$, the $g_{T'}(2300)$, and the $g_{T''}(2350)$, all

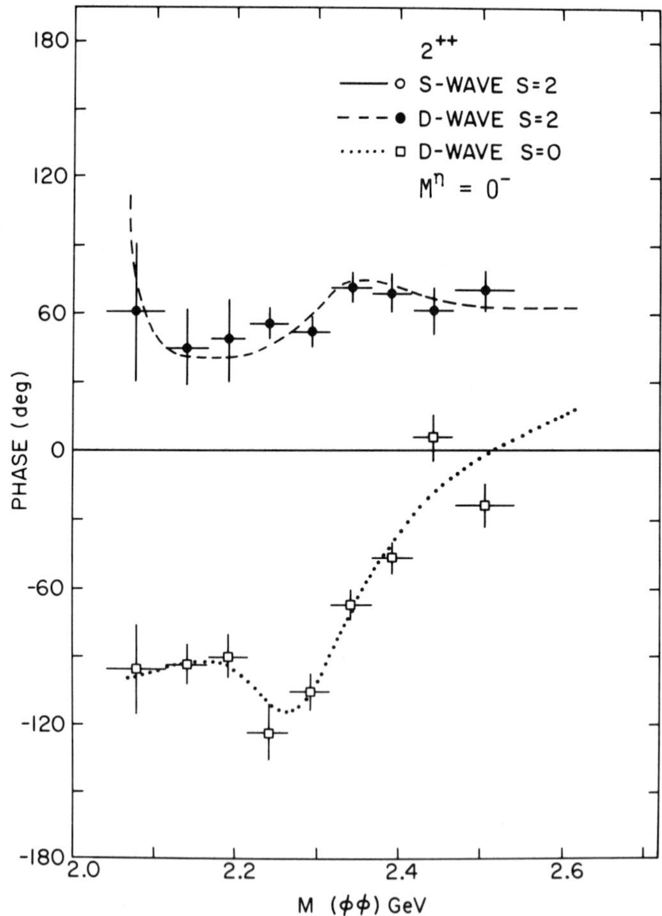

FIGURE 6b. *D-S* phase difference from the partial wave analysis versus $\phi\phi$ mass. The smooth curves are derived from a three-pole *K*-matrix fit.

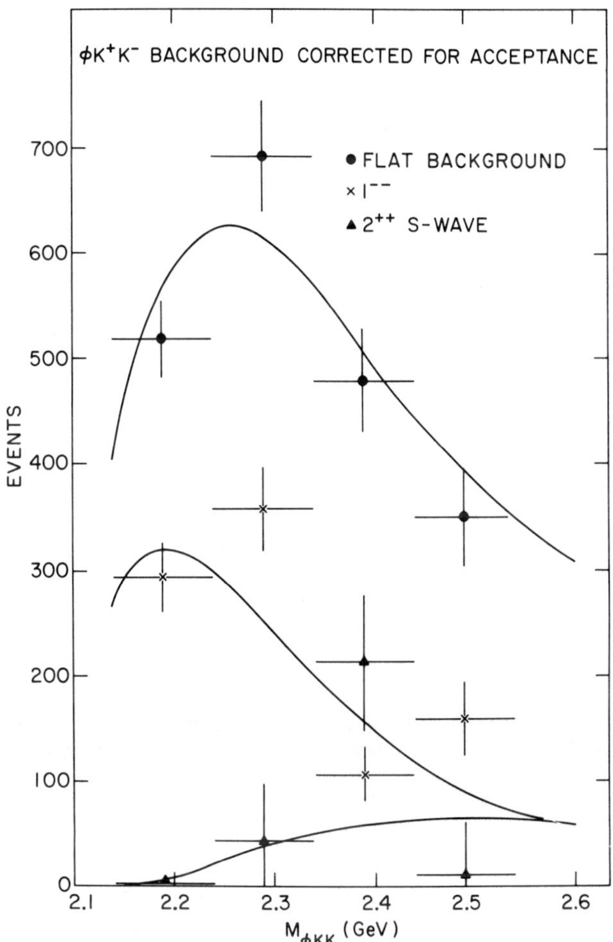

FIGURE 7. The partial wave intensities in the background reaction $\pi^- p \rightarrow \phi K^+ K^- n$.

with the same quantum numbers $I^G J^{PC} = 0^+ 2^{++}$. In contrast, hadronic reactions in other channels (e.g., $\pi^- p \rightarrow K_s^0 K_s^0 n$)[18,19] show no such selective quantum numbers or resonance phenomena.

(2) The OZI allowed background reaction $\pi^- p \rightarrow \phi K^+ K^- n$, which is unexpectedly only a few percent of the OZI forbidden $\pi^- p \rightarrow \phi\phi n$, consists mainly of a structureless, incoherent background that is flat in all angular distributions. Approximately 28% of the reaction has $J^{PC} = 1^{--}$, which are expected quantum numbers for a $\phi K^+ K^-$ system where all particles are in a relative S-wave (i.e., a kinematic effect). Only approximately 7% of the cross section is nonresonant 2^{++} S-wave.[d] Thus, it is entirely different than the $\phi\phi$ 2^{++}

---

[d] The $K^+ K^-$ system and the ϕ are in a relative S-wave. The $K^+ K^-$ pair are in a relative P-wave.

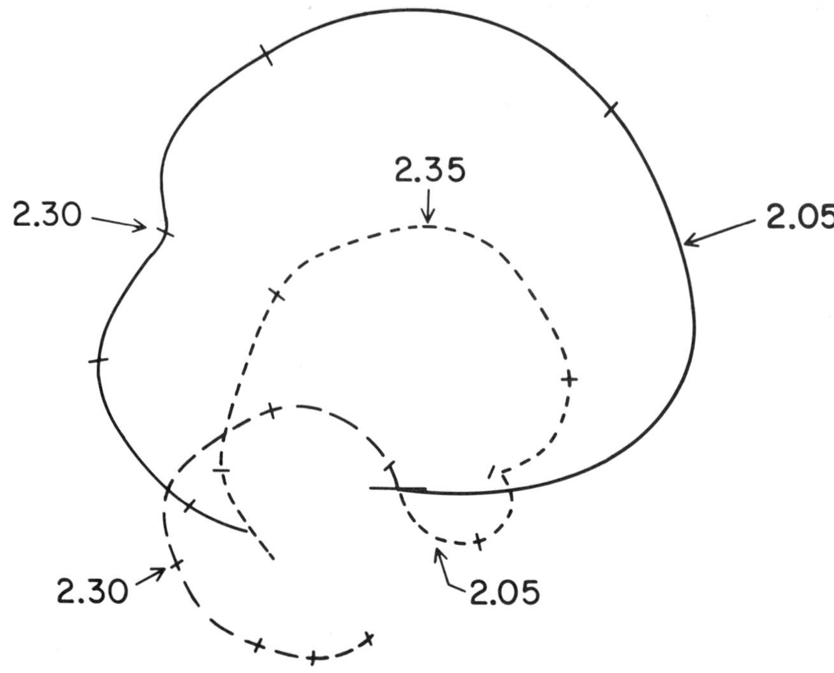

FIGURE 8. Argand plot from K-matrix fit. Solid line is S-wave, $S=2$; (———) is D-wave, $S=2$; (---) is D-wave, $S=0$.

TABLE 1. Breit-Wigner Resonance Parameters of the Three-Pole K-Matrix Fit[a]

$$M_1 = 2.050^{+0.090}_{-0.050} \quad \Gamma_1 = 0.200^{+0.160}_{-0.050} \approx 50\%^{+10\%}_{-10\%} \text{ data}$$

$$S\text{-wave, } S = 2 \approx 98\%^{+2\%}_{-70\%} \text{ coupling sign } (+) \text{ defined}$$

$$D\text{-wave, } S = 2 \approx 0\% + 50\% \text{ coupling sign } (-)$$

$$D\text{-wave, } S = 0 \approx 2\%^{+25\%}_{-2\%} \text{ coupling sign } (-)$$

$$M_2 = 2.300^{+0.020}_{-0.100} \quad \Gamma_2 = 0.200^{+0.060}_{-0.050} \approx 20\%^{+20\%}_{-10\%} \text{ data}$$

$$S\text{-wave, } S = 2 \approx 30\%^{+20\%}_{-20\%} \text{ coupling sign } (+)$$

$$D\text{-wave, } S = 2 \approx 50\%^{+20\%}_{-10\%} \text{ coupling sign } (+)$$

$$D\text{-wave, } S = 0 \approx 20\%^{+20\%}_{-20\%} \text{ coupling sign } (+)$$

$$M_3 = 2.350^{+0.020}_{-0.030} \quad \Gamma_3 = 0.270^{+0.090}_{-0.130} \approx 30\%^{+10\%}_{-20\%} \text{ data}$$

$$S\text{-wave, } S = 2 \approx 40\%^{+10\%}_{-20\%} \text{ coupling sign } (+)$$

$$D\text{-wave, } S = 2 \approx 5\%^{+15\%}_{-5\%} \text{ coupling sign } (-)$$

$$D\text{-wave, } S = 0 \approx 55\%^{+20\%}_{-15\%} \text{ coupling sign } (+)$$

[a]Masses and Γ are in GeV. $I^G J^{PC} = 0^+ 2^{++}$ for all three resonances.

amplitudes. This reaction has about the same threshold and similar kinematics to the $\pi^- p \rightarrow \phi\phi n$ reaction; therefore, threshold effects would be quite similar in the two reactions, and it is clear that the striking characteristics of the $\phi\phi$ data cannot be attributed to such a naive mechanism.

If one assumes that 1–3 primary glueballs with $J^{PC} = 2^{++}$ produce these states,[7,8,13,14] then the above characteristics of the data can be very well explained naturally within the context of QCD. One or two primary $J^{PC} = 2^{++}$ glueballs could break the OZI suppression and mix with nearby $q\bar{q}$ states with the same quantum numbers and similar masses. However, the simplest explanation is that we have a triplet of $J^{PC} = 2^{++}$ glueballs. These would be in the right mass range predicted from lattice gauge calculations[3,4,17] and would fit[14] the prediction of three distinct masses made by T. D. Lee.[20]

ALTERNATIVE EXPLANATIONS AND CRITICISMS

Recent differences[21] regarding the degree of OZI forbiddenness of the reaction $\pi^- p \rightarrow \phi\phi n$ observed by BNL/CCNY have been resolved.[22] It was concluded that these resonances would be OZI forbidden if they were of the $q\bar{q}$ type, and they thus constituted strong evidence for glueball(s).

In recent papers,[14,23] we have clearly shown that alternative explanations by Gomm,[24] Karl *et al.*,[25] and Donoghue[26] are incorrect and do not fit the data. Therefore, the glueball resonance hypothesis is the only published explanation of the data that is viable and it naturally arises in the context of QCD.

WHY HAVE THE g_T's NOT BEEN SEEN IN OTHER CHANNELS?

The MK III results observe $J/\psi \rightarrow \gamma\phi\phi$.[27] Their detection efficiency for $\phi\phi$ is very low in the mass region of the $g_T(2050)$, $g_{T'}(2300)$, and $g_{T''}(2350)$. Thus, they find only ~ 10 events in this mass region. However, if one corrects their $\phi\phi$ mass spectrum for the detection efficiency, it is not inconsistent with the shape of the mass spectrum seen by BNL/CCNY. One should note, though, that we are comparing $\approx 4{,}000$ observed events to ~ 10. It appears that the MK III can only observe strong-signal, narrow, high-mass $\phi\phi$ states such as the decay of the η_c and, thus, it is not likely to be able to observe the BNL/CCNY states. David Hitlin may have some newer results.[27b]

At the Bari conference, the DM2 group[28] has reported approximately 50 $\gamma\phi\phi$ events in the mass region of the BNL/CCNY experiment. At present, due to the limited statistics, they are unable to say whether this signal is related to the resonant structures (i.e., the g_T, $g_{T'}$, $g_{T''}$) observed by BNL/CCNY.

It should be noted that in a related experiment, $\pi^- \mathrm{Be} \rightarrow \phi\phi$ inclusive,[29] the data are found to be consistent with the $g_{T'}$ and $g_{T''}$, and two Breit-Wigner resonances are needed to explain the results. The acceptance discriminates against the g_T. This reaction would only be expected to be partially Zweig suppressed. If a $K\bar{K}$ or $K(^\Lambda_\Sigma)$ pair were created, the Zweig suppression would not apply. Booth *et al.* have recently studied[30] the case of central production of $\phi\phi$ in the exclusive reactions:

$$A\mathrm{p} \rightarrow A K^+ K^- K^+ K^- \mathrm{p},$$
$$A\mathrm{p} \rightarrow A \phi K^+ K^- \mathrm{p},$$
$$A\mathrm{p} \rightarrow A \phi\phi \mathrm{p} \ (\text{where } A = \pi^+ \text{ or p}).$$

Their results demonstrate that the $\phi\phi$ reaction does not exhibit the expected OZI suppression.

The other hadronic production experiments involve OZI allowed channels; therefore, one would expect the g_T's to be submerged in the many other hadronic states one could expect. Thus, their detection would likely require very large statistics, and even then it might be quite difficult to separate them from the other hadronic states.

FIGURE 9 shows the results of the analysis of a 23-GeV/c $\pi^-p \rightarrow K_s^0 K_s^0 n$ experiment.[18] In the mass region of the g_T's, the $J^{PC} = 2^{++}$ amplitude behavior is smooth and structureless and shows no phase motion. Furthermore, the 0^{++} and 4^{++} states are also populated, unlike the selection of only $J^{PC} = 2^{++}$ in $\pi^-p \rightarrow \phi\phi n$. This experiment has had its statistics raised by a factor of ~ 3 recently[19] and the results are the same. This is what I would expect when the effects of the OZI suppression filter action are eliminated, as they are in this reaction.

Incidentally, in this $\pi^-p \rightarrow K_s^0 K_s^0 n$ experiment, there is no evidence for the $\theta(1700)$.[27,31] In the analysis, we conclude that the coupling of $\theta \rightarrow \pi\pi$ is consistent with the OZI suppression expected if the θ is an $s\bar{s}$ pair.

As to why the g_T's have not yet been seen in the radiative J/ψ decay, I would suggest the following: We argue that the Zweig suppression in our channel (with a pure glue intermediate state) should filter out other hadronic states and give a highly enriched sample of glueballs. What we found in the data is certainly consistent— namely, we find three new states with the same quantum numbers and nothing else accompanied by no background (within errors).

In the J/ψ radiative decay, approximately 90% of the observed states are known conventional ones, and thus it is an inefficient filter for glueballs. As I discussed in reference 14, if glueballs were strongly coupled in the radiative decay of the J/ψ, then the perturbatively calculated decay ratio $J/\psi \rightarrow \gamma gg/ggg$ would be expected to be much larger than the observed rate,[32] whereas, experimentally, they agree. Furthermore, the width of the J/ψ would be expected to be broadened.

The evidence for the glueball candidates $\iota(1420)$ and $\theta(1700)$ are reviewed here by Hitlin[27b] and Meshkov,[33] and the reader is referred to their papers.

In regard to the g_T's, it is also worth noting that Chanowitz and Sharpe[34] have concluded that strange quarks may well be favored in glueball decay and, in particular, in the $\phi\phi$ S-wave.[e]

Furthermore, I would like to point out that except for color, the quantum numbers of a gluon and a ϕ are the same. Thus, one can imagine that gluons would like to go into ϕ mesons just as photons like to go into vector mesons (i.e., similar to VDM). Of course, the color must be changed to a singlet, but such color rearrangements might perhaps be accomplished by soft gluon exchanges in the final hadronization. Thus, this may also be another reason why the g_T's, if they are glueballs, are only seen in the $\phi\phi$ decay mode. If sufficient statistics are gathered in $J/\psi \rightarrow \gamma\phi\phi$, some evidence for the g_T, $g_{T'}$, and $g_{T''}$ states may be seen.

---

[e]They also have meikton states that break the OZI suppression and that are possibly associated with our states as well as glueballs. However, this argument depends on bag calculations, and the dynamical mechanisms are not clear.

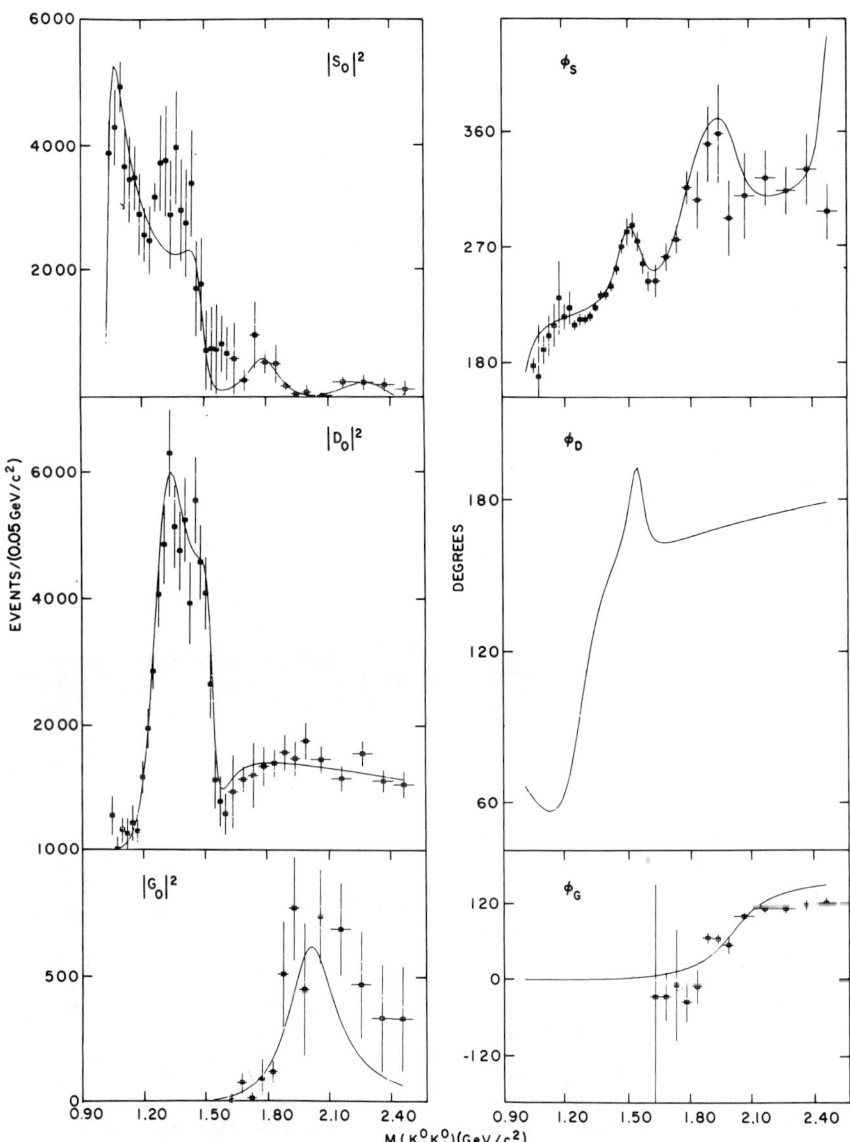

FIGURE 9. The square of the moduli of the S_0, D_0, and G_0 amplitudes together with their absolute phases from the best fit as functions of $K_s^0 K_s^0$ effective mass for $t' < 0.1$ $(\text{GeV}/c)^2$. The solid curves are the results of our preferred mass-dependent fit in the same t' interval (reference 31).

CONCLUSIONS ON THE STATUS OF THE g_T's AS GLUEBALL STATES

One can prove the g_T, $g_{T'}$, and $g_{T''}$ are glueballs with the appropriate input axioms. Then, as we concluded previously, the $g_T(2050)$, $g_{T'}(2300)$ and $g_{T''}(2350)$ are produced by 1–3 primary $J^{PC} = 2^{++}$ glueballs if you assume as input axioms:

(1) QCD is correct;
(2) The OZI rule is universal for weakly coupled glue in disconnected Zweig diagrams, where the disconnection is due to the creation or annihilation of new flavor(s) of quark(s), and $J \geq 1$ for the disconnected system (to avoid possible vacuum mixing effects).

We have previously stated that the BNL/CCNY $g_T(2050)$, $g_{T'}(2300)$, and $g_{T''}(2500)$ are naturally explained within the context of QCD by concluding that they are produced by 1–3 primary glueballs. One or two broad primary glueballs could in principle break down the OZI suppression and mix with one or two quark states that accidentally have the same quantum numbers and have close to the same mass. However, the simplest explanation of the rather unusual characteristics of our data is that we have found a triplet of $J^{PC} = 2^{++}$ glueball states.

Alternatives to the glueball resonance explanation have been discussed earlier and have been found to be incorrect or do not fit the data, or both.[14,23]

ACKNOWLEDGMENTS

The author wishes to thank Sidney Meshkov for presenting a summary of this paper at the conference when the author was not able to do so due to temporary illness. I also wish to thank Loyal Durand and Martin Block for their cooperation.

REFERENCES

1. (a) FRITZCH, H. & M. GELL-MANN. 1972. XVIth Int. Conf. on High Energy Physics, Chicago-Batavia. Vol. 2, p. 135. National Accelerator Laboratory. Batavia, Illinois; FRITZCH, H., M. GELL-MANN & H. LEUTWYLER. 1973. Phys. Lett. **47B:** 365; (b) WEINBERG, S. 1973. Phys. Rev. Lett. **31:** 49; Phys. Rev. **D8:** 4482; (c) GROSS, D. J. & F. WILCZEK. 1973. Phys. Rev. **D8:** 3633.
2. FRITZCH, H. & P. MINKOWSKI. 1975. Nuovo Cimento **30A:** 393; FREUND, R. P. & Y. NAMBU. 1975. Phys. Rev. Lett. **34:** 1645; JAFFE, R. & K. JOHNSON. 1976. Phys. Lett. **60B:** 201; KOGUT, SINCLAIR & SUSSKIND. 1975. Nucl. Phys. **B114:** 199; ROBSON, D. 1977. Nucl. Phys. **B130:** 328; BJORKEN, J. SLAC Pub. no. 2372.
3. MICHAEL, C. & I. TEASDALE. 1985 (March). Glueballs from asymmetric lattices. Liverpool Univ. preprint no. LTH-127; DE FORCRAND, PH., G. SCHIERHOLZ, H. SCHNEIDER & M. TEPER. 1985. Phys. Lett. **152B:** 107; OTTO, S. W. & P. STOLORZ. 1985. Phys. Lett. **151B:** 428; SCHEIRHOLZ, G. & M. TEPER. 1984. Phys. Lett. **136B:** 64; BERG, B. 1984 (February). The spectrum in lattice gauge theories. DESY preprint no. 84-012.
4. BERG, B. & A. BILLOIRE. 1983. Nucl. Phys. **B221:** 109; **B226:** 405; TEPER, M. Proc. 1983 Int. Europhysics Conf. on High Energy Physics, Brighton, U.K., July 20–27, 1983. J. Guy & C. Costain, Eds. Rutherford Appleton Laboratory; 1983. Preprint no. LAPP-TH-91.
5. ETKIN, A. *et al.* 1978. Phys. Rev. Lett. **40:** 422–425; **41:** 784–787.
6. LINDENBAUM, S. J. *et al.* A new higher statistics study of $\pi^-p \rightarrow \phi\phi n$ and evidence for glueballs. Proc. 21st Int. Conf. on High Energy Physics, Paris, France, 26–31 July 1982. Journal de Physique **43.** P. Petiau & M. Porneuf, Eds.: C3-87–C3-88. Editions Physique.

Les Ulis, France; ETKIN, A. *et al*. 1982. The reaction $\pi^- p \to \phi\phi n$ and evidence for glueballs. Phys. Rev. Lett. **49**: 1620–1623.

7. LINDENBAUM, S. J. 1983. Hadronic production of glueballs. Proc. 1983 Int. Europhysics Conf. on High Energy Physics, Brighton, U.K., July 20–27, 1983. J. Guy & C. Costain, Eds.: 351–360. Rutherford Appleton Laboratory.

8. LINDENBAUM, S. J. 1984. Production of glueballs. Comments Nucl. Part. Phys. **13** (no. 6): 285–311.

9. PARTICLE DATA GROUP. 1984 (April). Review of particle properties. Rev. Mod. Phys. **56** (no. 2): part II.

10. NOVIKOV, V. A. *et al*. 1981. Nucl. Phys. **B191**: 301.

11. OKUBO, S. 1963. Phys. Lett. **5**: 165; 1977. Phys. Rev. **D16**: 2336; ZWEIG, G. 1964. CERN report nos. TH.401 and TH.412; IIZUBA, J. 1966. Prog. Theor. Phys. Suppl. **37–38**: 21; IIZUBA, J., K. OKUDA & O. SHITO. 1966. Prog. Theor. Phys. **35**: 1061.

12. LINDENBAUM, S. J. Hadronic physics of $q\bar{q}$ light quark mesons, quark molecules and glueballs. Proc. of the XVIII Course of the Int. School of Subnuclear Physics, July 31–August 11, 1980, Erice, Trapani, Italy. Subnuclear Series, vol. 18, "High Energy Limit". A. Zichichi, Ed.: 509–562. Plenum. New York; 1981. Nuovo Cimento **65A**: 222–238.

13. LINDENBAUM, S. J. 1985. Status of glueballs (invited lecture). How Far Are We From the Gauge Forces (Proc. of the 21st Course of the International School of Subnuclear Physics on "How Far We Are From the Electronuclear Interactions and the Other Gauge Forces", Erice, Trapani-Sicily, 3–14 August 1983). A. Zichichi, Ed.: 631–672. Plenum. New York.

14. LINDENBAUM. S. J. 1986. Invited lecture. Proc. of the Summer Workshop in High Energy Physics and Cosmology, International Centre for Theoretical Physics, Trieste, Italy, 7–14 July 1985. The ICTP Series in Theoretical Physics–Vol. 2. Superstrings, Supergravity, and Unified Theories. World Scientific. Singapore.

15. ETKIN, A. *et al*. 1985. Phys. Lett. **165B**: 217–221.

16. MESHKOV, S. Proc. of the Seventh Int. Conf. on Experimental Meson Spectroscopy, April 14–16, 1983, Brookhaven National Laboratory. S. J. Lindenbaum, Ed: 125–156. AIP Conf. Proc. no. 113. Amer. Inst. Phys. New York; FISHBANE, P.M. & S. MESHKOV. 1984. Comments Nucl. Part. Phys. **13**: 325.

17. See reference 16 for a fuller discussion of this topic.

18. ETKIN, A. *et al*. 1982. Amplitude analysis of the $K_s^0 K_s^0$ system produced in the reaction $\pi^- p \to K_s^0 K_s^0 n$ at 23 GeV/c. Phys. Rev. **D25**: 1786–1802.

19. BRANDEIS/BNL/CCNY/DUKE/NOTRE DAME COLLABORATION. Private communication.

20. LEE, T.D. Time as a dynamical variable. Columbia Univ. preprint no. CU-TP-266; Talk at Shelter Island II Conf., June 2, 1983; Proc of the 21st Course of the International School of Subnuclear Physics on "How Far We Are From the Electronuclear Interactions and the Other Gauge Forces", Erice, Trapani-Sicily, 3–14 August 1983. To be published.

21. LIPKIN, H. J. 1983. Phys. Lett. **124B**: 509; 1984. Nucl. Phys. **B224**: 147; LINDENBAUM, S. J. 1983. Phys. Lett. **131B**: 221; see also reference 13.

22. LINDENBAUM, S. J. & H. J. LIPKIN. 1984. Phys. Lett. **149B**: 407.

23. LINDENBAUM, S. J. & R. S. LONGACRE. 1985. The glueball resonance and alternative explanations of the reaction $\pi^- p \to \phi\phi n$. Hadron Spectroscopy—1985 (International Conference, Univ. of Maryland). S. Oneda, Ed.: 51–66. AIP Conf. Proc. no. 132. Amer. Inst. Phys. New York; also in: 1985. Phys. Lett. **165B**: 202–204.

24. GOMM, H. 1984. Phys. Rev. **D30**: 1120.

25. KARL, G., W. ROBERTS & N. ZAGURY. 1984. Phys. Lett. **149B**: 403; KARL, G. *In* Hadron Spectroscopy—1985 (International Conference, Univ. of Maryland). S. Oneda, Ed.: 73. AIP Conf. Proc. no. 132. Amer. Inst. Phys. New York.

26. DONOGHUE, J. The status of unusual meson candidates. Proc. of the Yukon Advanced Study Institute: The Quark Structure of Matter, August 11–27, 1984. To be published; Univ. of Massachusetts preprint no. UMHEP-209; 1985. Theory summary. Hadron Spectroscopy—1985 (International Conference, Univ. of Maryland). S. Oneda, Ed.: 460. AIP Conf. Proc. no. 132. Amer. Inst. Phys. New York.

27. (a) EINSWEILER, K. Proc. 1983 Int. Europhysics Conf. on High Energy Physics, Brighton, U. K., July 20–27, 1983. J. Guy & C. Costain, Eds.: 348–350. Rutherford Appleton

Laboratory; (b) HITLIN, D. Radiative decays and glueball searches. Proc. of the 1983 Int. Symposium on Lepton and Photon Interactions at High Energies, Cornell University, August 4–9, 1983. David G. Cassel & David L. Kreinick, Eds.: 746–778; (c) HEUSCH, C. Proc. 22nd Course of the Int. School of Subnuclear Physics or "Quarks, Leptons, and Their Constituents", Erice, Trapani-Sicily, Italy, 5–15 August 1984. To be published.

28. JEAN MARIE, B. DM2 results on hadronic and radiative J/ψ decays. Proc of the Int. Europhysics Conf. on High Energy Physics, Bari, Italy, 18–24 July 1985. To be published; 1985. Preprint no. LAL 85–27. This paper contains other results relevant to the iota, θ. See also: AUGUSTIN *et al.* 1985. Preprint no. LAL 85–27.

29. BOOTH *et al.* A high statistics study of the $\phi\phi$ mass spectrum. To be published in Z. Phys.

30. ARMSTRONG, T. A. *et al.* 1986. Phys. Lett. **166B:** 245–248.

31. EDWARDS, C. *et al.* 1982. Phys. Rev. Lett. **48:** 458.

32. ABRAMS, G. S. *et al.* 1980. Phys. Rev. Lett. **44:** 114.

33. MESHKOV, S. 1987. Glueball and meson spectroscopy. These proceedings.

34. CHANOWITZ, M. & S. SHARPE. 1983. Phys. Lett. **132B:** 413.

A Survey of E/Iota Meson in Hadroproduction[a]

S. U. CHUNG

Brookhaven National Laboratory
Upton, New York 11973

The $E(1420)$ meson has been around with us since 1963, when its observation was first reported by R. Armenteros et al.[1] With the discovery of the iota(1460) in the same decay channel from J/ψ radiative decays[2] in 1980, there has been a resurgence of interest in this meson because it may turn out to be a pseudoscalar glueball. This review covers the E/iota states seen in hadroproduction, while an update of the iota meson from J/ψ radiative decays is covered by D. Hitlin in this volume. The present review represents an updated version of the previous reviews given by the author at Lund,[3] Leipzig,[4] and Santander.[5] For a report on the status as of summer 1985, see the rapporteur talk by S. Cooper given in Bari.[6]

P. Baillon et al.[7] carried out the first systematic study on the properties of $E(1420) \rightarrow K\bar{K}\pi$ as seen in $\bar{p}p$ annihilations at rest into $E\pi\pi$ (this represents the completed work of a preliminary report given earlier by R. Armenteros et al.). They determined that $I(E) = 0$ from a study of various charge combinations of $K\bar{K}\pi$, and from a symmetry under $K \leftrightarrow \bar{K}$, they determined that $G(E) = +1$ and hence $C(E) = +1$. In addition, they found that E couples to both $\delta(980)\pi$ and $K^*(890)\bar{K}$ (and its charge conjugate). In a fit to a series of one-dimensional distributions, they found that $J^{PC}(E) = 0^{-+}$ is favored, with $J^{PC}(E) = 1^{++}$ as a poor second. In fact, they have concluded in the paper that J^{PC} is 0^{-+} and that E couples to both $\delta(980)\pi$ and $K^*(890)\bar{K}$. In a later maximum-likelihood analysis by P. Baillon on the same data,[8] he confirmed their original conclusion that $J^{PC}(E) = 0^{-+}$. The difference in the logarithm of the likelihood ($\Delta\ln L$) is 68 in favor of 0^{-+} over 1^{++}.

In 1980, C. Dionisi et al.[9] published results of their Dalitz-plot analysis on the E events from $\pi^- p \rightarrow En$ at 3.95 GeV/c. They found that $I^G(E) = 0^+$ as before, but that the E appeared as a $J^{PC} = 1^{++}$ state coupling mainly to $K^*(980)\bar{K}$, which is contrary to the results of P. Baillon et al. The number of E events was about 150 on a background of 200, which is to be compared with ≈ 800 E events for P. Baillon et al. with less than 100 background events. After the Particle Data Group listed the E meson as having been measured as a $J^{PC} = 1^{++}$ state, the high-energy physics community seemed to take it as such (especially theorists). When, in 1982, a $J^{PC} = 0^{-+}$ $K\bar{K}\pi$ state at 1440 MeV was discovered in J/ψ radiative decays, it was henceforth given a new name—iota(1440)—distinct from the E meson.

However, T. A. Armstrong et al.[10] analyzed E events observed in a central production in the reactions $(\pi^+/p)p \rightarrow (\pi^+/p)Ep$ at 85 GeV/c. From an identical Dalitz-plot analysis as that of Dionisi et al., they concluded once again that $J^{PC}(E) =$

[a]This research was supported by the U.S. Department of Energy under Contract No. DE-AC02-76CH00016.

1^{++} $(K^*\overline{K})$. Assuming a single spin-parity state on a noncoherent background in the $K\overline{K}\pi$ mass region from 1.37 to 1.49 GeV, they found that 1^{++} $(K^*\overline{K})$ is favored with $\Delta \ln L = 91$ over 0^{-+}. In addition, in the same experiment, they observe a shoulder at the E-mass region in the $\eta\pi^+\pi^-$ channel;[11] in a similar Dalitz-plot analysis, they find hint of a 1^{++} bump $(\delta\pi)$ at the E mass, but a flat distribution in the 0^{-+} wave throughout the mass region.

A. Ando et al.[12] carried out an experiment at KEK on a reaction identical to that of Dionisi et al., but at higher energy: $\pi^- p \rightarrow En$ at 8.06 GeV/c with $E \rightarrow \eta\pi^+\pi^-$. Their full partial-wave analysis relies on a maximum-likelihood technique in five variables, that is, three Euler angles describing the $\eta\pi^+\pi^-$ orientations and two Dalitz-plot variables. From this analysis, they confirm existence of a 0^{-+} $\eta_r(1275)$ first reported by N. Stanton et al.[13] in the same reaction, but also a significant bump at the E mass in the $I = 0$, $J^{PC} = 0^{-+}$ $(\delta\pi)$ wave. What is most remarkable is that they state that the 0^{-+} $(\delta\pi)$ wave destructively interferes with a 0^{-+} $(\epsilon\eta)$ wave with the result that the E bump nearly disappears in the $\eta\pi^+\pi^-$ spectrum. This could also explain observations of a shoulder in the same channel by T. A. Armstrong et al. One of the most sensitive measures of a resonance behavior is its rapid phase motion with respect to a background wave. This can be readily examined in a full partial-wave analysis (but not in a Dalitz-plot analysis); A. Ando et al. claim that all three $\delta\pi$ states, 0^{-+} $\eta_r(1275)$, 0^{-+} $E(1420)$, and 1^{++} $D(1285)$, exhibit rapid phase motions that are consistent with resonance interpretations.

A high-statistics study of the E/iota meson is being carried out at BNL with the Multi-Particle Spectrometer. To date, we have accumulated some 4000 E/iota events on a background of about the same magnitude in the reaction $\pi^- p \rightarrow K^+ K_S \pi^- n$ at 8 GeV/c. Coincidentally, this happens to be the same reaction at the same energy as those of A. Ando et al. (performed at KEK) and N. Stanton et al. (performed at Argonne).

The trigger requires detection of a K^+ on-line through a random-access memory system, a K_S decay into $\pi^+\pi^-$ via an increase in the multiplicity downstream of a liquid-hydrogen target, and a veto in a lead-scintillation sandwich counter surrounding the target. Because we trigger on $K^+\overline{K}^0\pi^-$ and not on $K^-K^0\pi^+$, we can only measure the G parity and not the isospin, which we presume to be zero.

The data from our 1983 run have been completely analyzed and the results of our Dalitz-plot analysis have already been published.[14] These data contain about 2000 E events, as shown in FIGURE 1; still, it represents so far the highest published E/iota statistics in the world. It is seen that the $D(1285)/\eta_r(1275)$ bump is also prominent, representing about 1000 events.

We have performed on these data a complete partial-wave analysis using two isobars, $\delta(980) \rightarrow K\overline{K}$ and $K^*(890) \rightarrow K\pi$. We have used all allowed J^{PG} states with $J = 0$ and $J = 1$ for $K\overline{K}\pi$ mass up to 1.6 GeV. The generalized density matrix for all J^{PG} waves and their interferences has been constrained to rank two to reflect an arbitrary mixture of spin-nonflip/spin-flip amplitudes at the nucleon vertex, but its elements were allowed to have all possible values within the bounds of the positivity conditions.[15]

The results,[16] which are consistent with our previous Dalitz-plot analysis, show a sharp 0^{-+} bump in the channels $\delta\pi$ and $K^*\overline{K}$ at the E mass (see FIGURE 2). Note also that some 0^{-+} bump can be accommodated at the D. The behavior of the 1^{++} wave is

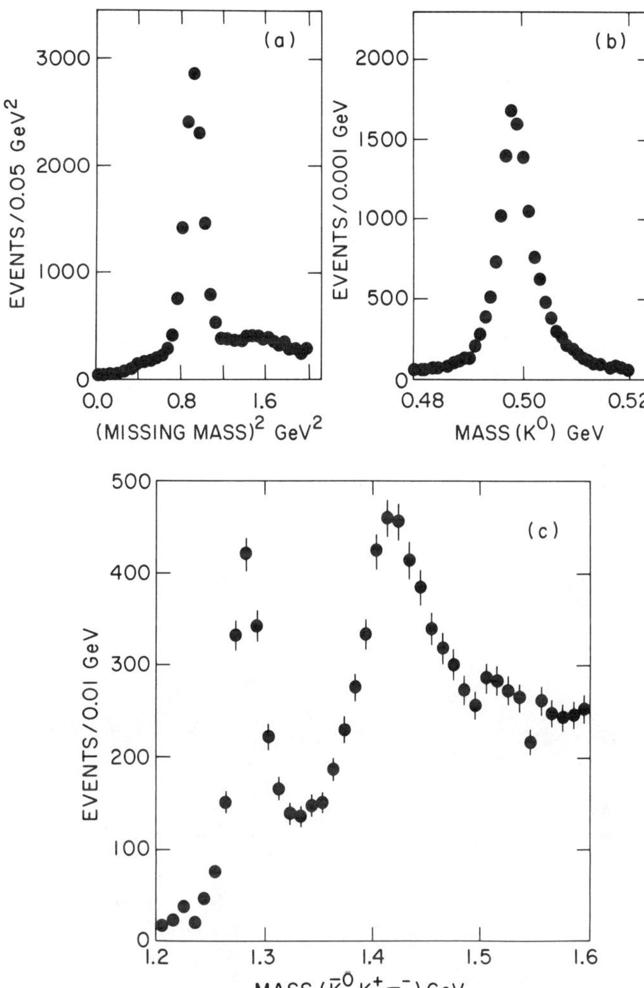

FIGURE 1. (a) Missing-mass squared spectrum showing neutron peak for the reaction $\pi^- p \rightarrow K^+ K^0_S \pi^- n$ at 8 GeV/c (reference 16); (b) $\pi^+\pi^-$ spectrum for decay vertex; (c) $K^+ K_S \pi^-$ spectrum with cuts on neutron mass, K_S mass, and $t < 1$ GeV^2.

remarkable in that the *E*-mass region contains a large 1^{++} wave, which is not inconsistent with those of C. Dionisi *et al.* and T. A. Armstrong *et al.*, but it exhibits merely a sharp rise at the $K^*\overline{K}$ threshold and thereafter a gentle plateau up to 1.6 GeV—this is not a resonant behavior. Indeed, in a 40-MeV bin for $KK\pi$-mass from 1.40 to 1.44 GeV, the predicted events for a 0^{-+} wave are about equal to those for a 1^{++} wave. It is determined in our analysis that the $D(1285)$ is mostly a $1^{++}(\delta\pi)$ state, but a sizable $0^{-+}(\delta\pi)$ state can also be accommodated that is consistent with A. Ando *et al.* and N. Stanton *et al.*

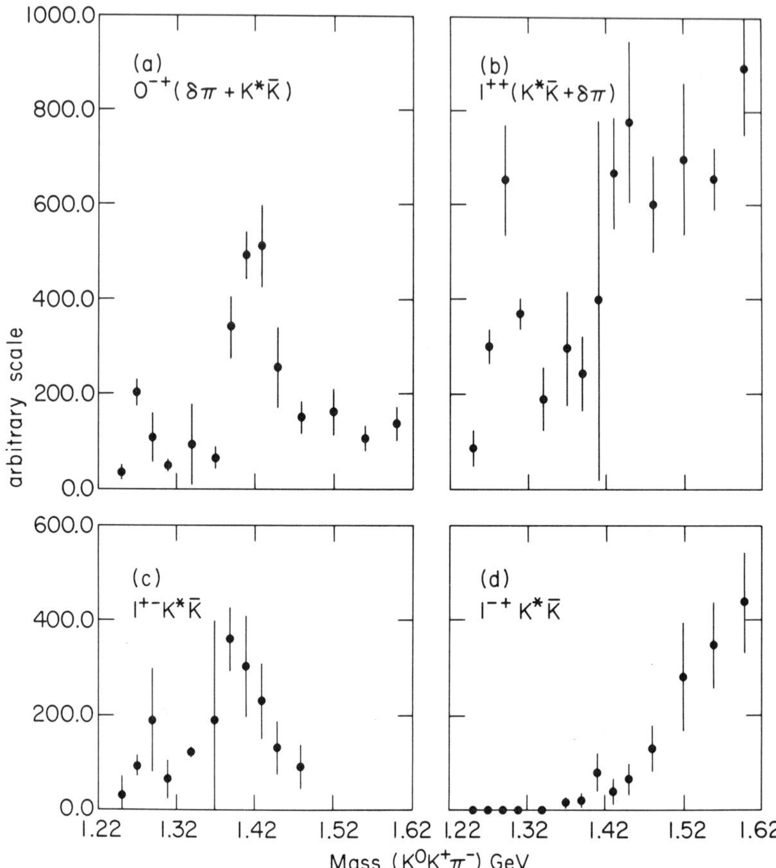

FIGURE 2. Results of partial-wave analysis: (a) $J^{PG} = 0^{-+}$ wave with $\delta\pi$, $K^*\overline{K}$, and their interferences combined together; (b) $J^{PG} = 1^{++}$ wave with δ and K^* isobars combined as above; (c) $J^{PG} = 1^{+-}(K^*\overline{K})$ wave; (d) $J^{PG} = 1^{-+}(K^*\overline{K})$ wave.

FIGURE 3 shows a few relevant phases from our fits. A rapid positive rise of the $0^{-+}(\delta\pi)$ relative to the $1^{++}(K^*\overline{K})0^+$ is consistent with the interpretations of a 0^{-+} resonance and a nonresonant $1^{++}(K^*K)0^+$ wave. The $1^{+-}(K^*K)$ wave gives a hint of a bump at the E; however, the $1^{+-}(K^*K)0^+$ state does not seem to exhibit a rapid phase motion with respect to the nonresonant $1^{++}(K^*K)0^+$ wave. No firm conclusions can be drawn: it is hoped that new 1985 data included in the analysis will help clarify the nature of the 1^{+-} wave.

In our BNL experiment, the E/iota is also observed in a $\overline{p}p$ inclusive process at 6.6 GeV/c. A Dalitz-plot analysis[17] of the $K\overline{K}\pi$ system again confirms that the E enhancement is mostly a $0^{-+}(\delta\pi + K^*\overline{K})$ state.

There are a number of experiments that report observation of E/iota in hadroproduction, but, as yet, no reported spin-parity analyses. The Lepton-F Spectrometer[18] at

Serpukhov finds a bump at the E in the reaction $K^-p \rightarrow K^+K^-\pi^0 Y$ at 32.5 GeV/c, but not in the reaction $\pi^-p \rightarrow K+K^-\pi^0 n$. The GAMS collaboration[19] sees E/iota in the reaction $\pi^-p \rightarrow \eta\pi^0\pi^0 n$ at 100 GeV/c. In contrast to the shoulders observed in $\eta\pi\pi$ channels by A. Ando *et al.*,[13] the GAMS experiment sees a prominent bump at the E mass. Finally, P. Chauvat *et al.*[20] have observed a sharp E enhancement in the diffractive dissociation of protons at an ISR experiment.

A succinct summary of all the E/iota observations in hadroproduction is given in TABLE 1. It is noteworthy that the measured mass and width of both $D(1285)$ and $E(1420)$ are all consistent with one another; in particular, the mass of the E does not seem to vary by more than 5 MeV from one experiment to another. This is in contrast to the iota seen in both Mark III and DM2 experiments, which observe $m = (1456 \pm 5 \pm 6)$ MeV, $\Gamma = (95 \pm 10 \pm 15)$ MeV and $m = (1460 \pm 3 \pm 8)$ MeV, $\Gamma = (100 \pm 12 \pm 15)$ MeV, respectively (see D. Hitlin's paper in these proceedings).

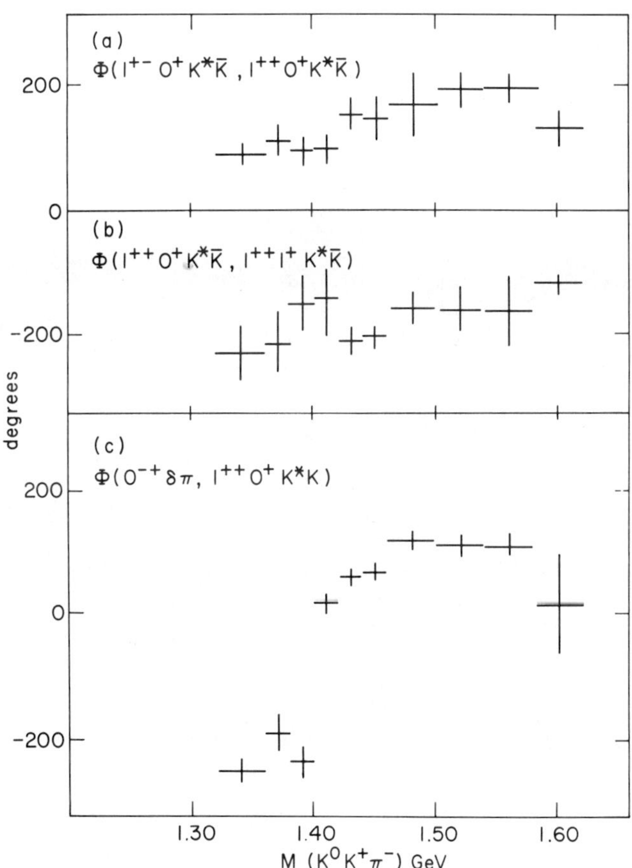

FIGURE 3. (a, b, c) Relative phases of a few relevant partial waves. Notation: $J^{PG} M^\epsilon$ Isobar, where M is the z-component of spin and ϵ is the reflectivity quantum number.

TABLE 1. *E*/Iota in Hadroproduction

Beam Momentum (GeV/c)	Reaction and E/Iota Decay	Mass/Width (MeV)	I	G	Spin Parity Analysis[a]	J^P	Isobars[b]	Ref.
0	$\bar{p}p \to E\pi\pi, E \to K\bar{K}\pi$	$(1425 \pm 7)/(80 \pm 10)$	0	+	PWA	0^-	δ, K^*	7, 8
3.95	$\pi^- p \to En, E \to K\bar{K}\pi$	$(1426 \pm 6)/(40 \pm 15)$	0	+	DPA	1^+	K^*	9
85	$(\pi^+/p)p \to (\pi^+/p)Ep$							
	$E \to K\bar{K}\pi$	$(1422 \pm 2)/(60 \pm 2)$		+	DPA	1^+	K^*	10
	$E \to \eta\pi\pi$	$(1423 \pm 20)/\text{---}$		+	DPA	1^+	δ	11
8.06	$\pi^- p \to En, E \to \eta\pi^+\pi^-$	$(1420 \pm 6)/(31 \pm 26)$	0	+	PWA	0^-	δ	12
8.0	$\pi^- p \to En, E \to K^+K_S\pi^-$	$(1421 \pm 2)/(60 \pm 10)$		+	PWA	0^-	δ, K^*	14, 16
					DPA			
6.6	$\bar{p}p \to EX, E \to K^+K_S\pi^-$	$(1416 \pm 5)/(80 \pm 30)$		+	DPA	0^-	δ, K^*	17
100	$\pi^- p \to En, E \to \eta\pi^0\pi^0$	$\approx 1420/\approx 40$						19
63^c	$p \to PE, E \to K\bar{K}\pi$	$(1422 \pm 3)/(47 \pm 10)$						20

[a]PWA stands for partial-wave analysis; DPA stands for Dalitz-plot analysis.
[b]Intermediate states: $\delta(980)\pi$ or $K^*(890)\bar{K}$ + c.c.
[c]$\sqrt{s}$ for ISR.

A study of a resonance coupling to $\delta(980)\pi$ necessarily involves the problem of parameterizing the $\delta(980)$ resonance, which is thought to decay into $K\bar{K}$ and $\pi\eta$ channels. A coupled-channel parameterization for the δ was given by S. Flatté,[21] and his formulation has been used by most of the spin-parity analyses so far performed on the *E*/iota in hadroproductions. As some of the finer points of the problem are not as well known, a short comment seems to be in order.

The best experimental information regarding the δ decay into $K\bar{K}$ and $\pi\eta$ may be that $D(1285)$ decays into both $K\bar{K}\pi$ and $\pi\pi\eta$ and that both require a δ-like intermediate state. In an elegant derivation, R. Cahn and P. Landshoff[22] showed that Flatté's coupled-channel formula is perhaps the simplest, which satisfies the two-channel unitarity constraints. Therefore, in the absence of firm theoretical prejudice for alternative formulations, it seems prudent to continue to use Flatté's formula for the δ parameterization.

We used the following amplitude in our analysis: for $i, j = 1, 2$,

$$T_{ij} = \frac{g_i g_j}{m_R^2 - m^2 - im_R \left(g_1^2 P_1 + g_2^2 P_2 \right) \dfrac{m_t}{m}}, \tag{1}$$

where m_t is the $K\bar{K}$-threshold mass, P_1 and P_2 are the breakup momentum of $\pi\eta$ and $K\bar{K}$ systems, respectively, and g_i stands for the corresponding coupling constant. With this definition, note that g_i is a unitless constant. The ratio g_1^2/g_2^2 is typically fixed at its $SU(3)$ value:[23] $\frac{2}{3}$. It should be noted that we take P/m as the relativistic phase-space factor, rather than just P, as was done by Flatté and by A. D. Martin *et al.*[24]

So far, the clearest observation of the δ is in the reaction $K^- p \rightarrow \delta^- \Sigma^+(1385), \delta^- \rightarrow \pi^- \eta$ at 4.2 GeV/c.[25] It was shown that the mass and width are $m_0 = (981 \pm 6)$ MeV and $\Gamma_0 = (55 \pm 5)$ MeV, respectively, from a simple s-wave Breit-Wigner form on a smooth noncoherent background.

If one demands that T_{ij} of equation 1 has a peak at m_0, one finds, after setting $P_2 = i|P_2|$,

$$m_R^2 = m_0^2 - r \frac{|P_2(m_0)|}{P_1(m_0)} m_0 \Gamma_0, \tag{2}$$

where we put $g_2^2 = rg_1^2$. We shall define an "intrinsic" $\pi\eta$ width through g_1 by

$$\Gamma_1(m) = g_1^2 P_1(m) \frac{m_t}{m}. \tag{3}$$

Then,

$$m_R \Gamma_1(m_0) = m_0 \Gamma_0. \tag{4}$$

Note that equations 2 and 4 have been obtained by equating the T_{ij} to $(m_0^2 - m^2 - im_0\Gamma_0)^{-1}$.

The shape of the resonance peak given by equation 1 at m_0 is highly asymmetric; there is a gentle rise below m_0, followed by a sharp drop at the $K\bar{K}$ threshold. Because of this, if g_1^2 is fixed through equations 3 and 4, one finds that the apparent width is somewhat smaller (e.g., 35 MeV for $\Gamma_0 = 55$ MeV and $r = \frac{3}{2}$). In order to restore the apparent width to the observed 55 MeV, one has to increase Γ_0 or decrease r, or both. The following display shows a few sets of Γ_0 and r along with other relevant parameters:

r	Γ_0(MeV)	m_R(MeV)	g_1^2	$\Gamma_1(m_R)$(MeV)
0.5	68	978	0.212	68
1.0	88	973	0.277	88
1.5	122	962	0.387	122

The ratio r also determines the $K\overline{K}/\pi\eta$ branching ratio for the δ, that is, 10%/90%, 17%/83%, and 22%/78% for $r = 0.5$, 1.0, and 1.5, respectively.

In a Dalitz-plot fit, we have found that small values of r are preferred; our published paper[14] of Dalitz-plot analysis on D and E corresponds to $r = 0.5$. On the other hand, we find that our fits with full partial-wave amplitudes are stable against this variation, with no significant difference in the likelihood between $r = 0.5$ and 1.5. Because Dalitz-plot analysis appears to be somewhat sensitive to δ parameterizations, one needs to be cautious about accepting quoted branching ratios for $\delta\pi/K^*\overline{K}$. Moreover, the two decay channels can interfere for a given partial wave, so it is not meaningful to quote the branching ratios without also giving the amount of interference effects.

It has recently been determined by K. L. Au, D. Morgan, and M. R. Pennington[26] that $I = 0$ s-wave system coupling to $\pi\pi$ and $K\overline{K}$ is extremely complicated; they find that at least four resonances are required with masses ranging between 0.99 to 1.48 GeV. Therefore, they argue that there must exist additional states, including a glueball or hybrid state beyond the two quarkonium states expected in this mass range. It may very well be that $I = 1$ $\pi\eta$ and $K\overline{K}$ channels are just as complicated. However, until further information is forthcoming, one must perhaps rely on the simplest, unitarized coupled-channel formula given by S. Flatté for the δ parameterization.

Turning now to the future prospects regarding the E/iota experiments, we can point out that our published data come from our 1983 run alone, with the data of 1985 to be added to our analysis sample shortly. We expect to run once more at the MPS in June 1986; with it, we should eventually have a total of quadruple the statistics of the 1983 data. In addition, one can expect spin-parity analyses from the GAMS experiment on their $\eta\pi^0\pi^0$ channel, from the ISR diffractive production of $K\overline{K}\pi$ systems, and from the KEK experiment on the $K\overline{K}\pi$ channel. The GAMS group is planning another experiment to look at the central production of neutral particles with the 300 GeV π^- beam: $\pi^-p \rightarrow \pi^-p + M^0$, with $m^0 \rightarrow \gamma\gamma$, $\eta\eta$, $\eta\pi^0\pi^0$, $\eta\eta'$, etc. The Ω' Spectrometer is likewise planning on a further study of the mesonic systems produced centrally in the reactions $pp \rightarrow pp + M^0$ and $\pi^-p \rightarrow \pi^-p + M^0$ at 340 GeV/c. These two experiments should be sensitive to glueballs and hybrids because of their double-diffractive production processes whose central systems are thought to be gluon-rich.

In summary, the E/iota meson is produced in a variety of hadron-induced reactions, of which most are from exclusive processes. The energies also vary widely, from $p\overline{p}$ annihilations at rest to π^- beams at 100 GeV/c. It is becoming increasingly clear that a pseudoscalar E/iota meson is a mixed system with one or more quarkonium states and perhaps a glueball. There are, in fact, theoretical papers[27,28] that show that the isoscalar $J^{PC} = 0^{-+}$ meson system is highly complex with radial excitations and/or mixed glueballs occurring at around 1.4 GeV.

In a systematic study of meson systems from π to Υ by S. Godfrey and N. Isgur,[29] we learn that there should exist a total of six *s*- and *p*-wave light-quark systems in the mass range between 1.36 to 1.53 GeV; not only the $J^{PC} = 0^{-+}$ *E*/iota, but also other meson states with $J^{PC} = 1^{+-}, 1^{--}, 0^{++}, 1^{++}$, and 2^{++}. The only way such a complex superposition of states can be disentangled is to apply isobar-model partial-wave techniques on high-statistics data.

A great deal of theoretical and experimental work also needs to be done on the decay modes of the *E*/iota meson. Very little is known regarding the $\delta(980)$ decays into $\eta\pi$ and $K\overline{K}$.[30] A coupled-channel analysis of $\eta\pi\pi$ and $K\overline{K}\pi$ systems in J/ψ radiative decays may shed new information on this problem. So far, there is little evidence of iota $\rightarrow \eta\pi\pi$ in J/ψ radiative decays. W. T. Palmer and S. S. Pinsky[31] pointed out that $\delta\pi$ ($\delta \rightarrow \eta\pi$) and $\epsilon\eta$ ($\epsilon \rightarrow \pi\pi$) decay modes may interfere destructively to obscure the iota signal in $\eta\pi\pi$; this mechanism certainly seems to hold for the *E*/iota seen by A. Ando *et al.*[12] Recently, M. Frank *et al.*[32] speculated that *E*/iota $\rightarrow K\overline{K}\pi$ could be explained by formation of the "$\overline{K}K$ molecule" and not the $\delta(980)$ intermediate state. However, their $\overline{K}K$ formulation gives a poor fit to our data in the upper half of the *E*-mass region where the $K^*\overline{K}$ states are strong. This presumably indicates that if "$K\overline{K}$ molecules" are formed at all in the *E*/iota decays, they must represent a small addition to the $\delta(980)$ intermediate state.

One indisputable fact that emerges in a survey of the *E*/iota in hadroproductions is that the measured masses and widths are consistent with one another, averaging around $m \simeq 1420$ MeV and $\Gamma \simeq 60$ MeV, respectively. In contrast, the iota of J/ψ radiative decays are measured to be of $m \simeq 1460$ and $\Gamma \simeq 100$ MeV, respectively. This may indicate that we are dealing with two or more objects of the same J^{PC} that are produced with very different cross sections and interferences between hadroproductions and J/ψ radiative decays.

ACKNOWLEDGMENTS

The author wishes to thank M. Block, L. Durand, and the other organizers of the 1986 Aspen Winter Conference for a most pleasant and rewarding American version of Rencontre de Moriond.

REFERENCES

1. ARMENTEROS, R. *et al.* 1963. Int. Conf. on Elem. Particle Physics, Siena, Italy, p. 287.
2. SCHARRE, D. L. *et al.* 1980. Phys. Lett. **97B**: 329.
3. CHUNG, S. U. 1984. XV Int. Symp. on Multiparticle Dynamics, Lund, Sweden, p. 186.
4. CHUNG, S. U. 1984. XXII Int. Conf. on High Energy Physics, Leipzig, East Germany, vol. II, p. 167.
5. CHUNG, S. U. 1984. XII Int. Winter Meeting on Fundamental Physics, Santander, Spain, p. 295.
6. COOPER, S. 1985. Rapporteur talk. Int. Europhysics Conf. on High Energy Physics, Bari, Italy, p. 947.
7. BAILLON, P. *et al.* 1967. Nuovo Cimento **50A**: 393.
8. BAILLON, P. 1983. Experimental Meson Spectroscopy, BNL, p. 78.

9. DIONISI, C. *et al.* 1980. Nucl. Phys. **B169:** 1.
10. ARMSTRONG, T. A. *et al.* 1984. Phys. Lett. **146B:** 273.
11. BAILLIE, O. V. *et al.* 1985. Int. Europhysics Conf. on High Energy Physics, Bari, Italy, p. 314.
12. ANDO, A. *et al.* 1985. KEK preprint no. 85-15.
13. STANTON, N. *et al.* 1979. Phys. Rev. Lett. **42:** 346.
14. CHUNG, S. U. *et al.* 1985. Phys. Rev. Lett. **55:** 779.
15. CHUNG, S. U. & T. L. TRUEMAN. 1975. Phys. Rev. **D11:** 633.
16. Results of our partial-wave analysis on BNL data have been presented by: ZIEMINSKA, D. 1985. Int. Conf. on Hadron Spectroscopy-1985, Univ. of Maryland, p. 27; DOWD, J. 1985. Int. Europhysics Conf. on High Energy Physics, Bari, Italy, p. 318; CHUNG, S. U. 1985. 20th Rencontre de Moriond, Les Arcs, France, p. 489; PROTOPOPESCU, S. 1985. Annual Meeting of the APS Div. of Particles and Fields, Eugene, Oregon. BNL 37234.
17. REEVES, D. 1985. Ph.D. thesis. Florida State Univ. preprint no. FSU-HEP-850801; see also reference 16.
18. BITUKOV, S. I. *et al.* 1985. Serpukhov preprint no. EFVE85-19; see also references 4 and 6.
19. MOUTHUY, TH. 1985. Int. Europhysics Conf. on High Energy Physics, Bari, Italy, p. 320.
20. CHAUVAT, P. *et al.* 1984. Phys. Lett. **148B:** 382.
21. FLATTÉ, S. M. 1976. Phys. Lett **63B:** 224.
22. CAHN, R. & P. LANDSHOFF. 1985. CERN-TH.4191.
23. MORGAN, D. 1974. Phys. Lett. **51B:** 71.
24. MARTIN, A. D. *et al.* 1977. Nucl. Phys. **121:** 514.
25. GAY, J. B. *et al.* 1976. Phys. Lett. **63B:** 220.
26. AU, K. L., D. MORGAN & M. R. PENNINGTON. 1986. Phys. Lett. **167B:** 229.
27. FRANK, M. & P. J. O'DONNELL. 1984. Phys. Rev. **D29:** 921; FRANK, M. & P. J. O'DONNELL. 1983. Phys. Lett. **133B:** 253.
28. LIPKIN, H. J. 1985. ANL-HEP-PR-85-DRAFT.
29. GODFREY, S. & N. ISGUR. 1985. Phys. Rev. **D31**.
30. REVIEW OF PARTICLE PROPERTIES. 1984. Rev. Mod. Phys. **56:** part II.
31. PALMER, W. T. & S. S. PINSKY. 1983. Phys. Rev. **D27:** 2219.
32. FRANK, M. *et al.* 1985. Phys. Lett. **158B:** 442.

Glueball and Meson Spectroscopy[a]

SYDNEY MESHKOV

Center for Radiation Research
National Bureau of Standards
Gaithersburg, Maryland 20899
and
Institute for Theoretical Physics
University of California
Santa Barbara, California 93106

INTRODUCTION

Glueballs[1]—colorless, flavorless composites of two or more gluons—should exist according to the basic tenets of quantum chromodynamics (QCD). From the preceding discussions of Hitlin,[2] Chung,[3] and Lindenbaum,[4] it is clear that the study of glueball spectroscopy is intertwined with the study of meson spectroscopy. In this brief review, I shall try to correlate the experimental results presented by the aforementioned authors with our current understanding of QCD. A major problem with attempting to offer any help is that we have no analogue of the quarkonia phenomenological potentials (see Cornell,[5] Richardson,[6] and Kaus[7]) to predict glueball spectra. Lattice Gauge Theory calculations are just approaching a postnatal stage.

SPECTROSCOPY

Level Order

The simplest models that give the order and J^{PC} of glueball states arise from potentials for confinement. Unfortunately, we cannot compute masses in this approach. Although this approach has been presented in detail before, we mention the model and results briefly.

The model for constructing the two gluon glueball states has two gluons moving with a relative orbital angular momentum L.[1,8,9] We require that the two gluon wave functions be totally symmetric under the interchange of the space, spin, and color coordinates of the gluons. Taking $J = L + S$, where S is the spin of the two gluons, the allowed 2g states are shown below in FIGURE 1. Note that an oddball,[10,11] 1^{-+}, occurs for the first time at the $L = 1$ level. Oddballs are glueballs with J^{PC} quantum numbers that do not occur in $q\bar{q}$; L mesons, that is, 0^{--}; $1^{-+}, 3^{-+}, \ldots ; 0^{+-}, 2^{+-}, \ldots$.

Oddball states are enclosed in boxes in FIGURE 1. Although two massless spin 1 states cannot couple to a 1^{-+} state, Muzinich[12] has shown that the physical 1^{-+} state

[a]This research was supported in part by the National Science Foundation under Grant No. PHY82-17853, supplemented by funds from the National Aeronautics and Space Administration.

87

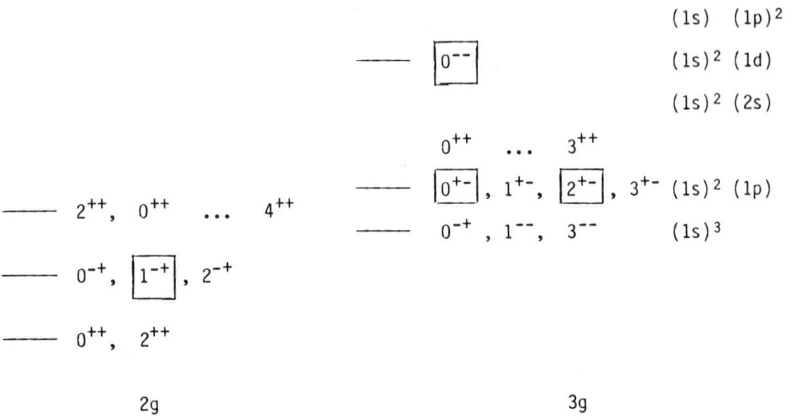

$$
\begin{array}{ll}
 & (1s)\ (1p)^2 \\
\boxed{0^{--}} & (1s)^2\ (1d) \\
 & (1s)^2\ (2s) \\
0^{++}\ \ldots\ 3^{++} & \\
\boxed{0^{+-}},\ 1^{+-},\ \boxed{2^{+-}},\ 3^{+-} & (1s)^2\ (1p) \\
0^{-+},\ 1^{--},\ 3^{--} & (1s)^3
\end{array}
$$

$$
2^{++},\ 0^{++}\ \ldots\ 4^{++}
$$

$$
0^{-+},\ \boxed{1^{-+}},\ 2^{-+}
$$

$$
0^{++},\ 2^{++}
$$

$$2g \qquad\qquad\qquad 3g$$

FIGURE 1. 2g and 3g levels.

may be considered to be a superposition of 2-gluon, 3-gluon, . . . , N-gluon 1^{-+} Fock states.

Also shown is the 3g spectrum obtained by putting three gluons in a potential well.[9b] No estimates for the masses have been made, so the exact normalization of the 3g sector with respect to the 2g sector is arbitrary. We note that the lowest 3g states are 0^{-+}, 1^{--}, 3^{--}. In addition we note that there are a large number of low-lying 2^{++} states, as well as 0^{-+} and 0^{++} states.

Masses

Three methods for estimating glueball masses continue to dominate the thinking of those interested in the field: Lattice Gauge Theories, Bag Models, and Massive Constituent Gluon Theories. They all produce spectra with some similarities; that is, a 0^{++} ground state and low-lying 0^{-+} and 2^{++} states. They are also all dependent on the choice of some parameter to set the scale. We shall see that this is not an academic point when we confront the ultimate arbiter, experimental results.

Lattice Gauge Theories

Lattice Gauge Theories (LGT) offer us the possibility of carrying out a well-defined fully controllable nonperturbative QCD calculation of the glueball spectrum. The Monte Carlo technique has been the vehicle for carrying out this program. After the workshop on Glueballs and Lattice Gauge Theory at the Aspen Center for Physics in August 1984, I was discouraged. At that time, about the only thing on which there was any agreement was that there was a mass gap; that is, a 0^{++} state existed. Even the existence of a 2^{++} state was in question. Since this workshop, the scene has brightened. De Forcrand, Schierholz, Schneider, and Teper[13] have done an "experiment" for the

0^{++} glueball. They find $m_g/\sqrt{K} = 280$, where $\sqrt{K}$ (the string tension) sets the mass scale, and $\sqrt{K} = 280\ (0.38\ \Lambda_L)$. Consequently, $\Lambda_L/\sqrt{K} = 9.4 \times 10^{-3}$, as compared to the old Creutz[14] value of 6×10^{-3}. This gives $m_g = \sqrt{K}/0.38$. As may be seen from TABLE 1, $m_g\ (0^{++}) = 1050{-}1100\ \mathrm{MeV}/c^2$ for usual values of the string tension.

In reference 13, de Forcrand *et al.* also find a 2^{++} state such that $m(2^{++})/m(0^{++}) \simeq 2.9$. They are unable to achieve a believable 1^{-+} state, and they do not try for a 0^{-+} state. Note: After the Conference, a preprint by Berg *et al.*[15] was published that claimed that $m(0^{++})/m(2^{++}) \simeq 1.2$. A telephone conversation with Berg indicates that $m(0^{++}) \simeq m(2^{++})$. This muddies the waters once again, but it is in agreement with the gg, $L = 0$ model in which $m(0^{++})$ is degenerate with $m(2^{++})$. We await a definitive Monte Carlo Lattice Gauge Theory calculation.

Bag Models and Constituent Gluon Models

These models have been reviewed extensively in the past[16] and will not be discussed further.

Glueball Solitons

Zachariasen[17] has discussed this approach, which produces both spherical and smokering glueballs. The calculations yield a wide range of 0^{++} masses (2–4 GeV/c^2) depending on parameter choices.

Conclusions

(a) We have no analogue of the potential models that are so successful for describing heavy quarkonia.
(b) Some progress is being made in Lattice Gauge Theories, but their predictive powers for describing glueball masses are minimal.
(c) $Q\bar{Q}$ potentials that correspond to the real world have been extracted from Lattice Gauge Theories,[18] so there is hope.

Hermaphrodites-$q\bar{q}g$

Excellent reviews of this system exist in the literature,[19,20] so we shall not pursue this subject further.

TABLE 1. $m_g(0^{++})$ as a Function of $\sqrt{K}$

$\sqrt{K}$ MeV/c^2	Λ_L MeV/c^2	$m_g(0^{++})$ MeV/c^2
400	3.75	1050
420	3.95	1100

GLUEBALL PRODUCTION

The two main ways to produce glueballs are via the radiative decays of vector mesons and as final state products in Zweig-forbidden processes. In addition, double pomeron exchange dominated processes yield final states with the quantum numbers of glueballs.

(a) The radiative decays of vector mesons (production in hard gluon channels) is the way in which the glueball candidates—the $\iota(1460)$[21,22] and the $\theta(1720)$[23,24]—were first discovered. [They were originally $\iota(1440)$ and $\theta(1640)$.] This "classic" method looks attractive because the estimate $\Gamma(J/\psi \rightarrow \gamma gg)/\Gamma(J/\psi \rightarrow ggg) = 16\alpha/5\alpha_s$ implies that the $B(J/\psi \rightarrow \gamma G) \simeq$ 6–10%. It is the method for producing the three glueball candidates:

$$\iota(1460)\,, \qquad \theta(1720)\,, \qquad \text{and} \quad \xi(2230) \quad .$$
$$0^{-+} \qquad\qquad 2^{++} \qquad\qquad 0^{++}, 2^{++}, 4^{++}$$

Many new results from Mark III have just been reported by Hitlin,[2] and I shall make extensive use of them in what follows. By now, so many resonances have been observed in the J/ψ radiative decay that the earlier penchant for labeling all such states as glueballs has abated. The relative amount of glue in the states X_i is given by the relative branching fractions $B(J/\psi \rightarrow \gamma X_i)$. I shall try to reconcile the radiative decay data with the often contradictory hadronic data summarized for us by Chung.[3]

(b) A powerful method for producing glueball candidates is to study final state products that arise from Zweig-forbidden hadronic processes.[4,25,26] The production of resonating $\phi\phi$ final states in $\pi^- p$-initiated reactions has led to three 2^{++} glueball candidates: $g_T(2050)$, $g_T(2300)$, and $g_T(2350)$.[4]

0^{-+} SYSTEM

We must treat the 0^{-+} system as a whole in order to understand the spectroscopy—both $q\bar{q}$ and glueball. The properties of many of the states of the 0^{-+} system are quite uncertain. All of the candidate 0^{-+} states, together with their production modes, are listed in TABLE 2.

In addition, in TABLE 3, we list a set of spectroscopic calculations for the

TABLE 2. 0^{-+} States and Their Production Modes

State	Produced Radiatively	Produced Hadronically	2γ
η (549)	√	√	√
η' (960)	√	√	√
ζ (1275)	?	√	
E (1420)	?	√	
ι (1460)	√	√	
X (1550)	√		
Y (1800)	√		

TABLE 3. 0^{-+}, $I = 0$ States (MeV/c^2) Spectroscopic Calculations

State	Godfrey-Isgur[28] $P2$ (1984)	Stanley-Robson[29] (1980)	Cohen-Lipkin[30] (1979, 1984/85)	Bender-Palmer-Pinsky[31] (1984)	Chao-Palmer-Pinsky[32] (1985/86)
1η	530	565	549	549	549
$1\eta'$	960	937	960	960	960
$2\eta_n$	1270	1250	1260	1275	1275
$2\eta_s$	1540	1550	1420	1700	1420
$3\eta_n$	1870	1850	1420		
$3\eta_s$	2120	2060			
G				1460	1460

pseudoscalar states. Inclusion in TABLE 3 depends on whether a radial excitation of the $\eta(548)$ is predicted to lie at $\simeq 1275$ MeV/c^2 because we take the experimental indications for such a state seriously. The excellent work of Frank and O'Donnell[27] is excluded from TABLE 3 only because they do not obtain a radial excitation at $\zeta(1275)$.

A few comments about the entries in TABLE 3 are in order. The earliest indication for a radial excitation at 1275 MeV/c^2 was in the first of a series of papers by Cohen and Lipkin[30] that added the other states piecemeal. Stanley and Robson[29] obtained the $n = 2$ and $n = 3$ radial excitations in 1980. The original Godfrey-Isgur calculation[28] ($P1$) had no state at 1270 MeV/c^2, but had $2\eta_n = 1430$ MeV/c^2 and $2\eta_s = 1640$ MeV/c^2. They added an additional annihilation term to accommodate the $\zeta(1275)$. Their values of $3\eta_n$ and $3\eta_s$ are taken to be as reported in the paper of Carlson and Peterson.[33] The papers of Palmer, Pinsky, and co-workers[31,32] are not true spectroscopic calculations, but instead are attempts to carry out the mixing suggested earlier, testing the resulting wave functions with various experimental production and decay rates. In their latest paper,[32] they find that mixing the five states shown in TABLE 3, where both the 1420 and 1460 MeV/c^2 are taken as 0^{-+}, does not work. O'Donnell[34] has informed me, after this conference, that all of the solutions presented in TABLE 3 have experimental difficulties associated with them.

Subsets of the 0^{-+} System

η,η' System

If the η and η' resonances are treated as an isolated mixed pair of flavor $SU(3)$ octet and singlet states, the mixing angle is found to be $11°$ from the Gell-Mann Okubo mass formula, $19°$ from $\gamma\gamma$ decays, and $18°$ from production experiments. Despite our earlier caution, most attempts to unravel the nature of the mixing of the η and η' have usually mixed them only with each other and/or a glueball. Using what has now become a standard notation, due to Rosner,[35]

$$|\eta> = X_\eta |N> + Y_\eta |S> + Z_\eta |G>,$$

$$|\eta'> = X_{\eta'} |N> + Y_{\eta'} |S> + Z_{\eta'} |G>.$$

It must be emphasized that G stands for a glueball and/or any radially excited quark states; that is, anything that is not made out of ground-state up, down, and strange quarks.

In a published analysis, Mark III finds[36]

$$X_\eta^2 + Y_\eta^2 = 1.1 \pm 0.2 \quad \text{and} \quad X_{\eta'}^2 + Y_{\eta'}^2 = 0.65 \pm 0.18.$$

This result implies that nothing else is needed for η, but that something else is needed for the η'. Note that Mark III used somewhat different (earlier) $\Gamma(\eta \rightarrow \gamma\gamma)$ values than those summarized by Kolanoski at Kyoto:[37]

	Mark III[36]	Kolanoski[37]
$\Gamma(\pi^\circ \rightarrow \gamma\gamma)$	7.8 ± 0.6 eV	7.33 ± 0.20 eV
$\Gamma(\eta \rightarrow \gamma\gamma)$	343 ± 47 eV	560 ± 40 eV
$\Gamma(\eta' \rightarrow \gamma\gamma)$	4.16 ± 0.43 keV	4.3 ± 0.1 keV.

The use of these new values in the Mark III analysis may alter the final numbers somewhat.

Palmer and Pinsky[38] disagree with the conclusion that nothing additional is needed for the η wave function. They claim that $B(J/\psi \rightarrow \gamma\eta)$ is large ($\sim 8 \times 10^{-4}$) and that therefore the glue component is large. In fact, they claim that if the mixing angle is 11°, then the admixture of singlet in the GMO formula must be all glue.

The implication of $X_{\eta'}^2 + Y_{\eta'}^2 = 0.65 + 0.18$ is that η' is mixed with states other than η. Such a mixing with the other 0^{-+} states should be calculable in a correct mixing model. This has been attempted in numerous mixing models in the past and at present without great success.

Structure of Pseudoscalar Radial Excitations

The unmixed η and η' are taken to be pure octet and singlet states, respectively, that is, $\eta \approx \eta_8$ and $\eta' \approx \eta_1$. An interesting question is whether $\eta_r \approx \eta_{r_8}$ and $\eta_r' \approx \eta_{r_1}$, or (as for the vector and tensor mesons) whether we take $\eta_r \approx \eta_n = (u\bar{u} + d\bar{d})/\sqrt{2}$ and $\eta_r' \approx \eta_s = s\bar{s}$. Palmer and Pinsky, doing unitary corrections à la Tornqvist[39] find that the latter choice is preferred.[38]

$\zeta(1275)$

The existence of this state is critical for the spectroscopy of the pseudoscalar system. It was originally seen in 1979 by Stanton et al.[40] as a 0^{-+} $\pi\pi\eta$ resonance in the reaction $\pi^- p \rightarrow \eta\pi\pi\eta$. This has been confirmed in an analogous experiment at KEK by Ando et al.[41] There are also hints of such a state (no partial wave analysis, however) in $J/\psi \rightarrow \gamma(\gamma\rho)$ by Crystal Ball,[42] DM2,[43] and Mark III,[2,44,45] and as a possible shoulder in the Mark III $J/\psi \rightarrow \gamma(\eta\pi\pi)$ results.[2,44]

Both Stanton and Ando find that $\delta\pi$ and "ϵ"η interfere destructively at $\zeta(1275)$. Stanton et al. find an excess in the 0^{-+} $\epsilon\eta$ channel at ~ 1400 MeV/c^2. Ando et al. also see a peak with some structure in the same channel. In addition, they find a sharp peak in the 0^{-+} $\delta\pi$ channel at 1420 MeV/c^2. Palmer and Pinsky[46] in 1983 suggested that the destructive interference observed at $\zeta(1275)$ might also hold in the ι region, thus giving a small $\eta\pi\pi$ yield in the decay $J/\psi \rightarrow \gamma\eta\pi\pi$.

E(1420) – $\iota(1460)$ Region

The Split Iota (hopefully, not shades of the split A_2) has had a checkered history, as emphasized by Hitlin[2] and Chung.[3] These states must be studied in conjunction with those below—η, η' and ζ— and with those above—1550 ($\rho\rho$), 1800 ($\rho\rho$ and $\omega\omega$), and possibly 1710 ($\eta\pi\pi$). Interest in the $\iota(1460)$ as a glueball candidate is due to the large value of B$[J/\psi \rightarrow \iota(1460)]$, that is, B = $(6.9 \pm 0.4 \pm 1) \times 10^{-3}$, the largest J/ψ branching fraction.

The overriding question that we must address is the following: Is $E(1420)$ the same particle as $\iota(1460)$, or are they two distinct particles? As we just heard from Hitlin[2] and Chung,[3] E is produced hadronically, and possibly radiatively, and ι is now produced hadronically, as well as being produced radiatively. If both E and ι are different 0^{-+} particles, they are mixed with each other and their 0^{-+} companions.

$\gamma\rho$ and $\gamma\gamma$ Decays. $\gamma\rho$ is an enigma. Does $\gamma\rho$ decay from E, ι, or both? Experimentally, $J/\psi \rightarrow \gamma X$, $X \rightarrow \gamma\rho^0$. $M(X)$ and $\Gamma(X)$ are listed in TABLE 4 for the DM2[43] and Mark III[44] experiments. If we take $X \equiv \iota$ and if we assume B$(K\overline{K}\pi) = 1$, then $\Gamma(\iota \rightarrow \gamma\rho) = 1.9 \pm 0.7$ MeV. This value is consistent with many theoretical treatments, for example, 0.4–1.5 MeV obtained by Rosenzweig[47a] and 3.5 MeV obtained by Milton, Palmer, and Pinsky.[47b]

TABLE 4. Masses, Widths, and Products of Branching Fractions for $J/\psi \rightarrow \gamma X$, $X \rightarrow \gamma\rho^0$

	$M(X)$	$\Gamma(X)$	B$(J/\psi \rightarrow \gamma X)B(X \rightarrow \gamma\rho^0)$
Mark III	$1420 \pm 15 \pm 20$	$133 \pm 55 \pm 30$	$(1.0 \pm 0.2 \pm 0.2) \times 10^{-4}$
DM2	1401 ± 18	174 ± 44	$(0.9 \pm 0.2 \pm 0.14) \times 10^{-4}$

We note that $M(X)$ for both experiments (and for the older Crystal Ball experiment as well) lies near the E meson, not the ι. If we perversely say, "it looks like a skunk, smells like a skunk, so it is a skunk," then because X is near the E mass, it is the E. Thus, it is the E that is radiatively produced, and both $\Gamma(E \rightarrow \gamma\rho)$ and $\Gamma(\iota \rightarrow \gamma\rho)$ must be redetermined. In a discussion of this point with Hitlin[48] before this session, he pointed out that if $\gamma\rho$ comes from the E meson, then J/ψ should decay into the E radiatively in the mode $J/\psi \rightarrow \gamma K\overline{K}\pi$, with $K\overline{K}\pi$ in the E mass region. This leads us to suggest that the decay $J/\psi \rightarrow \gamma K\overline{K}\pi$ should be analyzed by fitting with two Breit-Wigner resonances. In addition, the process $J/\psi \rightarrow \gamma X$, $X \rightarrow \gamma\rho$ should be fitted with two Breit-Wigners as well.

The $\iota \rightarrow \gamma\gamma$ decay width is also difficult to understand at present, particularly with respect to the $\iota \rightarrow \gamma\rho$ decay. The most recent result from TPC/2γ is a limit $\Gamma(\iota \rightarrow \gamma\gamma)$ B $(\iota \rightarrow K\overline{K}\pi) < 1.9$ KeV.[49] If B$(K\overline{K}\pi) = 1$, $\Gamma(\iota \rightarrow \gamma\gamma) < 1.9$ keV, which is less than the 5 keV expected from the application of vector dominance to the $\iota \rightarrow \gamma\rho$ decay width.[47b] Why is this so? At the moment there is no good answer. [After this conference, Meshkov, Palmer, and Pinsky[50] solved this dilemma by showing that the combination $u\bar{u} + d\bar{d} - 5s\bar{s}$ has $\Gamma(\iota \to \gamma\gamma) = 0$, yet has a finite $\gamma\rho$ decay width. This result was also known to M. Chanowitz and D. Caldwell.[49]]

It may well be that we have two conditions—$\Gamma(\iota \to \gamma\gamma)B(\iota \to K\bar{K}\pi) < 1.9$ keV and $\Gamma(E \to \gamma\gamma)B(E \to K\bar{K}\pi) < 1.9$ keV—in keeping with the spirit of the reanalysis of the $\gamma\rho$ decay suggested above.

Mixing Calculations

A complete mixing calculation for the 0^{-+} system still has not been carried out. A summary of earlier attempts has been given previously, so only three of the more recent attempts are discussed here:

(1) Mark III coupled channel fit for $J/\psi \to \gamma X$ for 0^{-+} X.[2,51] Four channels for X are $K\bar{K}\pi$, $\rho\rho$, $\omega\omega$, and $\gamma\rho$. They assumed that $\iota(1460)$ is the 1550 MeV $\rho\rho$ peak below threshold. Their coupled channel fit couples ι to all four channels; in addition, a second resonance is introduced to fit the 1800-MeV peak in the $\rho\rho$ distribution. They do not include η, η', ζ, or E in their fit. The first three channels are fit fairly well, but the fit to $\gamma\rho$ is poor.

(2) Chao, Palmer, and Pinsky[32] mix five bare states: 1η, $1\eta'$, $2\eta_n$, $2\eta_s$, and G. Taking the eigenstates of this 5×5 mass matrix to be the physical $\eta(549)$, $\eta'(960)$, $\zeta(1275)$, $E(1420)$, and $\iota(1460)$ particles gave no satisfactory solution. Future calculations should include $3\eta_n$ and $3\eta_s$. They found extreme sensitivity in their calculations to the exact mass of the radially excited pion, labeled as 2π. In particular, they found that $m(2\pi)$ had to be less than $m[\zeta(1275)]$.

(3) Lipkin[30] has proposed a degenerate pair $2\eta_s$ (which couples to hadrons) and $3\eta_n$ (which couples to J/ψ radiatively) that mix. He lists various ways of determining the mixing angle, but does not include mixing with other states.

0^{-+} Conclusions

(1) We must decide whether $E(1420)$ and $\iota(1460)$ are really two different particles.
(2) $\gamma\rho$ and $\gamma\gamma$ decays pose a problem.
(3) The coupling of both E and ι to J/ψ via radiative decays and by hadronic production implies mixing.
(4) Mixing calculations that include 1η, $1\eta'$, $2\eta_n$, $2\eta_s$, $3\eta_n$, $3\eta_s$, at least one glueball G, and possibly $q\bar{q}g$ seem required.
(5) It is still premature to label $\iota(1460)$ as a glueball.

2^{++} ISOSCALAR SYSTEM

We treat the 2^{++} system in analogy with the 0^{-+} states. All of the candidate 2^{++} states, together with their production modes, are listed in TABLE 5. In addition, in TABLE 6, we list a spectroscopic calculation for the 2^{++} $q\bar{q}$ states by Godfrey and Isgur,[28] together with a possible set of assignments to these and to the lowest 2-gluon 2^{++} glueball states.

TABLE 5. 2^{++} States and Their Production Modes

State	Produced Radiatively	Produced Hadronically	2γ
$f(1273)$	$\surd$	$\surd$	$\surd$
$f'(1516)$	$\surd$	$\surd$	$\surd$
$\theta(1720)$	$\surd$	$\surd$	$\surd$ (limit)
$f(1810)$		$\surd$	
$\xi(2230)?(0^{++}, 4^{++})$	$\surd$		$\surd$
$g_T(2050)$		$\surd$	
$g_{T'}(2300)$		$\surd$	
$g_{T''}(2350)$		$\surd$	

The existing 2^{++} resonances are the $f(1273)$, $f'(1516)$, $\theta(1720)$, $f(1810)$, $g_T(2050)$, $g_{T'}(2300)$, $g_{T''}(2350)$, and possibly $\zeta(2230)$. The J^{PC} of $\xi(2230)$ has not been determined, but because of its decays into K^+K^- and into $K_s^0 K_s^0$,[2] it can be 0^{++}, 2^{++}, or 4^{++}. Of these states, the possible glueball candidates are the $\theta(1720)$, $g_T(2050)$, $g_{T'}(2300)$, $g_{T''}(2350)$, and $\xi(2230)$. The 2^{++} system has been discussed in detail before,[1] but a few comments to update the situation are in order. The assignments of the $f(1273)$ and $f'(1516)$ as 1^3P_{2n} and 1^3P_{2s} are standard. Our state of knowledge about the rest of the system is so poor that even the first radial excitations are uncertain. I tentatively assign the resonance that we call $f(1810)$ and that has been seen in only two experiments[52,53] to the 2^3P_{2n} slot, in agreement with the Godfrey-Isgur[28] prediction of 1790 MeV/c^2. Its 2^3P_{2s} companion, predicted to be at 2030 MeV/c^2, has yet to be found, although it is conceivable that the mysterious $\theta(1720)$ might be assigned that role (discussed below).

Subsets of the 2^{++} System

f(1273) and f'(1516)

These are a pair of ideally mixed $q\bar{q}$, $L = 1$, $^3P_2$ states. For such an assignment, the radiative widths should satisfy the condition $\Gamma(J/\psi \to \gamma f')/\Gamma(J/\psi \to \gamma f) \leq 0.5$.[54]

TABLE 6. 2^{++}, $I = 0$ States (MeV/c^2)

State	Spectroscopic Calculation (Godfrey-Isgur)	Assignments (Meshkov)
1^3P_{2n}	1280	$f(1273)$
1^3P_{2s}	1510	$f'(1516)$
2^3P_{2n}	1790	$f(1810)$
2^3P_{2s}	2030	$\theta(1720)$?
1^3F_{2n}	2070	
1^3F_{2s}	2270	$\xi(2230)$?
1^3F_{4s}	2210	
$G(2g)\ L = 0, S = 2$		$g_T(2050)$
$G(2g)\ L = 2, S = 0$		$g_T(2300)$
$G(2g)\ L = 2, S = 2$		$g_{T''}(2350)$

Experimentally, this condition is satisfied by the Mark III result that the ratio is $\geq 0.35 \pm 0.14$. This removes any problem about an apparent violation of ideal mixing originally due to a small Mark II value for $B(J/\psi \rightarrow \gamma f')$. There is currently no need to mix glue or anything else into the f, f' system.

$\theta(1720)$

A test for the candidacy of a resonance to be a glueball is that it must be a flavor singlet.[1] As such, the decays of a 2^{++} glueball into two pseudoscalar mesons stand in the simple ratios shown in TABLE 7. A comparison with the data, listed in TABLE 7, indicates a discrepancy between the predicted ratios and those obtained experimentally.[51] (In order to compare, I have divided the experimental branching ratio by the phase-space for each mode.) The ratios do not look like those that correspond to a flavor singlet, so it seems that the $\theta(1720)$ is not a glueball. There are some caveats, of course. The $\eta\eta$ data is taken from Crystal Ball experiments, whereas the other two modes are from Mark III.[51] In a theoretical vein, Chanowitz and Sharpe[19b] have shown in a Bag Model calculation that the $s\bar{s}$ contribution may be enhanced for glueballs and meiktons containing TM gluons. (In addition, see reference 10.)

TABLE 7. Comparison of $\theta(1720)$ Decays to Two 0^{-+} States[51] with Flavor Singlet Predictions

	BR(Exp)	BR/β^5	$SU(3)$
$\pi\pi$	$(1.6 \pm 0.4 \pm 0.3) \times 10^{-4}$	1.7	3
$K\bar{K}$	$(9.6 \pm 1.2 \pm 1.8) \times 10^{-4}$	27	4
$\eta\eta$	$(3.8 \pm 1.6) \times 10^{-4}$	14	1

Contour plots of polarization fits to f, f', and θ in an analysis of Mark III production data show qualitatively different behavior for the production of quarklike f and f' on the one hand, and θ on the other.[54] The exact meaning of this difference is not understood, but it serves as an indication that there might be something different about θ.

$\theta(1720)$ was first discovered in the radiative decay channel $J/\psi \rightarrow \gamma\eta\eta$. DM2 has now produced θ hadronically[43] in the direct decay $J/\psi \rightarrow \phi K_s^0 K_s^0$. The experimental product of branching ratios is $B(J/\psi \rightarrow \phi\theta)B(\theta \rightarrow K\bar{K}) = (3.6 \pm 0.7 \pm 0.7) \times 10^{-4}$. However, Lindenbaum[4] reports that in $\pi^- p \rightarrow K_s^0 K_s^0 n$, a partial wave analysis in the 2^{++} partial wave amplitude of $K_s^0 K_s^0$ shows no evidence for the θ.

Therefore, what is $\theta(1720)$?

(1) $\pi\pi$, $K\bar{K}$, $\eta\eta$ decay ratios are not consistent with a flavor singlet assignment.

(2) The hadronic production of θ is consistent with its being an $s\bar{s}$ 2^3P_2 radial excitation of f'.

(3) The decays of θ into two pseudoscalars is consistent with its being primarily $s\bar{s}$ 2^3P_2 with some nonstrange quark admixture.

(4) The problem with comments 2 and 3 (above) is that $\theta(1720)$ lies 70 MeV below the nonstrange 2^3P_{2n} [or the $f(1810)$] in the Godfrey-Isgur calculation. A possible, but noncompelling explanation for how an $s\bar{s}$ state could lie below its

nonstrange counterpart is the following: perhaps the $s\bar{s}$ state originally does lie higher than the $f(1810)$, but it is preferentially depressed due to its interaction with numerous glueball states above it. This preferential interaction could come about if glueballs couple more strongly to strange quarks than to nonstrange quarks.[10]

$\xi(2230)$ $J^{PC} = 0^{++}, 2^{++}, 4^{++}$

For more information on this, see references 2 and 55.

(a) It is produced in $J/\psi \to \gamma K\bar{K}$ with $\Gamma = 18^{+23}_{-15} \pm 10$ MeV/c^2. This narrow width is consistent with the $\sqrt{OZI}$ rule estimate for a glueball decay width:[10]

$$\Gamma_G = \sqrt{\underset{\sim\ 100\ \text{MeV}}{\Gamma_{\text{ordinary}}} \quad \underset{1\ \text{MeV}/c^2}{\Gamma_{\text{OZI-violating}}}} \simeq 10\ \text{MeV}.$$

We have taken 30 MeV/c^2 as a standard.

(b) $\xi(2230)$ is consistent with the estimates of 1^3F_2, $q\bar{q}$, $L = 3$ at 2270 MeV/c^2, or 1^3F_4, $q\bar{q}$, $L = 3$ at 2210 MeV/c^2.

(c) ξ has possibly been seen in the DM2 experiment $J/\psi \to \gamma\Phi\Phi$.

(d) ξ conceivably could still be a Higgs particle.

$g_T(2050)$, $g_{T'}(2300)$, and $g_{T''}(2350)$ $\phi\phi$ Resonances

These three unusually broad states are the glueball candidates produced in the Zweig-forbidden reaction $\pi^- p \to \phi\phi n \to K^+K^-K^+K^- n$.[4,25,26] Although apparently Zweig-forbidden, the cross section is comparable to Zweig-allowed processes such as $K^- p \to \phi\phi(\Sigma^0, \Lambda)$ and $\pi^- p \to \phi K^+ K^- n$. The strength of the Zweig-forbidden process is attributed to the formation and subsequent decay into two ϕ's of a glueball. The most recent determination of the properties of the g_T states is given in TABLE 8. Note that the g_T masses have changed somewhat, that is, $g_T(2120) \to g_T(2050)$, $g_{T'}(2220) \to g_{T'}(2300)$, and $g_{T''}(2360) \to g_{T''}(2350)$.

What assignments can be made for these states?

(a) They could correspond to the three lowest $G(2g)$ states shown in FIGURE 1, that is,

$$g_T(2050) \qquad L = 0, S = 2$$
$$g_{T'}(2300) \qquad L = 2, S = 0$$
$$g_{T''}(2350) \qquad L = 2, S = 2.$$

(b) They could correspond to a mixed $q\bar{q}$, $L = 1$ and two $G(2g)$ states, or various permutations and combinations of $q\bar{q}$ and $G(2g)$ states. An example of such a system is:

$$g_T(2050) \qquad 2^3P_{2s} \text{ (Godfrey-Isgur 2030)}$$
$$g_{T'}(2300) \qquad G(2g)$$
$$g_{T''}(2350) \qquad G(2g).$$

We note that Mark III has seen a new $\pi^+\pi^-$ state (no J^{PC} determination) with a mass

TABLE 8. Quantum Numbers and Parameters of the Breit-Wigner Resonances Extracted from the Three-Pole K-Matrix Fit[4]

% Data		$I^G J^{PC}$	Mass (GeV)	Γ_{tot} (GeV)	Branching Ratio (%)		
					$L = 0, S = 2$ S_2	$L = 2, S = 2$ D_2	$L = 2, S = 0$ D_0
50^{+10}_{-10}	g_T (2050)	0^+2^{++}	$2.05^{+0.09}_{-0.05}$	$0.20^{+0.16}_{-0.05}$	98^{+2}_{-70}	0^{+50}_{-0}	2^{+25}_{-2}
20^{+20}_{-10}	$g_{T'}$ (2300)	0^+2^{++}	$2.30^{+0.02}_{-0.10}$	$0.20^{+0.06}_{-0.05}$	30 ± 20	50^{+20}_{-10}	20 ± 20
30^{+10}_{-20}	$g_{T''}$ (2350)	0^+2^{++}	$2.35^{+0.02}_{-0.03}$	$0.27^{+0.09}_{-0.13}$	40^{+10}_{-20}	5^{+15}_{-5}	55^{+20}_{-15}

$M_x = 2086 \pm 15$ MeV$/c$ and a width $\Gamma_x = 210 \pm 63$ MeV$/c^2$. Is this state related to $g_T(2050)$?

2^{++} *Conclusions*

(1) The g_T states are still glueball candidates. To confirm this assignment, we need: (a) a measurement of B($J/\psi \to \gamma g_T$) and (b) a measurement of their branching ratios for decay into the two pseudoscalar modes, $\pi\pi$, $K\overline{K}$, and $\eta\eta$ to verify that they correspond to those of flavor singlets.

(2) $\xi(2230)$ is a candidate for glueball status because of its $\sqrt{\text{OZI}}$ width. We need to study its two pseudoscalar decay branching ratios as discussed above.

(3) $\theta(1720)$ does not seem to be a viable glueball candidate due to $\pi\pi$, $K\overline{K}$, and $\eta\eta$ branching ratios that do not correspond to decays coming from a flavor singlet.

(4) A good 2^{++} $q\bar{q}$ QCD spectroscopic calculation is needed. This is easier to do than for the 0^{-+} case.

$\gamma\gamma$ PRODUCTION OF VECTOR MESON PAIRS

In the two-photon energy interval, $W\gamma\gamma \simeq 1.5$–2.5 GeV, it is found experimentally that $\sigma(\gamma\gamma \to \rho^0\rho^0) \gg \sigma(\gamma\gamma \to \rho^+\rho^-)$. An ordinary $q\bar{q}$ $I = 0$ resonance has $\rho^+\rho^-/\rho^0\rho^0 = 2$. In order to explain this effect, ingenious but rather farfetched explanations (for example, the interference of two isospin amplitudes for a four-quark resonance with a width $\Gamma \simeq 1$ GeV$/c^2$) have been invoked.[37] It seems very unreasonable to invoke an exotic 1 GeV$/c^2$ wide resonance when the energy region $W_{\gamma\gamma} = 1.5$–2.5 GeV is filled with many hadronic resonances of standard widths (see TABLES 2 and 5, for example) that can decay to $\rho^0\rho^0$ and $\rho^+\rho^-$. In an attempt to see whether standard physics can account for this effect, S. J. Brodsky and S. Meshkov[56] have used factorization and vector dominance to yield a major piece of the $\gamma\gamma \to \rho^0\rho^0$ cross section. The rest comes from ordinary resonance decay.

One method for doing this[37,57] relates $d\sigma/dt(\gamma\gamma \to \rho^0\rho^0)$, $d\sigma/dt(\gamma p \to \rho^0 p)$, and $d\sigma/dt(pp \to pp)$, with comparisons carried out at the same Q – final state kinetic energy. Agreement with experiment is obtained for $W_{\gamma\gamma} > 2$ GeV for $\gamma\gamma \to \rho^0\rho^0$, $\gamma\gamma \to \omega\omega$, and other vector meson processes. A lack of low energy photoproduction data and proton-proton data is a problem at lower $W_{\gamma\gamma}$ energies. Further work continues.

SUMMARY

(A) $g_T(2050)$, $g_{T'}(2300)$, and $g_{T''}(2350)$ are glueball candidates.

(B) $\xi(2230)$ is also a glueball candidate, but it could be assigned to a $q\bar{q}$, $L = 3$ state.

(C) The 0^{-+} system is very interesting and complicated.

(D) $\iota(1460)$ cannot be confirmed as a glueball until the 0^{-+} system is unscrambled.

ACKNOWLEDGMENTS

The author is indebted to D. O. Caldwell, P. M. Fishbane, D. Hitlin, S. J. Lindenbaum, W. F. Palmer, and S. S. Pinsky for countless useful and informative discussions about the data and their interpretation. Also, thanks are due to S. J. Brodsky, S. U. Chung, I. J. Muzinich, P. J. O'Donnell, and P. Zerwas for important comments.

REFERENCES

1. See, for example, a recent review by: FISHBANE, P. M. & S. MESHKOV. 1984. Comments Nucl. Part. Phys. **13:** 325, and references cited therein.
2. HITLIN, D. 1987. These proceedings.
3. CHUNG, S. U. 1987. These proceedings.
4. LINDENBAUM, S. J. 1987. These proceedings.
5. EICHTEN, E. *et al.* 1979. Phys. Rev. **D17:** 3090; 1980. Phys. Rev. **D21:** 203; EICHTEN, E. 1985. *In* The Sixth Quark. Proceedings of the Twelfth SLAC Summer Institute on Particle Physics, 1984. P. M. McDonough, Ed. SLAC. Stanford; and references therein.
6. RICHARDSON, J. 1979. Phys. Lett. **82B:** 272; BUCHMULLER, W. & S-H. H. TYE. 1981. Phys. Rev. **D24:** 132.
7. BEAVIS, D., S-Y. CHU, B. R. DESAI & P. KAUS. 1979. Phys. Rev. **D20:** 2345; SCHMITZ, S., D. BEAVIS & P. KAUS. 1985. University of California, Riverside report no. UCR-TH-85-3. Unpublished.
8. MESHKOV, S. 1982. Proceedings of the Johns Hopkins Workshop on Current Problems in Particle Theory, Florence, Italy, June 2–4, 1982.
9. (a) ROBSON, D. 1977. Nucl. Phys. **B130:** 328; (b) COYNE, J., P. FISHBANE & S. MESHKOV. 1980. Phys. Lett. **91B:** 259; (c) BJORKEN, J. D. 1979. SLAC Summer Institute on Particle Physics. SLAC-PUB-2372.
10. CARLSON, C. E., J. J. COYNE, P. M. FISHBANE, F. GROSS & S. MESHKOV. 1981. Phys. Lett. **99B:** 353.
11. GELL-MANN, M. 1984. The meson $\kappa(725)$ and the eightfold way. *In* The Eightfold Way. M. Gell-Mann & Y. Ne'eman, Eds.: 98. Benjamin. New York.
12. MUZINICH, I. J. Private communication; MUZINICH, I. J. & V. P. NAIR. General results on glueball masses in QCD. ITP-UCSB preprint. Unpublished.
13. DE FORCRAND, PH. *et al.* 1985. Phys. Lett. **152B:** 107; **160B:** 137; CERN preprint no. CERN-TH.4167/85 (April).
14. CREUTZ, M. 1979. Phys. Rev. Lett. **43:** 553; 1980. Phys. Rev. **D21:** 2308.
15. BERG, B. & A. BILLOIRE. 1986. Phys. Lett. **166B:** 203; BERG, B. *et al.* 1986. Florida State University preprint no. FSU-SCRI-86-06.
16. CARLSON, C. E., T. H. HANSSON & C. PETERSON. 1983. Phys. Rev. **D27:** 1556, and references cited therein; CORNWALL, J. M. & A. SONI. 1983. Phys. Lett. **120B:** 431.
17. ZACHARIASEN, F. 1987. These proceedings.
18. OTTO, S. & J. D. STACK. 1984. Phys. Rev. Lett. **52:** 2328; BARKAI, D., K. J. M. MORIARTY & C. REBBI. Phys. Rev. **D30:** 1293.
19. CHANOWITZ, M. & S. SHARPE. 1983. Nucl. Phys. **B222:** 211; Phys. Lett. **132B:** 413.
20. BARNES, T. 1985. The Bag Model and Hybrid Mesons. SIN Spring School on Strong Interactions, Zuoz (Engadin), Switzerland, April 9–17, 1985.
21. SCHARRE, D. L. *et al.* 1980. Phys. Lett. **B97:** 329.
22. EDWARDS, C. *et al.* 1982. Phys. Rev. **D49:** 259.
23. EDWARDS, C. *et al.* **D48:** 458.
24. FRANKLIN, M. E. 1982. Preprint no. SLAC-254.
25. ETKIN, A. *et al.* 1978. Phys. Rev. Lett. **40:** 422; **41:** 784.
26. LINDENBAUM, S. J. 1984. Comments Nucl. Part. Phys. **13:** 285.
27. FRANK, M. & P. J. O'DONNELL. 1984. Phys. Lett. **144B:** 451; Phys. Rev. **D29:** 921.
28. GODFREY, S. & N. ISGUR. 1985. Phys. Rev. **D32:** 189.

29. STANLEY, D. P. & D. ROBSON. 1980. Phys. Rev. **D21:** 3180.
30. COHEN, I. & H. J. LIPKIN. 1979. Nucl. Phys. **B151:** 16; LIPKIN, H. J. & I. COHEN. 1984. Phys. Lett. **135B:** 215; LIPKIN, H. J. 1985. ANL-HEP-PR-85 (October) preprint.
31. BENDER, C., W. F. PALMER & S. S. PINSKY. 1984. Phys. Rev. **D30:** 1002.
32. CHAO, S. C., W. F. PALMER & S. S. PINSKY. 1986. Phys. Lett. **172B:** 172.
33. CARLSON, C. E. & C. PETERSON. 1985. Phys. Rev. Lett. **55:** 355.
34. O'DONNELL, P. J. Private communication.
35. ROSNER, J. L. 1983. Phys. Rev. **D27:** 1101.
36. BALTRUSAITIS, R. M. *et al.* 1985. Phys. Rev. **D32:** 2883.
37. KOLANOSKI, H. 1985. Proceedings of the 1985 International Symposium on Lepton and Photon Interactions at High Energies, Kyoto, Japan, August 19–24, 1985, p. 89–116.
38. PALMER, W. F. & S. S. PINSKY. Private communication.
39. TORNQVIST, N. A. 1979. Ann. Phys. (N.Y.) **123:** 1.
40. STANTON, N. *et al.* 1979. Phys. Rev. Lett. **42:** 346.
41. ANDO, A. *et al.* 1985. KEK 85-15 (June) preprint.
42. EDWARDS, C. 1984. Caltech Ph.D. thesis no. CALT-68-1165.
43. AUGUSTIN, J. E. *et al.* 1985. LAL 85/27 (July) Orsay preprint, and references cited therein.
44. RICHMAN, J. D. 1985. Caltech Ph.D. thesis no. CALT-68-1231.
45. See also review by: COOPER, S. 1985. SLAC-PUB-3819 (October); Proceedings of the International Europhysics Conference on High Energy Physics, Bari, Italy, July 18–24, 1985.
46. PALMER, W. F. & S. S. PALMER. 1983. Phys. Rev. **D27:** 2219.
47. (a) ROSENZWEIG, C. 1983. Syracuse preprint; (b) MILTON, K. A., W. F. PALMER & S. S. PINSKY. 1983. Phys. Rev. **D27:** 202.
48. HITLIN, D. Private communication.
49. CALDWELL, D. O. Private communication.
50. MESHKOV, S., W. F. PALMER & S. S. PINSKY. 1986. VIIth International Workshop on Photon-Photon Collisions, College de France, Paris, France, April 1–5, 1986.
51. WERMES, N. 1985. SLAC-PUB-3730. Invited talk at Physics in Collision V Conference, Autun, France, July 3–5, 1985.
52. CASON, N. M. *et al.* 1982. Phys. Rev. Lett. **48:** 1316.
53. COSTA, G. *et al.* 1980. Nucl. Phys. **B175:** 402.
54. EINSWEILER, K. F. 1984. Stanford Ph.D. thesis no. SLAC-272 (May).
55. TOKI, W. 1985. SLAC-PUB-3852 (November); Proceedings of SLAC Summer Institute on Particle Physics, Stanford, California, July 29–August 6, 1985.
56. BRODSKY, S. J. & S. MESHKOV. To be published.
57. ALEXANDER, G., A. LEVY & U. MAOR. 1986. Z. Phys. C. Part. Fields **30:** 65, and references cited therein.

Collider Physics: A Theorist's View

STEPHEN D. ELLIS

Department of Physics
University of Washington
Seattle, Washington 98195

INTRODUCTION

The following is (one) theorist's view of what has been learned or may soon be learned from the experimental results from the CERN $\bar{p}p$ Collider. The bad news is that nothing "new" has appeared yet. The good news is that the standard model (SM) is being quite precisely and beautifully tested by the results from the collider.[1] In particular, all three essential ingredients of the standard 1-2-3 model are readily observed at the collider. There are good data on the observation of photons [=(1)], W's and Z's [=(2) and some Nobel prizes], and QCD jets [=(3)]. We shall divide the following necessarily brief discussion into three main sections: (*i*) Present tests of the SM; (*ii*) Possible tests of the SM; (*iii*) Yet to be explained results that (almost surely) lie within the realm of the SM. For the sake of brevity, only a few examples will be discussed. The topics appear in order of decreasing detail, decreasing rigor, and decreasing expertise on the part of the author.

PRESENT TESTS OF THE STANDARD MODEL

A particularly interesting (and exciting last year) example in this category is the study of events containing a W or Z plus 0, 1, or 2 jets. This study not only turns out to afford an impressive test of the standard model (by normal hadronic machine standards), but it also yields a description of many of the standard model sources of "backgrounds" to more exciting processes. An example of this latter category is the production of supersymmetric partner particles, which is a process that has captured considerable attention at CERN and elsewhere for the last few years. Sadly, it now seems clear, as will be illustrated below, that all events at the CERN collider are well described[2] by the standard model when studied carefully.

The analysis[3-5] to be described briefly here was performed in collaboration with W. J. Stirling and R. Kleiss. There are several features of the analysis that, at least in combination, are new to this sort of work and should be helpful for subsequent work. The perturbative calculations of the various parton subprocesses were carried out utilizing the "spinor" techniques of CALKUL,[6] which deal directly with helicity amplitudes rather than squares of amplitudes and traces over spin. This approach not only simplifies the evaluation of Feynman graphs and expedites the inclusion of the matrix elements in Monte Carlo programs that perform the full phase-space integrals, but, more importantly for the present issues, it leads naturally to the evaluation of complete processes including the final state decay of the vector bosons. Thus, the amplitudes calculated describe quark and gluon scatterings into quarks, gluons, and leptons with the vector bosons appearing only as intermediate states. This leads to the second important new feature. In the perturbative contact described above, it is natural to impose the various experimental cuts directly on the perturbative final state, both on the partons ($\equiv$ JETS) and the leptons. The relative simplicity of this perturbative analysis is an important complement to the more complete and (necessarily) more opaque analyses[7] being carried out by the experimental groups themselves. These latter analyses include the full simulation of the final state, along with detector resolution and efficiency effects. It is interesting to note that even in the present simpler analysis, the calculation is essentially carried out event-by-event. Because the multibody phase-space integrals are performed by Monte Carlo techniques, one is essentially summing over points in phase points with subsequent weighting by the matrix element. This approach is similar, in many respects, to the more complete event–by–full event generation procedure. In particular, one implements the above cuts by keeping or discarding a given Monte Carlo generated phase-space point (event) depending on whether it passes the cuts just as one does in the more complete simulations.

Because these cuts will play such an important role here and below, it is helpful to review their basic form. Our (somewhat simplified theorist's) detector has a cylindrical geometry as illustrated in FIGURE 1. We are led both experimentally and theoretically to impose three types of cuts on the final states detected in this detector. First, there is a minimum P_T cut for both jets and leptons as illustrated in FIGURE 1. This simply guarantees that the jet or lepton has enough momentum in the direction of the detector to be detected and identified. As indicated in FIGURE 1, the second cut is on the rapidity and is intended to ensure that the corresponding particle is not lost out of the end of the detector (always a possibility at large enough rapidity for collider experiments). The final variety of cut, as illustrated in FIGURE 1, controls the angular separation between jets and between jets and detected leptons. This cut serves to distinguish between 0, 1, 2, or more jet events and to ensure that leptons are not "lost" in jets. This last cut should not be necessary for μ detection, but in the present study, we shall focus on electrons where it does play an essential role. For the essentially cylindrical structure of the UA1 experiment, the natural angular separation variable is $\Delta r = \sqrt{\Delta\phi^2 + \Delta\eta^2}$, where $\Delta\phi$ is the azimuthal separation and $\Delta\eta$ is the rapidity separation. For the more spherical geometry of the UA2 experiment, the natural angular separation variable is a polar angle separation $\Delta\theta$. The precise values of the various cut parameters depend, of course, on the particle type and the particular detector (and time). For the following comparison to data, we shall use cut parameters:

$$P_T^{\text{Jet}} \geq P_T^{\min,J} = 6 \text{ GeV}/c \text{ [UA1 and UA2]};$$

$$P_T^e \geq P_T^{\min,e} = 15 \text{ GeV}/c \text{ [UA1] or } 17 \text{ GeV}/c \text{ [UA2]};$$

$$P_T^\nu \geq P_T^{\min,\nu} = 15 \text{ GeV}/c \text{ [UA1] or } 25 \text{ GeV}/c \text{ [UA2]};$$

$$|\eta^{e,J}| \leq \eta_0 = 2.5 \text{ [UA1] or } 1.7 \text{ [UA2]};$$

$$\Delta r(e, J), \Delta r(J, J) \geq r_0 = 1.0 \text{ [UA1]};$$

$$\Delta\theta(e, J), \Delta\theta(J, J) \geq \theta_0 = 45° \text{ [UA2]}.$$

It is important to note that these cuts not only define the situation experimentally, but also theoretically. The $P_T^{\min}$ cut serves to ensure that there are no infrared

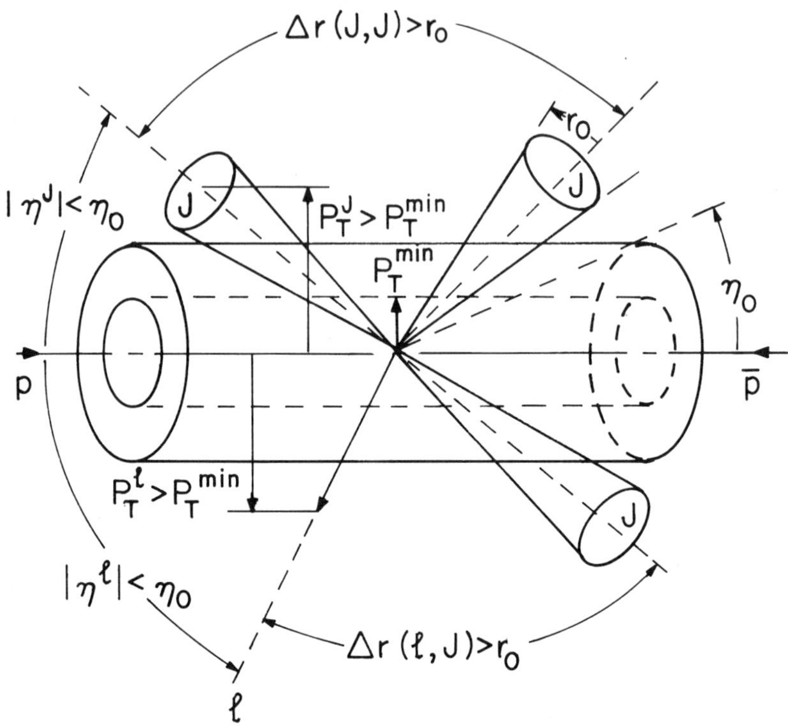

FIGURE 1. Idealized collider-style detector illustrating the various types of cuts discussed in the text. The astute reader will note that in order to limit the number of arrows in the center of this figure, the arrows indicating rapidities actually point in the direction of decreasing rapidity.

singularities. The rapidity cut, η_0, eliminates the possibility of collinear singularities due to initial state bremsstrahlung along the beam direction, while r_0 or θ_0 eliminates final state collinear singularities. The overall effects of the cuts in controlling potential singularities operate similarly to those introduced by Sterman and Weinberg[8] in the context of e^+e^- physics some time ago.

Before proceeding to a sampling of comparisons with the data, we should briefly review the major theoretical uncertainties in the present analysis. They fall into three major classes: (1) the choice of structure functions describing the incident flux of patrons; (2) the choice of scale Q characterizing the hard scattering process to be used in both the structure functions and the strong coupling constant, $\alpha_s(Q)$; (3) the interplay of the cuts on the jets with possible higher-order corrections and fragmentation effects. This latter issue is essentially the question of "K" factors,[9] which, due to the explicit cuts, will depend on the cut values. The first two questions can and will be directly addressed by calculating with two standard but different structure function sets (with different choices for Λ_{QCD}) and with two characteristic (effectively, maximum and minimum) values for the scale Q: $Q = M_V$ (the vector boson mass) and $Q = P_T^V$ (the vector boson transverse momentum). In particular, we will use structure functions from Duke and Owens[10] with $\Lambda_{QCD} = 0.2$ GeV and $Q = M_V$ (labeled DO1 and hereafter referred to as Mode 1) and those of Gluck, Hoffman, and Reya[11] with $\Lambda_{\mathrm{QCD}} =$ 0.4 GeV and $Q = P_T^V$ (labeled GHR and hereafter referred to as Mode 2). The differences between the resulting cross sections will yield a fair estimate of the corresponding theoretical uncertainty. The third issue is more complex and can be settled only by considerably more work being done both perturbatively and in utilizing Monte Carlo fragmentation studies. In the following analysis, we shall simply ignore any further corrections; that is, we set all K factors to unity. It is reassuring to note that the normalizations that result compare extremely well with the experimental results; this suggests that for the cut values chosen (and the chosen Q values), the K values are indeed (perhaps serendipitously) near unity. In any case, it is probably fair to assume that the uncertainties in the K factors are no larger than those directly illustrated by varying the structure functions and Q values.

Let us now turn to just a few illustrations of how well this simple description of the data works. More complete discussions can be found in the literature. We shall focus on $W^\pm$ production where the statistics are somewhat better. First, consider the total rate of produced and detected $W^\pm$'s that are detected via $e^\pm$'s plus the missing P_T

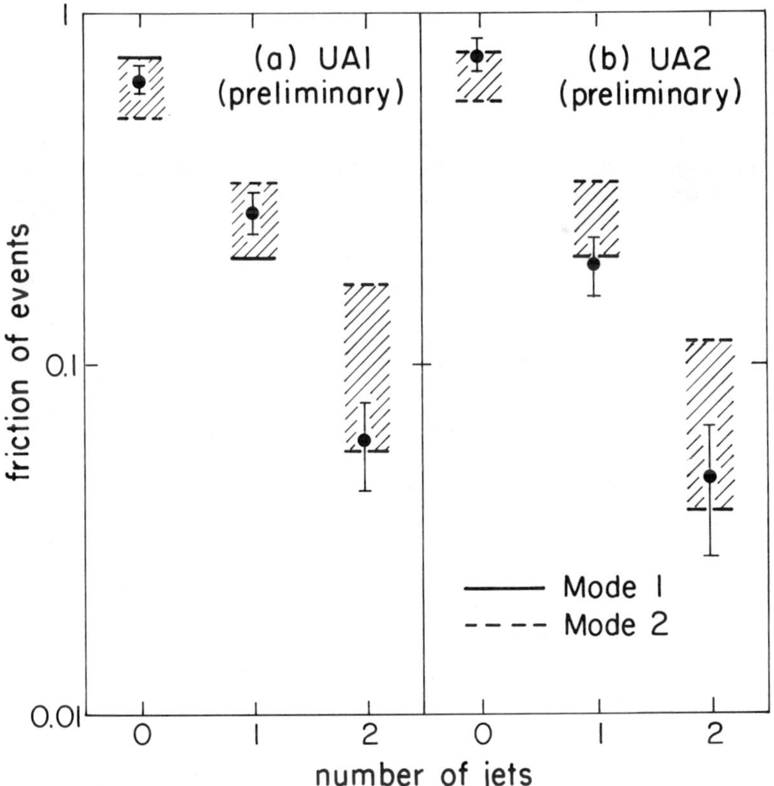

FIGURE 2. Fractions of events with various numbers of jets accompanying a detected W. The data are from reference 12. The Modes of the various theoretical results are defined in the text.

corresponding to the undetected ν's. For the UA1 sample, M_W is taken to be 81 GeV/c^2 with an integrated luminosity of 136 nb^{-1} at 540 GeV and 263 nb^{-1} at 630 GeV. The corresponding numbers for the UA2 sample are 83 GeV, 142 nb^{-1}, and 310 nb^{-1}. This results[3] in a predicted number of events of 174 for UA1 and 137 for UA2, with an uncertainty of about ± 3 events estimated by varying the structure function and Q. The observed numbers of events are 172 and 126 for UA1 and UA2, respectively. Now that we have established that the overall rate is about right, consider next how these events are distributed according to the number of accompanying jets (defined by the cuts above). The results[3] are illustrated in FIGURES 2a and 2b. The data[12] indicated are preliminary UA1 and UA2 results from the 1983 and 1984 runs. The theoretical results are indicated as the hatched regions, with the boundaries corresponding to the two sets of structure functions discussed above. Mode 1 gives the lower bound for 1 and 2 jets, while (due to the overall normalization) it gives the upper bound for 0 jets. Again, the agreement is good with no particular attempt to "tune" parameters.

Finally, we turn to the issue of the production of $W^{\pm}$'s at large P_T where we may expect our perturbative analysis to be most reliable and which will be of interest to us below. FIGURES 3a and 3b indicate the UA1 and UA2 data[13] as a function of the P_T^W again for the 1983 and 1984 runs. The upper curve (long dashes) in both cases is Mode 2 with Mode 1 (solid line) yielding a somewhat smaller large P_T^W tail. This is in agreement with FIGURE 2 and indicates that Mode 2 yields more jets and larger P_T^W on average. The primary reason for this is the choice of the scale $Q = P_T^W$ in Mode 2 instead of the choice $Q = M_W$ of Mode 1. This feature is also illustrated by the short dashed curve in FIGURE 3a, which is the result for Mode 2, but with $Q = M_W$, and which is essentially identical to the Mode 1 result with the same Q, but with different Λ_{QCD} and structure functions. The major theoretical uncertainty thus corresponds to the differences in the various Q values. A thorough understanding of this issue will follow only from a more complete (higher order) perturbative analysis including the virtual loop contributions (a rather unlikely prospect at the moment). It is also interesting to note that the theoretical uncertainty (the difference between the two curves) is of the order of the experimental uncertainty (the difference between the two experimental results). The integral under the UA1 results gives 10 events in good agreement with the Mode 1 result of about 9 events, while the Mode 2 result for UA2 of about 11 events (recall that the cut parameters are slightly different for the two detectors) agrees well with the observed 10 events. More refined analyses therefore must await further data, but the situation is already very impressive. For our analysis below, it is useful to summarize our results for the UA1 cuts with 136 nb^{-1} at 540 GeV and 263 nb^{-1} at 630 GeV and with the refinement that here Mode 2' corresponds to Mode 2 above with $Q = M_W$, which we saw gives a better description of the large P_T^W tail. Thus, the small differences between Modes arise entirely from the different Λ_{QCD} values and different structure functions. The corresponding numbers[5,13] of events are:

Channel	UA1	Mode 1	Mode 2'
$W^{\pm} \rightarrow e^{\pm}\nu$			
Total	172	177.6	171.0
$P_T^W \geq 20$ Gev/c	10	9.9	10.9
$Z^0 \rightarrow e^+ e^-$			
Total	18	18.9	19.8
$P_T^Z \geq 20$ GeV/c	0	0.9	1.1

This leads us to the following conclusions:

(A) The techniques used in the above analysis to calculate amplitudes directly work well and it seems reasonable to assume that one may reliably proceed to calculate tree-level amplitudes involving larger numbers of jets.

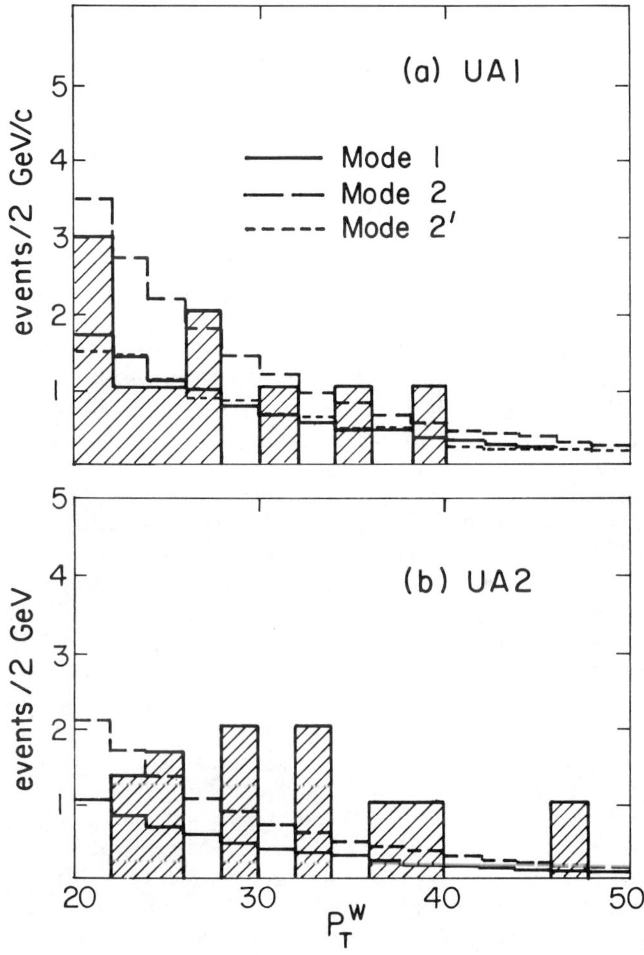

FIGURE 3. Large P_T distribution of detected W's including data from reference 13. The Modes characterizing each of the theoretical results are defined in the text.

(B) The technique of applying the jet-defining cuts directly on the partons with the accompanying simplifications also seems to work well.

(C) For reasons of brevity, we have not discussed here two other conclusions that arise from these analyses; for completeness, I will list them here:

(1) backgrounds to the hadronic decays of $W^\pm$ and Z^0 at the SSC are likely to play an important role;[14]

(2) processes involving two distinct hard scattering reactions can probably be rather simply separated from multijet single scattering processes using the ϕ correlation structure.[3]

(D) Finally, we can conclude that QCD can and will be carefully checked at hadronic colliders via the study of events involving the production of both jets and $W^{\pm}$'s and Z^0's.

However, we are left with one question. Because by directly applying detection and identification cuts to our perturbative final states we can quantitatively describe the observed rate of detected vector bosons, what happens to the unidentified ones? Because our analysis is "absolutely" normalized relative to the data via the detected bosons, the same analysis predicts with high reliability the rate of unidentified bosons. They must be somewhere. As the simplest example, consider the production of Z^0's that then decay via ν pairs, which are always undetected. The rate of such events is a straightforward factor times the detected rate of decays in e^+e^-. This ratio involves a factor of six to account for the ratio of branching ratios plus a factor (often ignored, but that can actually be of order 1.4) to account for the fact that (for real detectors) the acceptance to not detect ν's is larger that the acceptance to detect $e^{\pm}$'s. If the Z^0 carries sizable P_T, such events will appear in the very interesting category of "missing E_T" or "monojet" events. At one time, such events were considered as promising evidence[15] for the presence of physics beyond the Standard Model. However, as will be described below, they now appear instead to offer another, perhaps surprisingly precise, test of the Standard Model.

An important feature of Standard Model sources of monojet events is that there are so many distinct channels that can contribute. In particular, besides the trivial source mentioned above, there are many ways in which a $W^{\pm}$ can decay without being identified while leading to undetected ν's (instead of, say, photinos) with sizable E_T. The charged lepton produced in the $W^{\pm}$ decay can be undetected for reasons that exactly match the various cuts defined above. It can be "lost" out the end of the detector, $|\eta^l| > \eta_0^l$, be too "soft" to be identified, $P_T^l < P_T^{\min,l}$, or be "hidden" in a jet, $\Delta r(l, J) < r_0(l, J)$. While each of these channels makes a small contribution to the monojet signal, their sum can be physically significant. (In the early days of the consideration[2,15] of this signal, it was often the case that the various contributions were treated individually and discarded individually as being insignificant *individually*.)

Before we consider the magnitude of these various SM contributions to the monojet signal, there are three technical details to be considered. First, it is important to recognize that the various identification criteria (cuts) serve to divide vector boson events into three categories. These categories correspond to events with definitely identified vector bosons: that is, those events that pass the cuts defined above; those events that do not pass the above cuts, but in which there is some evidence of a vector boson and which are therefore not included in either the vector boson or the monojet sample; and finally those events that give no detectable indication of the vector boson and unavoidably appear in the monojet sample. Thus, the cuts for not detecting the vector boson, which define the third category, are not simply the complement of the cuts defining the first category, but are rather more restrictive (and are a source of possible debate). The cuts used in the present analysis[5] to define undetectable lepton events in the UA1 detector are:

Lost	$\|\eta^e\| \geq 3.0$	$\|\eta^\mu\| \geq 3.0$
Soft	$P_T^e \leq 10 \text{ GeV}/c$	$P_T^\mu \leq 3 \text{ GeV}/c$
Hidden	$\Delta r(e, J) \leq 0.4$	$\Delta r(\mu, J) \leq 0.0$ (no cut).

The second technical point concerns the treatment of events involving the production of $W^{\pm}$'s that subsequently decay into τ's. Such events are considered as unidentified $W^{\pm}$ events if the τ is not identified. The τ can avoid identification in several ways. It can decay purely leptonically into either an e or μ that can go undetected for the same reasons as for a primary e or μ as described above (plus 2 ν's that always escape detection). In the present analysis, the momentum of the secondary e or μ is obtained by convoluting the calculated momentum distribution of the τ's with the known leptonic decay spectrum. The hadronic decay $\tau \rightarrow \nu + \overline{Q}Q$ is always considered as an unidentified τ event. (This is different from earlier analyses[4] where the τ jet was considered to be identifiable, but it is consistent with present UA1 practice; see the further discussion below.) The resulting final state is treated as an unseen ν plus a single hadronic jet with a fractional momentum distribution relative to the τ momentum contructed[16] from the observed exclusive hadronic τ decays plus a 10% continuum contribution. This last contribution ensures momentum conservation, and the corresponding momentum distribution can be evaluated perturbatively. The jet from a τ decay is treated somewhat differently from other jets. If the jet passes the jet-defining cuts above, it is included in the jet counting for that event. However, even if the τ jet is missed because the jet is too soft or lost out of the end caps or hidden in another jet, the event is still kept as a potential monojet event depending on the energy carried off by the missing ν's and the presence of other jets. Note that for the case of a hidden τ jet, the total momentum of the resulting single jet is the sum of the QCD jet plus τ jet.

The third technical point involves the requirement that we define a theoretical monojet sample in a fashion commensurate with the UA1 definition. In particular, the experimental missing E_T is inferred from the measured transverse energy in the event that arises from both the hard scattering process (the perturbative jets) and the remainder of the "underlying event." Because this measurement is performed with limited resolution, a violation of transverse momentum conservation (that is, missing $E_T > 0$) can be observed even in the absence of ν's. For the UA1 detector, it is observed[17] that this resolution smearing is approximately described by a Gaussian distribution of width

$$\sigma = 0.7 \sqrt{E_T^{\text{TOT}}}. \tag{1.1}$$

We can simulate this effect by smearing our theoretical missing E_T distribution with the use of such a Gaussian distribution (i.e., with each event, we associate an extra contribution to the missing E_T according to such a Gaussian distribution). To define the E_T^{TOT} associated with a given perturbative event, we use a simple two-component picture:

$$E_T^{\text{TOT}} = E_T^{\text{Hard}} + E_T^{\text{ue}}, \tag{1.2}$$

where E_T^{Hard} is the transverse energy associated with the perturbatively calculated hard scattering process and E_T^{ue} is interpreted as the transverse energy associated with the underlying event (taken as incoherent with the hard process). Values for E_T^{ue} are generated for each event according to the empirical distribution

$$\frac{dN}{dE_T^{\text{ue}}} = [E_T^{\text{ue}}]^2 \, e^{-bE_T^{\text{ue}}}, \tag{1.3a}$$

where the average value of E_T^{ue} is taken to be

$$\langle E_T^{\text{ue}} \rangle = \frac{3}{b} = 30 \pm 10 \text{ GeV}. \tag{1.3b}$$

Within the indicated uncertainty, this distribution offers a reasonable description of the observed E_T^{TOT} distribution[18] of (low P_T) $W^{\pm} + 0$ jet events, as is illustrated in

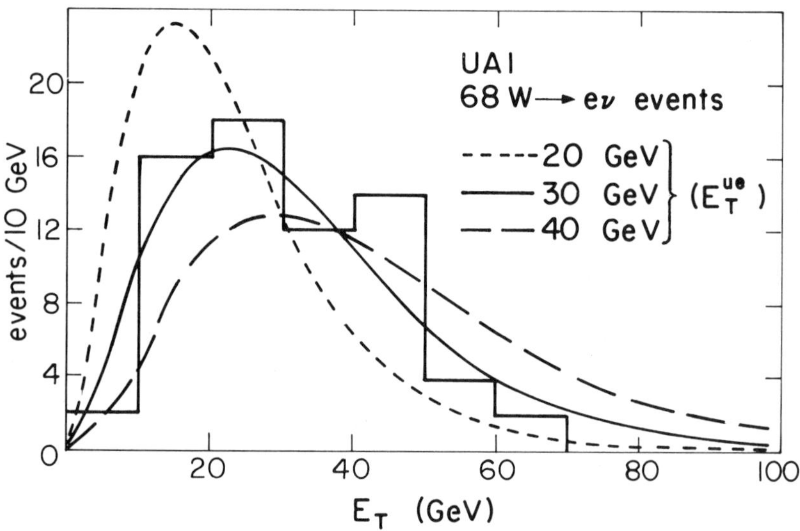

FIGURE 4. Scalar E_T accompanying detected W bosons. The data are from reference 18. The three theoretical distributions correspond to the three values for the average underlying event E_T, as described in the text.

FIGURE 4. As indicated, the three lines correspond to $\langle E_T^{\text{ue}} \rangle = 30$ GeV (solid), 20 GeV (short dashes), and 40 GeV (long dashes). We will see below that the implied uncertainty in this procedure has very little influence on the predicted cross section for large missing E_T events. This missing E_T smearing procedure also serves to effectively smear the P_T distribution of the W for $\tau + 0$ jet events where there is no perturbatively generated P_T. The corresponding missing E_T distribution then extends beyond the naive limit of $M_W/2$, as is indicated by the results below.

With these technical details behind us, we can proceed to compare the results of the theoretical analysis[5] outlined above with the 1984 UA1 monojet sample.[7] The event selection process, both experimentally and theoretically, involves a basic trigger or "on-line" event selection (discussed in more detail below) plus three specific selection criteria. The first cut is on the missing E_T itself. We require that

$$E_T^{\text{miss}} \geq \text{Max}\ [4\sigma,\ 15\ \text{GeV}], \tag{1.4a}$$

where σ is defined in equation 1.1, and

$$\phi(P_T^{\text{miss}},\ \text{vertical}) > 20°, \tag{1.4b}$$

which guarantees that the vector direction of the missing E_T (P_T^{miss} in the transverse plane) is not within 20° of the "cracks" that run longitudinally along the top and bottom of the UA1 detector. Cuts are also placed on the jets present in the event. The primary large E_T (mono-)jet, which serves to define the bulk of the missing E_T vector by momentum conservation, is required to satisfy

$$E_T^{\text{JET}} \geq 12\ \text{GeV} \tag{1.5a}$$

and

$$|\eta^{\text{JET}}| \leq 2.5. \tag{1.5b}$$

Any additional candidate jets in the event are required to satisfy the same rapidity cut plus

$$12\ \text{GeV} \geq E_T^{\text{jet}}. \tag{1.5c}$$

If

$$E_T^{\text{jet}} \geq 8\ \text{GeV}, \tag{1.5d}$$

then it is required that

$$\phi(P_T^{\text{miss}},\ \text{jet}) \geq 30°, \tag{1.5e}$$

$$\phi(\text{JET},\ \text{jet}) \leq 150° \tag{1.5f}$$

in order to ensure that the E_T^{miss} vector did not arise from a mismeasured jet P_T.

We can now apply these cuts to $W^{\pm}$ and Z^0 events generated with the Mode 1 structure functions and Λ_{QCD} values assuming 236 nb^{-1} at 630 GeV. The resulting expected numbers of events in each channel are:

CHANNEL	$E_T^{\text{miss}} \geq 4\sigma$	$\geq 30\ \text{GeV}$	$\geq 38\ \text{GeV}$
$\tau + 0$ jets	22.6	8.2	1.9
$\tau + 1$ jet	5.1	2.4	0.9
$\tau + 2$ jets	0.6	0.4	0.2
$\nu\bar{\nu} + 1$ jet	3.8	1.4	0.7
$\nu\bar{\nu} + 2$ jets	0.9	0.4	0.2
$\ell_{\text{miss}}^{\pm} + 1$ jet	1.1	0.7	0.4
$\ell_{\text{miss}}^{\pm} + 2$ jets	0.2	0.2	0.1
Total	34.3	13.6	4.4

CHANNEL	$E_T^{\text{miss}} \geq 4\sigma$	$\geq 30\,\text{GeV}$	$\geq 38\,\text{GeV}$
UA1	23	12	5
Total/Mode 2′	34.2	13.7	4.6
Total/Mode 1/$\langle E_T^{\text{ue}} \rangle = 20\,\text{GeV}$	39.3	13.4	4.2
Total/Mode 1/$\langle E_T^{\text{ue}} \rangle = 40\,\text{GeV}$	30.1	13.6	4.5

where the explicitly labeled τ contributions arise from hadronic τ decays, while unidentified leptonic τ decays are included in the missed lepton ($\ell_{\text{miss}}^{\pm}$) category. The

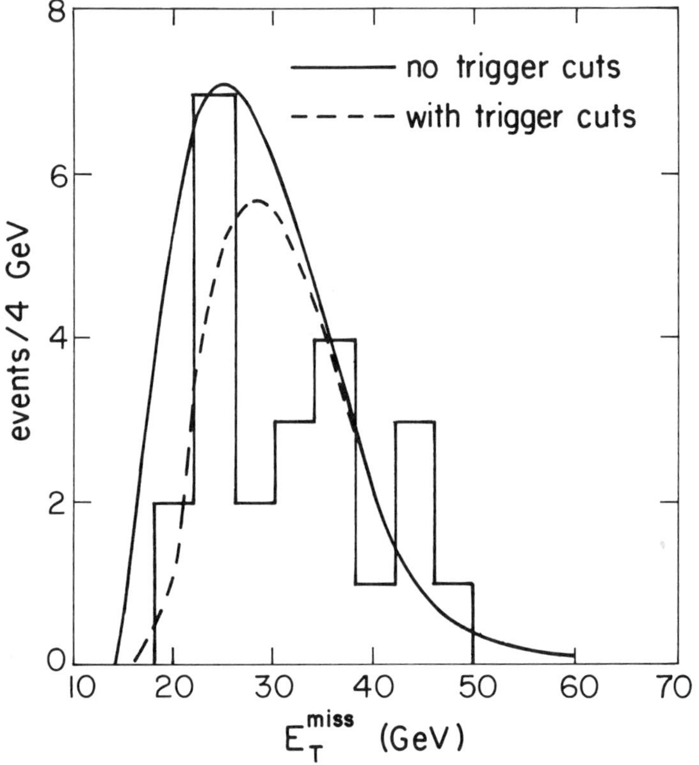

FIGURE 5. The distribution of missing E_T events in the 1984 monojet sample from reference 7. The theoretical curves are the SM predictions with (dashed line) and without (solid line) the trigger cut emulation described in the text.

last three lines offer some measure of the theoretical systematic uncertainty in the previous entries. These results are illustrated by the histogram (data) and solid line (SM theory) in FIGURE 5. At the level of ± 1 event, we can conclude that the results are reasonably reliable except near the edge of the 4σ cut. More surprisingly, except at the largest missing E_T values, we conclude that the SM predicts more monojet events than seen by UA1.

To see how such a situation can be understood, we must return to the issue of the "on-line" event selection process mentioned briefly above. The point is simply that some trigger criterion must be used on the initial crude experimental measurement of E_T and on the jet properties of a given event in order to decide whether to save it for a later careful examination. In the current data,[7] the trigger cut required either

(1) at least one hard jet, $E_T \geq 25$ GeV,

or

(2) a "left/right" E_T imbalance ≥ 17 GeV plus one jet with $E_T \geq 15$ GeV.

Due to real detector inefficiencies and resolution effects in the initial jet P_T measurements, it is likely that some of the events that would pass the final cuts are already rejected by these initial cuts. On the other hand, our theoretical analysis assumes that all acceptable events are included in the sample. Thus, for a more careful comparison with the experimental sample, we must simulate this trigger selection effect. Note that we are not concerned with unacceptable events that erroneously pass the initial cuts because these will presumably be eliminated by the later careful measurements and cuts. The issue here is "good" events that fail the trigger selection process. To estimate the effect, we follow previous efforts[19] in this direction and proceed by reducing (recall again that we are only concerned about losing events) all jet E_T by 20% and then smearing them further with a Gaussian distribution of width $E_T/5$. The above trigger selection criteria are then applied to these smeared E_T's to see if the event is discarded. This procedure was applied to all the events that had passed the "careful" analysis described above to see which events were "lost" due to the trigger. The corresponding revised line in the above table plus the data line again are

CHANNEL	$E_T^{\text{miss}} \geq 4\sigma$	≥ 30 GeV	≥ 38 GeV
Total/with Trigger	24.9	13.0	4.3
UA1	23	12	5

which is also indicated in FIGURE 5 by the dashed line. At this point, one must be almost embarrassed by how well the SM describes the monojet events. On the other hand, we must remember that the analysis has been "tuned" to the extent of choosing distributions, etc., (that is, the choice of Mode) that describe well the detected vector boson signal.

Before leaving this subject, there is one more point that can be usefully pursued somewhat more fully. It is clear from the above table that the SM predicts that monojet events generated by the hadronic decays of τ's are playing a sizable role. It is thus informative to try to identify this contribution in the data. An analysis[7] by UA1 has employed a combination of measures to attempt to determine the "tau-ness" of the jets in their monojet event sample. These include the angular size (τ's are more narrow), the charged multiplicity (τ's yield lower multiplicities), and the angular correlation between the most energetic charged particle and the neutral energy direction (τ's are more correlated). By testing these measures on Monte Carlo simulations of τ decays compared to the standard QCD jet sample in the data, a cut is determined that selects a sample from the monojet events that is thought to be essentially totally τ jets (defined by "likelihood of being τ" $\equiv L \geq 2$). Unfortunately, because our simple theoretical analysis treats the fragmentation of neither the QCD nor τ jets in detail, we cannot reproduce the effects of this cut on our theoretical sample. On the other hand, we know precisely how many τ jets are included. By considering the UA1 τ simulations[7] in detail, one can determine that only about one-half of all τ hadronic decays pass this "tau-ness" cut. Thus, we can make the following comparison of SM $\tau \rightarrow \nu$ + hadrons monojet events with data

CHANNEL	$E_T^{\text{miss}} \geq 4\sigma$	≥ 30 GeV	≥ 38 GeV
SM/with Trigger	18.7	9.2	2.2
SM/with Trigger $\times$ 0.5	9.4	4.6	1.1
UA1 ($L \geq 2$)	9	5	1

where again the agreement is strikingly good.

We close this section with some general conclusions about the SM and collider physics. It seems that the measurement of $W^{\pm}$'s and Z^0's plus jets is a fine arena for testing our perturbative understanding of QCD. Furthermore, we seem to be able to reliably proceed by applying the experimentally relevant cuts directly to the perturbative final states. This procedure is considerably simpler than and a helpful complement to a more complete Monte Carlo analysis. Finally, it appears that the SM offers a more than adequate explanation of the entire missing E_T signal with essentially no room for a non-SM contribution. At the same time, we have seen that the "cleanest" region to look for such new physics is for $E_T^{\text{miss}} \geq 30$ GeV (or more), which is where the SM predictions are apparently most insensitive to the systematic theoretical uncertainties.

POSSIBLE TESTS OF THE STANDARD MODEL

In this section, we will consider briefly the question of whether even more detailed tests of the SM are possible with collider data. For example, it has been suggested[20] by UA1 that a careful study of the events with two well-defined QCD jets (with the two-jet mass in the range of 150–250 GeV) yields evidence of nonscaling effects. In particular, the data indicate some deviation from the naive QCD result based on a fixed scale Q. These deviations are observed to vary with both the polar angle with respect to the beam direction of the two (essentially back-to-back) jets and with the total longitudinal momentum of the pair. Evidently, such deviations can either indicate a failure of perturbative QCD or offer the possibility of a more detailed test of QCD via a more complex variation of the "effective" scale Q_{eff} in the lowest-order expression. From the theoretical perspective, this issue can be resolved only by the careful evaluation of the higher-order corrections. Recall that a change in the definition of Q_{eff} serves only to change the structure of higher-order contributions. "Knowledge" of the Q_{eff} in the lowest-order terms can arise only from calculating the higher terms and then redefining Q_{eff} to absorb the bulk of the higher-order contributions into the first-order term. Evaluating these further contributions is a demanding task, but, surprisingly, giant steps in this direction have already been made.[21] The basic $2 \rightarrow 2$ and $2 \rightarrow 3$ processes have been evaluated. However, considerable effort remains to be invested in order to be able to apply these results to the experimental data on jets with all the associated cuts and constraints. The current theoretical results[21] are presented in the language of moments of single particle distributions. In order to encourage further activity in this field, I will make the only partially supported claim that the theoretical results seem to point in the opposite direction from the data.

One can characterize the data[20] for the θ dependence of the two-jet cross section by the suggestion that (in order to describe the data) Q_{eff} in the lowest-order $\alpha_s(Q_{\text{eff}})$ must relatively increase as θ approaches 90° (so that α_s decreases), while Q_{eff} is to decrease for $\theta \rightarrow 0°$ (so that α_s increases). On the other hand, I interpret the results of reference 21 to suggest that the next-order corrections are minimized if the lowest-order Q_{eff} is defined so as to increase as θ varies from 90° to 0°. Clearly, this area needs much more work, but the good news is that the preliminary work (both theoretical and experimental) described above suggests that important results can be obtained.

RESULTS YET TO BE EXPLAINED IN THE STANDARD MODEL

In this category, I would place the issue of understanding with some degree of detail and rigor the structure of hadronic multiplicities in the context of the SM. (One might also characterize this problem as: "Have you seen a Pomeron lately?".) It is important to appreciate that this issue has been with us since the results of the first accelerators, but it has been out of the limelight for some time. However, the situation has changed considerably in very recent times. We now have what we claim is the theory of strong interactions, plus, with the collider data, we have results that cover a tremendous range of energies. This situation is to some extent reflected by the large number of recent papers on this issue (approximately one paper per produced particle; a complete list will not be given here). The present brief discussion will include only my view of what the most recent data is telling us and an admonition to all of us to attempt to connect possible explanations[22] of these observations to the known structure of QCD.

I conclude that the collider data from UA5[23] are saying clearly that multiplicities are not displaying the property of KNO scaling[24] in the collider energy range, that is, the multiplicity distribution $P_n(s)$ is not characterized by the property

$$\langle n(s) \rangle P_n(s) = \Psi\left(z = \frac{n}{\langle n \rangle}\right), \tag{3.1}$$

where $\langle n \rangle$ is the average multiplicity and $\Psi(z)$ is a constant distribution if KNO scaling holds. Rather, it is observed that the distribution broadens as s increases. Furthermore, if the rapidity range $\Delta\eta$ over which particles are detected is varied, one observes that P_n broadens with increasing $\Delta\eta$, while Ψ becomes narrower.

The UA1 detector is able to collect data on hadronic multiplicities in the range $\Delta\eta \leq 2.5$ and to look for correlations with the presence or absence of jets in the corresponding events. Using a jet definition corresponding to the discussion above, but requiring only $E_T^{\text{Jet}} \geq E_T^{\text{min},J} = 5$ GeV, events can be divided into two categories: (1) no jets present or (2) ≥ 1 jets present in the event. It is observed[25] that these two components seem to separately display KNO scaling with characteristic properties:

$$\langle n_{\text{CH}} \rangle_{(2)} \approx 2 \cdot \langle n_{\text{CH}} \rangle_{(1)}, \tag{3.2a}$$

$$\langle P_T \rangle_{(2)} \approx {}^5\!/_4 \cdot \langle P_T \rangle_{(1)}. \tag{3.2b}$$

As indicated, these are only approximate properties. In particular, $\langle P_T \rangle_{(2)}$ is observed to be approximately independent of the charged particle multiplicity n_{CH}, while $\langle P_T \rangle_{(1)}$ seems to grow with n_{CH}. The distributions themselves can be approximately (but adequately) parameterized as single parameter "Gamma" distributions,

$$\Psi(z) \approx \frac{k^k z^{k-1}}{\Gamma(k)} e^{-zk}, \tag{3.3}$$

which appear to arise "naturally" in both QCD[26] and non-QCD[22] contexts. The "jet" sample (2) is narrower than the "no jet" sample (1) with

$$k_{(2)} \approx 2 \cdot k_{(1)}. \tag{3.4}$$

Furthermore, the sample (2) is essentially identical to a more restricted "hard jet" sample where every event has at least one jet with $E_T^{Jet} \geq 20$ GeV. As a fraction of the total signal, the jet sample is seen to increase with $\sqrt{s}$. The ratio $\sigma_{(2)}/\sigma_{TOT}$ is about 6% at 200 GeV and has increased to about 17% at 900 GeV.

How are we to understand all this new information? An intuitively attractive picture invokes two basic components[27] (e.g., a "hard" component leading to jets and a "soft" component with no associated jets), where each one is assumed to be "simple" and where each exhibits KNO scaling. Then, the energy dependence of the multiplicity data can be described by an energy-dependent ratio of the two components; for example, the "soft" component could be energy independent, while the "hard" component could grow with energy. Such a simple picture (with several tunable parameters) does seem to describe the data adequately.[27] Unfortunately, there are several points that suggest that there is much work still to be done:

(1) Are there really just two components or are there several?

There has been no serious discussion of this issue so far.

(2) Is the apparent KNO scaling of the two components real or only temporary in the current $\sqrt{s}$ range?

In this context, it is interesting to ask whether it might not be more informative to focus on the P_n instead of the Ψ's. Perhaps the apparently narrower $\Psi_{(2)}$ is really a comparable P_n, but scaled by a larger $\langle n \rangle$ (recall the matching factors of two above $k_{(1)}/\langle n \rangle_{(1)} \approx k_{(2)}/\langle n \rangle_{(2)}$).

In such two-component pictures, it is also important to recall that the energy independence of the "soft" component is only assumed. Furthermore, the separation of the components involves an arbitrary parameter, $E_T^{min,J}$. It is known[28] that for any fixed value of $E_T^{min,J}$, the "hard" component will eventually dominate as $\sqrt{s} \rightarrow \infty$, that is, $\sigma_{(1)} \rightarrow 0$. In this limit, the partons required to generate $E_T^{min,J}$ will correspond to steadily decreasing momentum fraction values; eventually, all events will exhibit such jets.

It will be very informative to make efforts to connect these data and models to our understanding of QCD. Unfortunately, work in this direction[26] suggests that, at present energies, the poorly understood recombination physics seems to be more important than the more controllable perturbative structure. On the other hand, such efforts seem mandatory if we are ever to claim to understand the strong interactions. A QCD-based connection[29] to the "old" ideas about the Pomeron and ladders diagrams would be very helpful. For those who are too young to remember such ideas, consider the simple picture of FIGURE 6 (which is included largely for its nostalgia value). This picture is intended to represent the scattering of proton, antiproton, or two protons via the scattering of quarks who exchange "ladders" of softer partons (qualitatively related to the concept of the ocean partons). One is then meant to sum over all the various ways to "cut" the ladders (i.e, sum over all contributions to the imaginary part of this forward elastic amplitude) to yield the multiparticle final states of the inelastic process. Large fluctuations in n then arise from cutting different numbers of ladders, while the ladders themselves are presumed to suffer only small fluctuations in the number of "rungs." Hard scattering events arise when there are "hot" (large energy) rungs at the center of one of the ladders; this is a rare occurrence. It is plausible to expect that such rare hot ladders (when cut) display a larger contribution to $\langle n \rangle$ (and $\langle P_T \rangle$) due to increased "beam-strahlung." Note that the presence of multiple ladders does not effect[30] the perturbative result for the rate of hard scattering processes, but

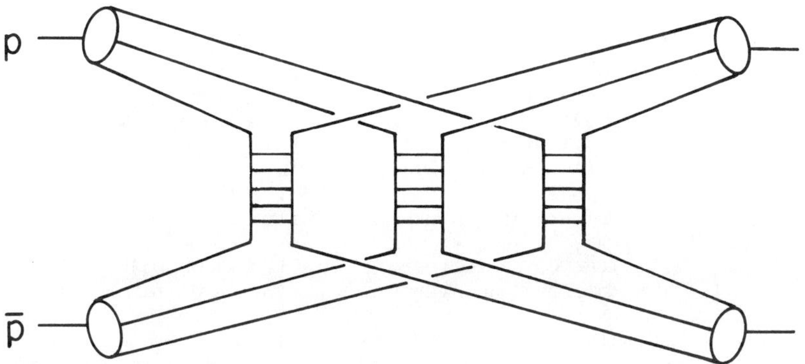

FIGURE 6. Idealized picture of inelastic pp̄ scattering representing multiple scattering as multiple "ladder" exchange.

does influence the associated multiplicity. Such a simple picture would seem to suggest that it is the P_n's that are of interest and not the Ψ's as was already suggested above. Finally, note that such pictures seem to bear at least a naive resemblance to the structure of recent Monte Carlo studies[31] of the multiplicity question. This work indicates that some sort of multiple soft scattering structure is required to adequately describe the data. Are these multiple ladders? Again, there is much to do both theoretically and experimentally, although the former is probably the really tough part.

SUMMARY

In summary, the collider seems to be observing an energy regime that is nearly ideal for studying the SM (and perhaps nothing else). The existing tests of 1-2-3 are more detailed than anyone would have anticipated two years ago and even more detail seems possible in the near future. Simple perturbative analyses constitute a useful complement to the more detailed (but less transparent) Monte Carlo work. There is also some possibility that collider results will give us important insight into the structure of soft physics and the Pomeron, but it is extremely important that such studies be connected to our existing understanding of QCD.

ACKNOWLEDGMENTS

The author thanks his collaborators, R. Kleiss and W. J. Stirling, without whose contributions much of the work reported on here would not exist. Helpful discussions with J. C. Collins, B. Durand, and R. K. Ellis are also gratefully acknowledged. Finally, thanks go to all those, especially L. Durand, who worked so hard to put this school and these proceedings together.

REFERENCES

1. CLINE, D. 1987. These proceedings.
2. Previous estimates of Standard Model backgrounds include: AURENCHE, P. & R. KINNUNEN. 1984. Annecy preprint no. LAPP-TH-108; BALLESTRERO, A. & G. PASSARINO. 1984. Phys. Lett. **148B:** 373, 378; CUDELL, J. R., F. HALZEN & K. HIKASA. 1985. Phys. Lett. **157B:** 447; GLOVER, E. W. N. & A. D. MARTIN. 1985. Z. Phys. **C29:** 399; HIKASA, K. 1985. University of Wisconsin preprint no. MAD/PH/261.
3. ELLIS, S. D., R. KLEISS & W. J. STIRLING. 1985. Phys. Lett. **154B:** 435; Proceedings of the 5th Topical Workshop on Proton-Antiproton Collider Physics. M. Greco, Ed.: 348. World Scientific. Singapore.
4. ELLIS, S. D., R. KLEISS & W. J. STIRLING. 1985. Phys. Lett. **158B:** 341; Proceedings of the 5th Topical Workshop on Proton-Antiproton Collider Physics. M. Greco, Ed.: 364. World Scientific. Singapore.
5. ELLIS, S. D., R. KLEISS & W. J. STIRLING. 1986. Phys. Lett. **167B:** 464.
6. THE CALKUL COLLABORATION: DE CAUSMAECKER, P. *et al.* 1981. Phys. Lett. **105B:** 215; BERENDS, F. *et al.* 1982. Nucl. Phys. **B206:** 61; 1984. Nucl. Phys. **B239:** 382; **B239:** 395; KLEISS, R. 1984. Nucl. Phys. **B241:** 61. See also: KLEISS, R. & W. J. STIRLING. 1985. Nucl. Phys. **B262:** 235.
7. UA1 COLLABORATION. 1985. Presented by J. Rohlf at the American Physical Society Division of Particles and Fields, Eugene, Oregon; CERN preprint no. CERN-EP/ 85-160.
8. STERMAN, G. & S. WEINBERG. 1977. Phys. Rev. Lett. **39:** 1439.
9. See, for example: STIRLING, W. J. 1982. Proceedings of the Drell-Yan Workshop, Fermilab, p. 131.
10. DUKE, D. W. & J. F. OWENS. 1984. Phys. Rev. **D30:** 49.
11. GLUCK, M., E. HOFFMAN & E. REYA. 1982. Z. Phys. **C13:** 119.
12. UA1 COLLABORATION. 1985. Presented by M. Levi at the Proceedings of the 5th Topical Workshop on Proton-Antiproton Collider Physics. M. Greco, Ed.: 144. World Scientific. Singapore; UA2 COLLABORATION. 1985. Presented by H. Plothow-Besch. Ibid., p. 57.
13. UA1 COLLABORATION. 1985. Presented by S. Geer at the Theoretical Advanced Study Institute in Elementary Particle Physics, Yale University; CERN preprint no. CERN-EP/85-163. See also the UA2 Collaboration in reference 12.
14. ELLIS, S. D., R. KLEISS & W. J. STIRLING. 1985. Phys. Lett. **163B:** 261.
15. See, for example: HALL, L. J., R. L. JAFFE & J. L. ROSNER. 1985. Phys. Rep. **125:** 103.
16. AURENCHE, P. & R. KINNUNEN. 1984. Annecy preprint no. LAPP-TH-108; GLOVER, E. W. N. & A. D. MARTIN. Z. Phys. **C29:** 399; GILMAN, F. J. & S. H. RHIE. Phys. Rev. **D31:** 1066.
17. UA1 COLLABORATION. 1985. Presented by M. Mohammadi at the Conference on Collider

Physics at Ultra-High Energies, Aspen, Colorado; CERN preprint no. CERN-EP/ 85-52.

18. UA1 COLLABORATION. 1984. Presented by J. Rohlf at the Proceedings of the XXII International Conference on High Energy Physics, Leipzig, vol. II, p. 16.

19. BARNETT, R. M., H. E. HABER & G. L. KANE. 1986. Nucl. Phys. **B267:** 625.

20. UA1 COLLABORATION. 1985. Presented by E. K. Buckley at the Proceedings of the 5th Topical Workshop on Proton-Antiproton Collider Physics. M. Greco, Ed.: 96. World Scientific. Singapore.

21. ELLIS, R. K. & J. C. SEXTON. 1986. Nucl. Phys. **B269:** 445.

22. For a "non-SM" approach, see: CARRUTHERS, P. 1986. Los Alamos preprint no. LA-UR-86-1540; CARRUTHERS, P. & C. C. SHIH. 1983. Phys. Lett. **127B:** 242; 1984. Phys. Lett. **137B:** 425.

23. UA5 COLLABORATION. 1985. Phys. Lett. **160B:** 193; Phys. Lett. **160B:** 199; Presentation by G. Ekspong at the Proceedings of the 5th Topical Workshop on Proton-Antiproton Collider Physics. M. Greco, Ed.: 478. World Scientific. Singapore.

24. KOBA, Z., H. B. NIELSEN & P. OLESEN. 1972. Nucl. Phys. **B40:** 317.

25. UA1 COLLABORATION. 1985. Presented by F. Ceradini at the International Europhysics Conference on High Energy Physics, Bari, Italy; CERN preprint no. CERN-EP/ 85-196.

26. DURAND, B. & I. SARCEVIC. 1986. Phys. Lett. **172B:** 104; 1985. University of Wisconsin preprint no. MAD/TH-85-7. See also: DURAND, B. & S. D. ELLIS. 1984. Proceedings of the 1984 Summer Study on the Design and Utilization of the SSC. R. Donaldson & J. G. Morfin, Eds.: 234. Amer. Phys. Soc. Division of Particles and Fields.

27. See, for example: GAISSER, T. K., F. HALZEN & A. D. MARTIN. 1986. Phys. Lett. **166B:** 219; MARTIN, A. D. & C. J. MAXWELL. 1986. Phys. Lett. **172B:** 248.

28. See, for example: ELLIS, S. D. 1984. Proceedings of the 1984 Summer Study on the Design and Utilization of the SSC. R. Donaldson & J. G. Morfin, Eds.: 221. Amer. Phys. Soc. Division of Particles and Fields.

29. For recent work in this direction using hard scattering/heavy flavor production as a flag, see: BERGER, E. L., J. C. COLLINS, D. E. SOPER & G. STERMAN. 1986. Argonne preprint nos. ANL-HEP-CP-86-83, ANL-HEP-CP-86-52, ANL-HEP-PR-86-14, ANL-HEP-CP-86-19, and ANL-HEP-CP-86-12.

30. See, for example: ELLIS, S. D., C. E. DETAR & P. V. LANDSHOFF. 1975. Nucl. Phys. **B87:** 176.

31. See, for example: SJÖSTAND, T. 1985. Fermilab preprint no. FERMILAB-PUB-85/119-T; PAIGE, F. & S. D. PROTOPOPESCU. 1986. Brookhaven preprint no. BNL-38034.

Some Recent Results from the UA1 Experiment at CERN

DAVID B. CLINE

Department of Physics
University of Wisconsin-Madison
Madison, Wisconsin 53706

INTRODUCTION: PHYSICS AT THE S$\bar{\text{p}}$pS

The S$\bar{\text{p}}$pS collider has provided a remarkable number of physics objectives. The most important physics goals of the present and of the next few years can be summarized as

$$W \rightarrow L + \nu_L$$
$$ \llcorner\!\!\rightarrow q + \bar{q} + \bar{\nu}_L$$

$$\bar{\text{p}}\text{p} \rightarrow \tilde{q} + \tilde{q} + \cdots$$
$$\llcorner\!\!\rightarrow \bar{q} + \tilde{\gamma}$$
$$\llcorner\!\!\rightarrow \bar{q} + \tilde{\gamma}$$

$$\bar{\text{p}}\text{p} \rightarrow B\bar{B} + \cdots$$
$$\llcorner\!\!\rightarrow \mu^+$$
$$\llcorner\!\!\rightarrow \mu^+.$$

We will present evidence for $\bar{B} - \bar{B}^o$ mixing in this report and we will discuss the search for heavy leptons.

THE UA1 DETECTOR UPGRADE

The detector upgrade consists of three parts:

(1) The μ system upgrade with limited streamer tubes and a muon wall spectrometer.[1]
(2) Installation of a microvertex detector.
(3) Upgrade of the calorimeter to a TMP-U plate system.

We will only discuss the first and third results here because the microvertex detector has been described elsewhere.[2,3]

FIGURE 1 shows a sideview of the UA1 detector indicating the μ wall magnetized iron spectrometer.[1] FIGURE 2 shows the details of the limited streamer tube spectrometer inside the magnetized wall. This system was built by the CERN–Padova–

Wisconsin group and the wall was working with high efficiency in the 1985 run. As an example, we show a candidate for a trimuon event observed with this system in the 1985 run in FIGURE 3.

The TMP calorimeter is schematically illustrated in FIGURE 4.[2,3] The EM calorimeter of the UA1 detector is being replaced by this density–high resolution warm liquid drift device. TABLE 1 gives some of properties of the uranium plate system. It is hoped that the improved detector will be ready to operate after the luminosity upgrade at CERN in 1988. FIGURE 5 shows a supergondola module with the new tower structure shown.

The new calorimeter is expected to increase the resolution of the UA1 detector for jets and for the measurement of the W, Z mass difference.

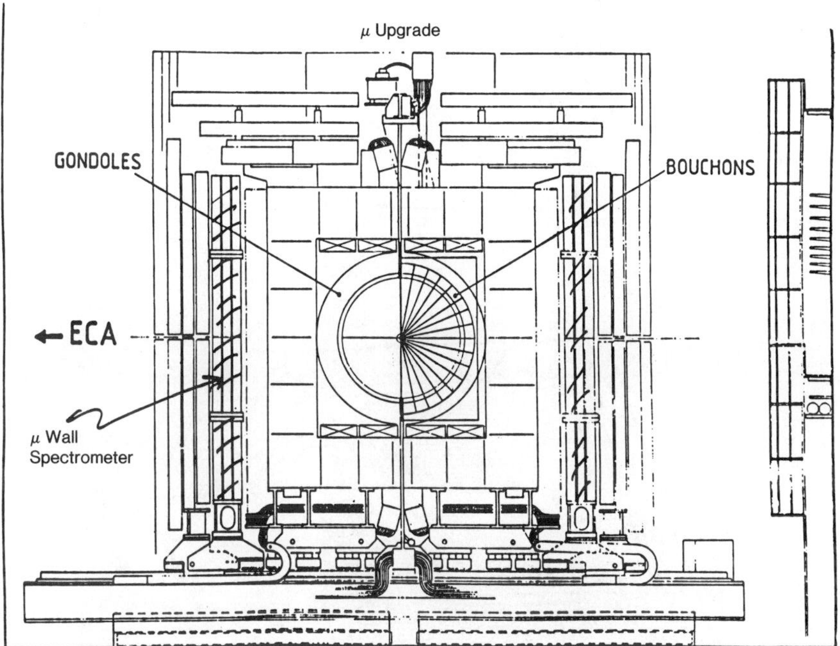

FIGURE 1. A sideview of the UA1 detector indicating the μ wall magnetized iron spectrometer.

STATUS OF THE DETECTION OF W AND Z PARTICLES AND THEIR PROPERTIES

During the 1985 run, the number of events observed was nearly equal to all of the events observed in 1983–84. TABLE 2 gives a list of the number of $W \to e\nu$ and $Z \to e^+e^-$ events observed up to the end of 1985, along with the projected number for 1986. Note that the $\bar{S}p\bar{p}S$ collider has given nearly 1 pB^{-1} and will provide ~600 $W \to e\nu$ for UA1 (~$X2$ for UA1 and UA2, thus giving nearly one thousand detected $W \to e\nu$

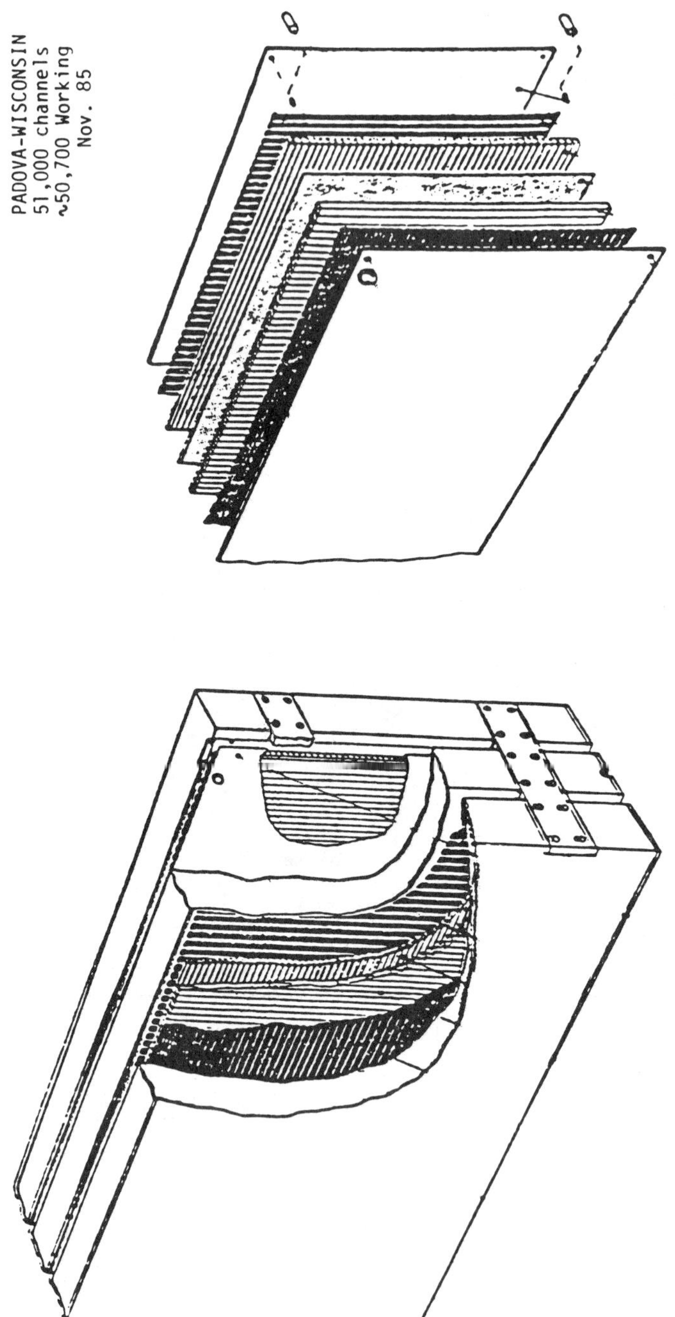

FIGURE 2. The details of the limited streamer tube spectrometer inside the magnetized wall. The left-hand diagram is a cut through a side absorber wall that shows the arrangement of the detectors in the iron; the right-hand one is a detailed layout of "Iarocci" tubes and readout strips.

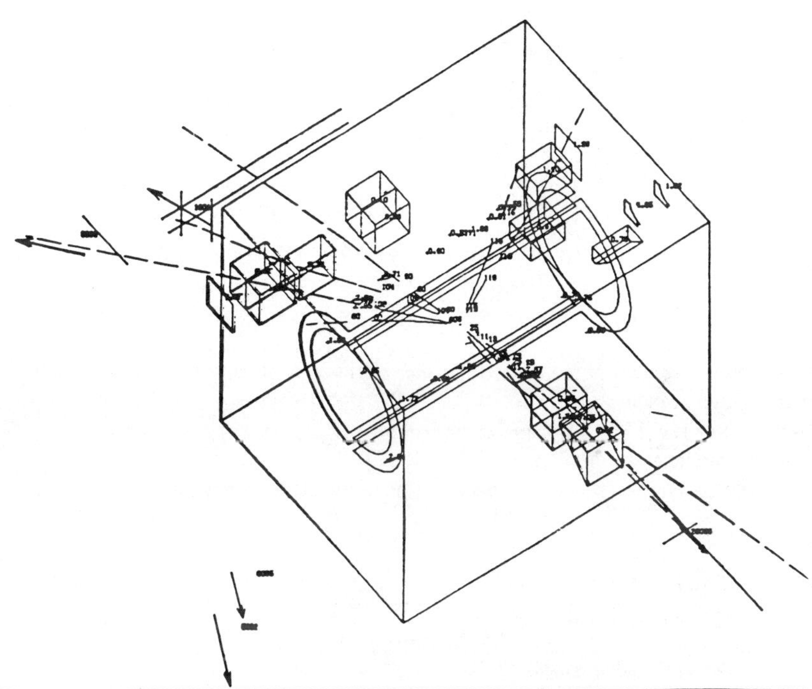

FIGURE 3. A candidate for a trimuon event.

events). Note also that the ratio of $Z \rightarrow e^+e^-$ to $W \rightarrow e\nu$ is stable at the value of[4]

$$\frac{N_{Z \rightarrow e^+e^-}}{N_{W \rightarrow e\nu}} = \frac{37}{359} \simeq 0.1,$$

which implies that if this trend continues, the number of neutrino families will hardly exceed the three families now observed. In FIGURES 6a and 6b, we show the mass determination of the W and Z from the properties of the $e\nu$ or e^+e^- system. These events are only from the 1983–84 samples of data.[5] In FIGURE 7, the implication of the

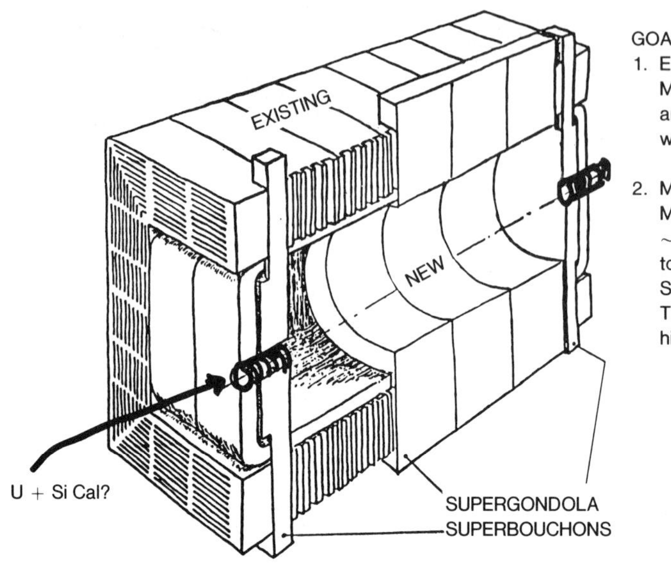

FIGURE 4. The TMP calorimeter.

present measurement of the mass difference of the $Z - W$ and the prospects for future observations are shown. The small dot at the right-hand corner shows the hoped-for resolution of the TMP calorimeter and indicates that it will be adequate to test the standard model at an important level.

FIGURE 8a shows the angular distribution of the electron or positron in the W center

TABLE 1. Calorimeter Parameters

Gap between DU plates	3 mm
Sensitive gap	1.25 mm
Number of plates per module	71
Number of gaps per module	70
Volume of liquid per module	70 liters
Weight per module	1.7 tons
Number of modules	160
Size of readout plates	$10 \times 12 \times 0.02$ cm^3
Number of electrons per gap in TMP (at 25 kV/cm for min. ionizing particle)	2000 electrons
Noise per gap	800 electrons

Sample	DU Plates	Capacitance	Electrons/MIP (25 kV/cm)
1	5	1.77 nF	16,000
2	10	4.10 nF	36,000
3	15	6.73 nF	56,000
4	10	4.66 nF	36,000
5	15	30.31 nF	56,000
6	15	33.03 nF	56,000

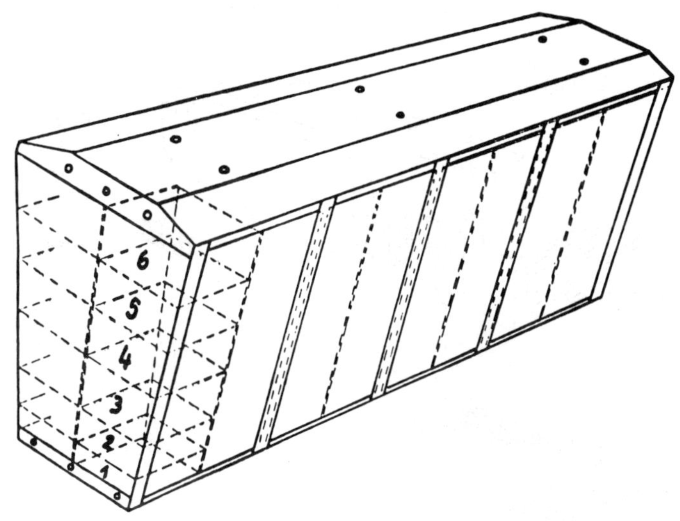

FIGURE 5. A supergondola module with the new tower structure. Six samples = one tower.

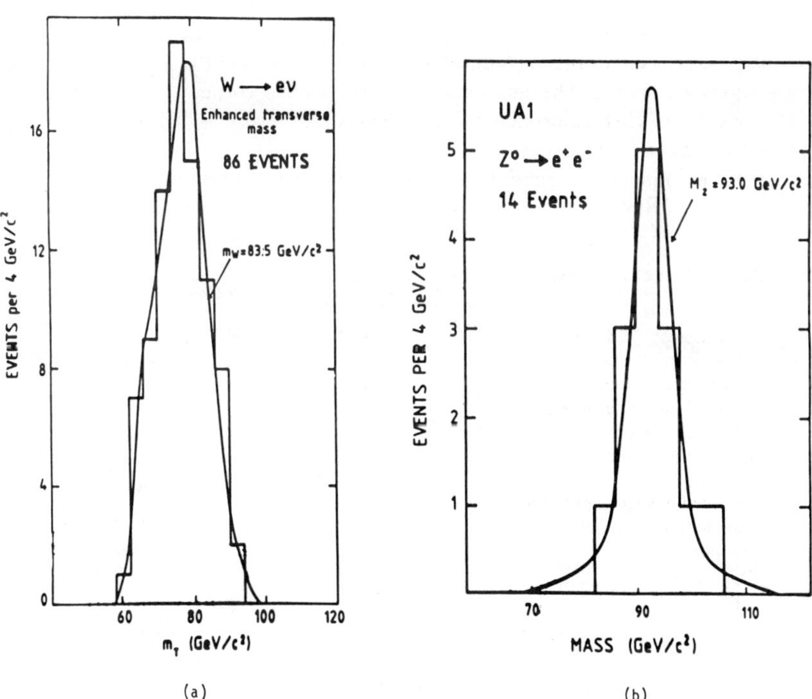

FIGURE 6. The mass determination of the (a) W or (b) Z from the properties of the (a) $e\nu$ or (b) e^+e^- system (1983–84 data).

TABLE 2. W-Z Events and $\int Ldt$ for the 1983–86 Runs

| | | Number of | |
Year	$\int Ldt$	$W \rightarrow e\nu$	$Z \rightarrow e^+e^-$
83	136 nb^{-1}	} 172	} 18
84	263 nb^{-1}		
85	374 nb^{-1}	187^a	19^a
	773 nb^{-1}	359^a	37^a
86	~300 (microvertex)	~180	~18
	>1 Pb^{-1}	>600	>60
87	Shutdown for TMP Calorimeter and ACOL		
88	Operate again with $L \simeq 5 \times 10^{30}\,\mathrm{cm}^{-2}\,\mathrm{sec}^{-1}$		

^{a}From the analysis of 274 nb^{-1}.

of mass.[5] The effects of right-handed currents would show up as a deviation at $\cos \theta^* = -1$. No such deviation is apparent in the data. FIGURE 8b shows a comparison of the ratio of $W \rightarrow e\nu$ and $Z \rightarrow e^+e^-$ rates with the exceptions for a large number of neutrino families in nature.[4,6] To 90% confidence level, the number of additional families is less than seven. It appears possible to reduce this to one or two before the systemetrics uncertainty of this technique becomes large.

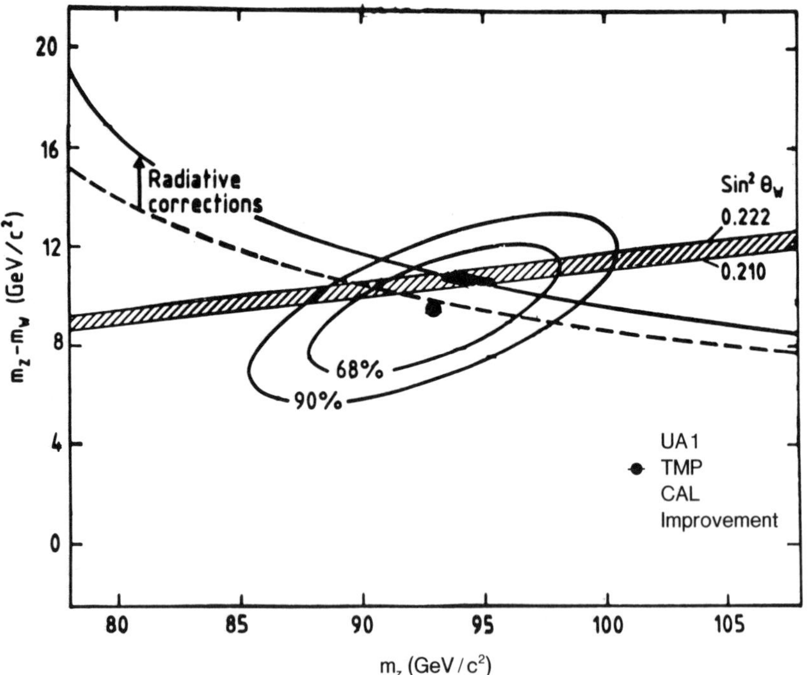

FIGURE 7. The implication of the present measurement of the mass difference of the $Z - W$ and the prospects for future observations.

SAME-SIGN DIMUONS AND B_s^o MIXING

We first detected low mass dimuon events in the 1983 data (reported in G. Bauer's thesis, University of Wisconsin, 1985). The properties of the early events (large ratio of same-sign dimuons, many events with nearby jet activity) have been reproduced in the

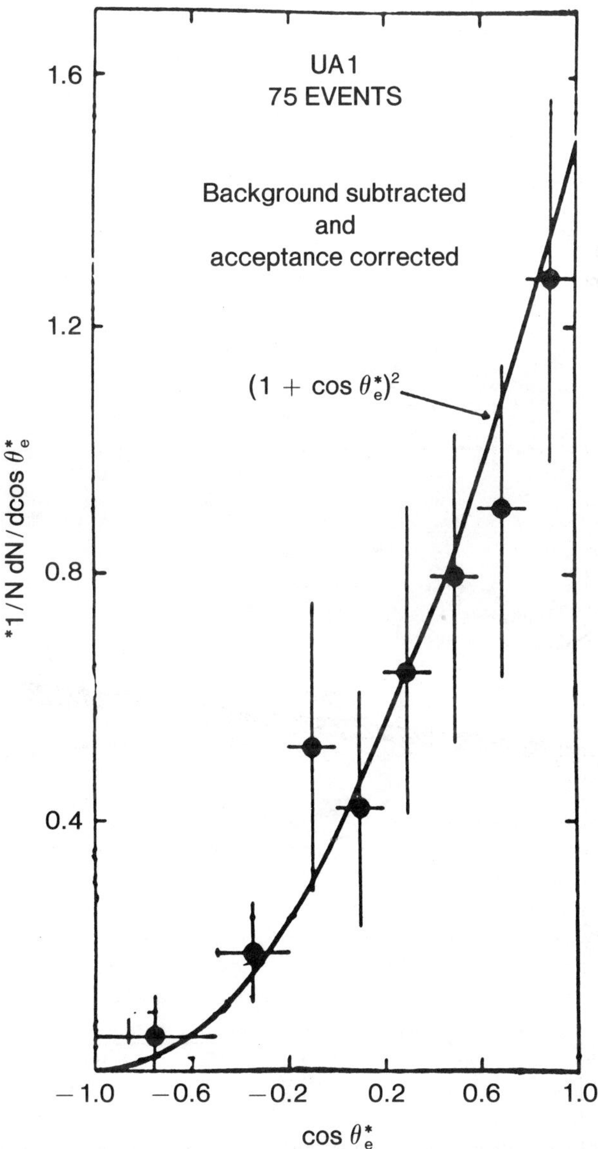

FIGURE 8a. The angular distribution of the electron or positron of the W center of mass.

1984 and 1985 data.[7–9] It is now possible to carry out a search for the mixing of B^o mesons as described below. FIGURE 9 shows the $\mu\mu$ mass distribution for four different classes of dimuons from the 1983–84 data. The background for these events has been "measured" as described in Bauer's thesis. There are two kinds of background as given in TABLE 3. FIGURES 10 and 11 show the data that are used to "measure" the (π, K)-decay background and the number of events expected in a sample taken from 100 nb^{-1} integrated luminosity, respectively. The second type of background for the B^o mixing search (described in TABLE 3) comes from cascade b decays. This is illustrated in FIGURE 12, where the possibility of B^o_s mixing is also shown.

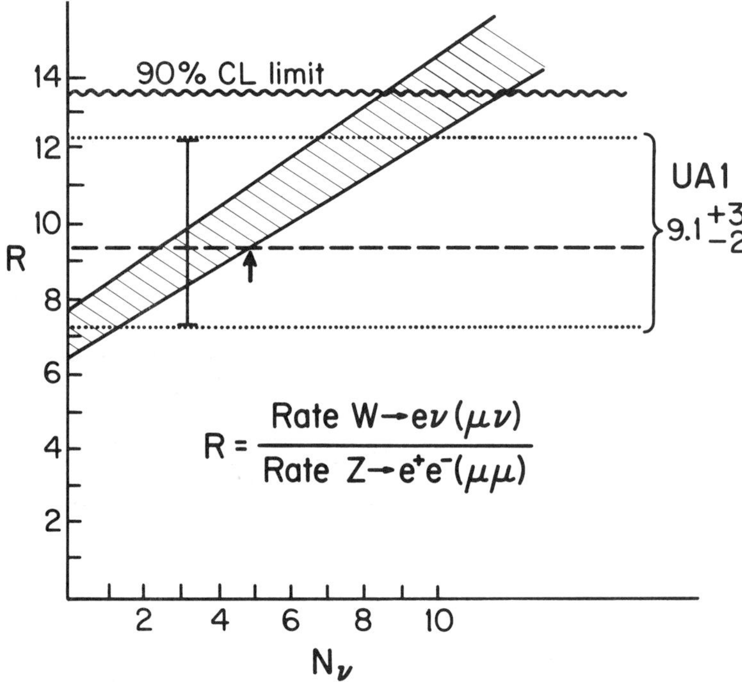

FIGURE 8b. Neutrino counting at the collider (1984 data). See text.

TABLE 4 gives the ratio of the nonisolated same-sign event rate to the opposite-sign rate for the 1983, 1984, and initial (preliminary) 1985 data.[9]

The prospects for detecting mixing of the B^o system are given in TABLE 5, where it is shown that the B^o_s system is more likely (theoretically) to mix with a large amplitude than the B^o_d. The reason for this is shown in FIGURE 13, where the various diagrams that contribute to the mixing of B^o_d and B^o_s mesons are shown. The B^o_s diagrams are almost Cabbibo-allowed processes.

In FIGURE 14, we show very preliminary data from 1983, 1984, and 1985 that

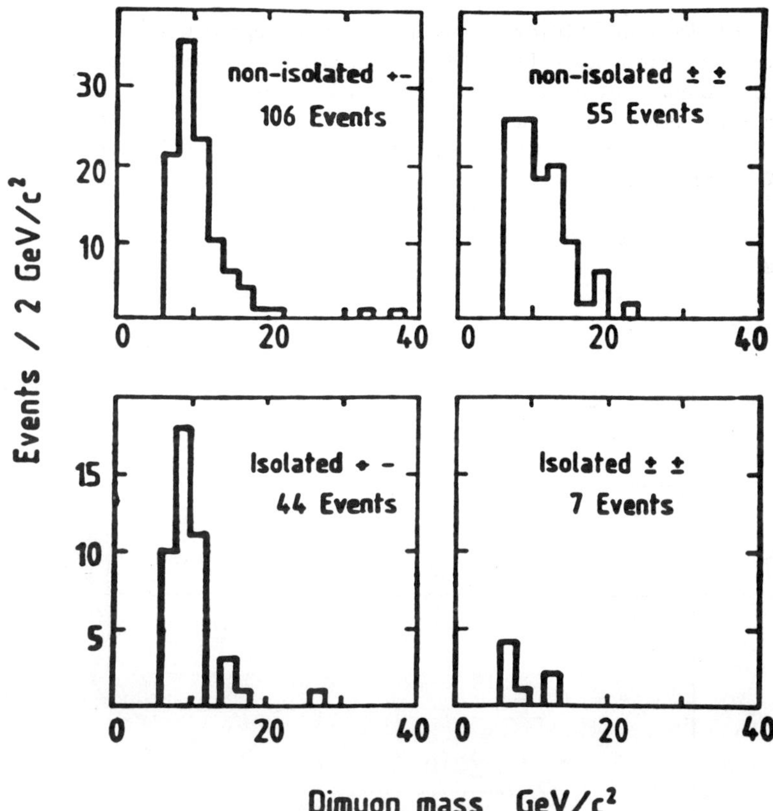

FIGURE 9. The $\mu\mu$ mass distribution for the four different classes of dimuons from the 1983–84 data.

TABLE 3. Backgrounds for $\mu^-\mu^-/\mu^+\mu^+$ Events

(1) π, k decay—Calculated from the data using $\mu + X$ events $\rightarrow \pi \dots$ included in the final result.

(2) Second generation decays

$$B\bar{B}$$

$$
\begin{array}{l}
\quad \llcorner\!\!\rightarrow \mu^- + c^+ \dots \rrbracket \text{ cascade} \\
\quad \llcorner\!\!\!\longrightarrow \bar{c} + \dots \rrbracket \text{ decays} \\
\qquad\quad \llcorner\!\!\rightarrow \mu^-\rrbracket
\end{array}
$$

Calculate ratio $R = \mu^\mp\mu^\mp/\mu^+\mu^-$ with ISAJET

(A) $P_{\perp\mu} \geq 3\,\mathrm{GeV}/c,\ R = 0.25 \pm 0.2$

(Cline/Rhoades)

(B) $P_{\perp\mu} \geq 3\,\mathrm{GeV}/c,\ R = 0.16 \pm 0.03$

$m_{\mu\mu} > 6\,\mathrm{GeV},\ R = 0.16 \pm 0.3$

(Ellis).

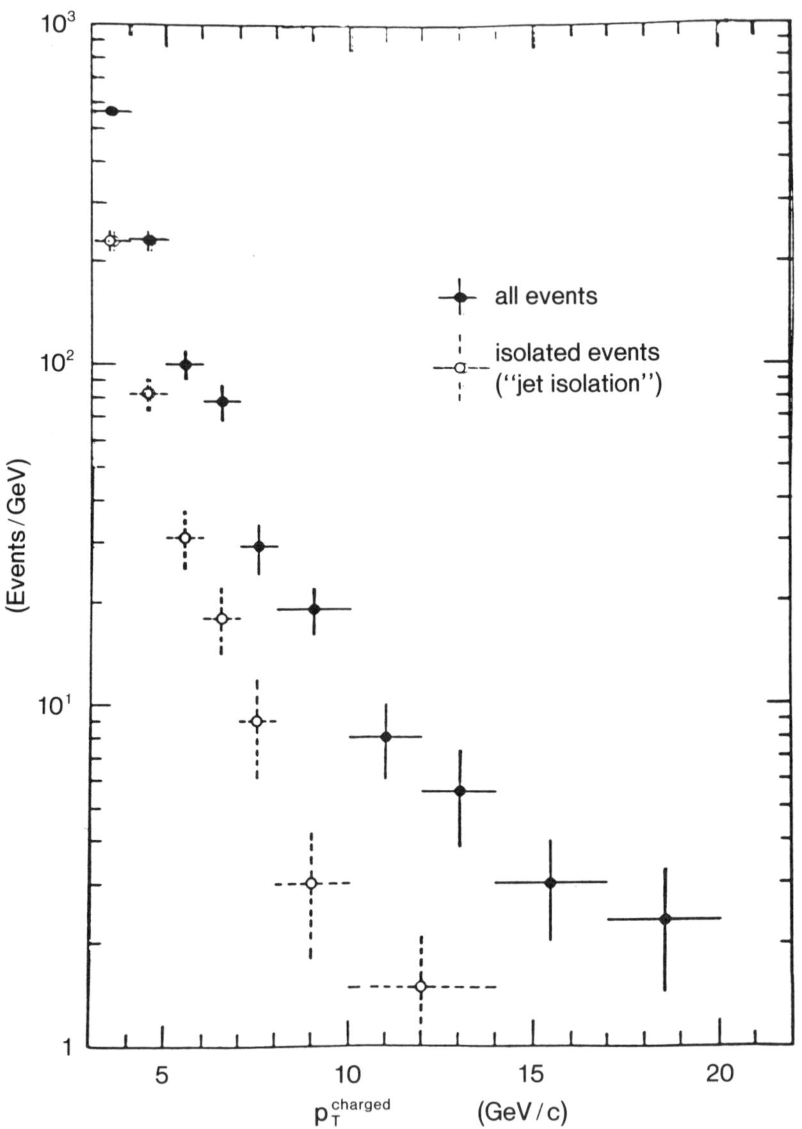

FIGURE 10. The spectrum of accepted charged particles (in events with a muon, $p_T^\mu > 5$ GeV/c).

indicate strong evidence for B^o mixing.[9] We have plotted the same-sign over opposite-sign data (nonisolated events) as a function of the $P_\perp$ cutoff of the two muons. If the same-sign events were due to cascade decays, the data should follow the lower curve (Cline and Rhoades, 1984). The data are in excellent agreement with a large mixing effect. Other possible origins of this ratio are now being studied by the UA1 group. If no other explanation can be found, the B_s^o mixing hypothesis will be strongly favored. This is the first example of a mixed system since the discovery of the K_1^o, K_2^o systems in the 1950s.

THE RISE OF σ_T AND MINIJETS

The origin of the rise in the total cross section of hadronic interactions is not well known. Data from the UA1 experiment may have given us a clue of great significance. As early as 1978, it was proposed that the rise of σ_T was due to hard scattering

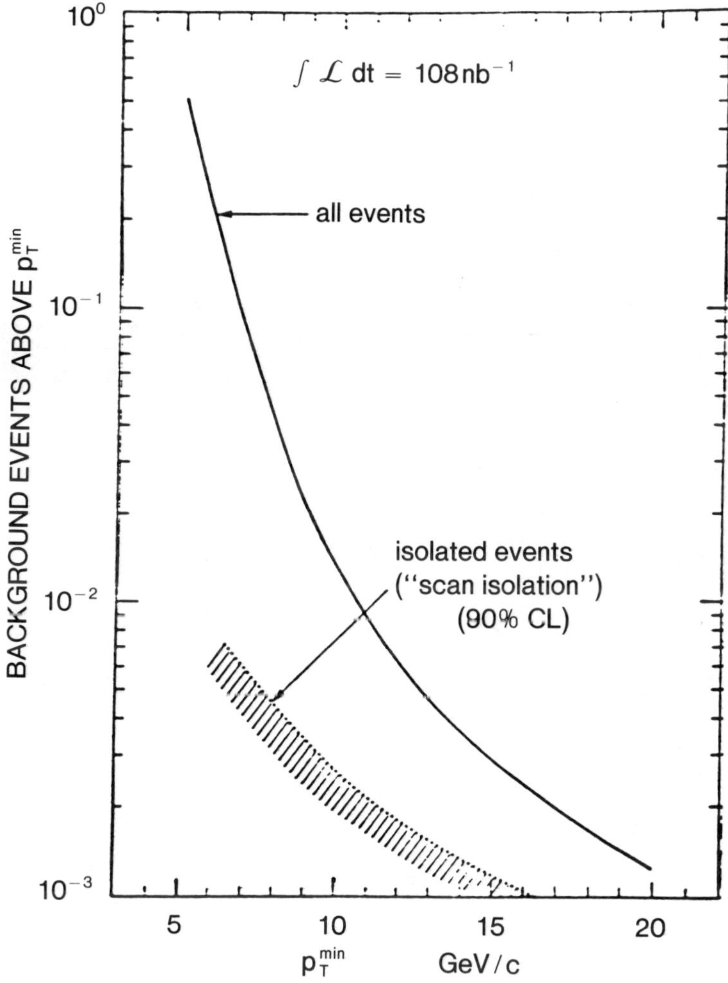

FIGURE 11. Integrated number of background events above p_t^{min}.

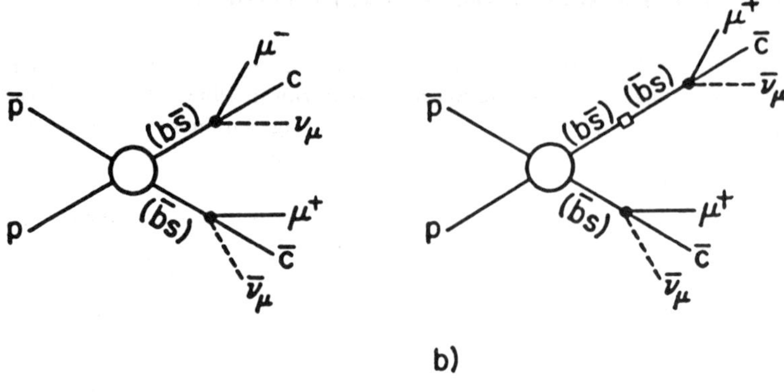

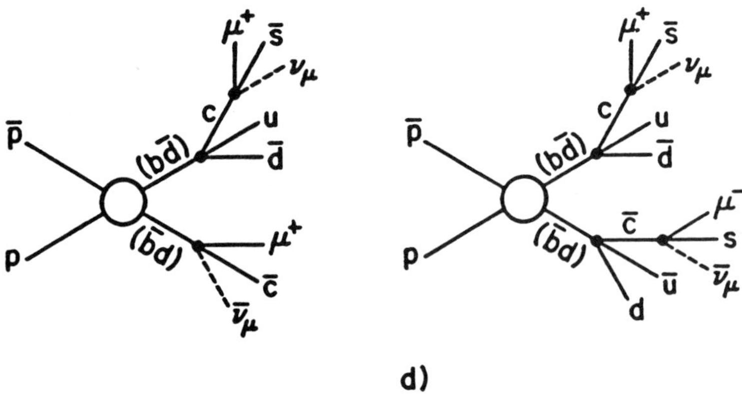

FIGURE 12. (a) First generation decay—B_s^o ($\mu^+\mu^-$) events; (b) first generation decay with B_s^o mixing ($\mu^\pm\mu^\pm$) events; (c) first and second generation decay—($\mu^\pm\mu^\pm$) events; (d) second generation decay ($\mu^+\mu^-$) events.

TABLE 4. Ratio of Nonisolated Same-Sign Event Rate to Opposite-Sign Event Rate

1985:		$+-$	$\pm\pm$
$\int Ldt = 190$ nb^{-1}	isolated μ's	31	5
160 events	nonisolated	76	48
	$R = 48/76 = 0.63 \pm 0.11$ (stat)		
	[raw numbers, not background corrected]		
1983–84:		$+-$	$\pm\pm$
$\int Ldt = 362$ nb^{-1}	isolated μ's	44	7
212 events	nonisolated	55	106
	$R = 55/106 = 0.52 \pm 0.09$ (stat) [raw numbers]		
	$\longrightarrow 0.46 \pm 0.10$ corrected		
Combined:			
	$R = 103/182 = 0.57 \pm 0.07$ (stat)		
	[raw numbers, not background corrected]		

processes and QCD (see Cline, Halzen, and Roth, 1973).[10] This possibility has now been tested using data from the $\overline{\text{S}}p\text{pS}$ ramping run in 1985.[11] FIGURE 15 shows the shape of the ramping cycle and the kinds of average luminosity obtained at various center-of-mass energies. Using these data that range from a center-of-mass energy of 200 GeV to 900 GeV, it has been possible to measure jet production down to very low $P_\perp$ ($\sim$5 GeV/c) and to estimate the fraction of the lower $P_\perp$ region that has been missed due to detector cuts. TABLE 6 gives the number of events analyzed at each energy. FIGURE 16 shows the jet transverse energy distribution at 540 GeV c.m. energy. For comparison, the E_T distribution for 900 GeV center-of-mass energy is shown in FIGURE

TABLE 5. Search for B^o Mixing

$$\underline{\overset{\text{b}}{\overline{\text{d}}}}\bigcirc B^o_d \qquad \underline{\overset{\text{b}}{\overline{\text{s}}}}\bigcirc B^o_s$$

$$\text{Define } \Delta_{d,s} = \frac{(\delta m_{d,s}/\Gamma_{d,s})^2}{2 + \left(\dfrac{\delta m_{d,s}}{\Gamma_{d,s}^{\,2}}\right)},$$

$$\text{where } \delta m = m_+ - m_- \qquad M+, \Gamma+,$$
$$\Gamma = \frac{\Gamma_+ + \Gamma_-}{2} \qquad M_-\Gamma_-$$

[two states for B^o system (like K^o_L, K^o_s)].

$$\text{Measurement } R = \frac{N(\mu^+\mu^+) + N(\mu^-\mu^-)}{N(\mu^+\mu^-) + N(\mu^-\mu^+)} = \frac{2\Delta}{1 + \Delta^2};$$

$$\text{for } \Delta \to 1, R \to 1,$$
$$\Delta \to 0, R \to 0.$$

$$\text{Theory expectations:} \qquad \frac{\delta m_s}{\Gamma_s} > 2\Delta \to \text{large},$$
$$\frac{\delta m_d}{\Gamma_d} \ll 1\Delta \to \text{small}.$$

17. (These data have been largely analyzed by the Rome group in the UA1 collaboration.), FIGURE 18 shows the estimated cross section for the production of these minijet events, indicating a large rise between 200 and 900 GeV center-of-mass energies. In FIGURE 19, we show the total cross section with the measured minijet cross section subtracted. Note that the "remainder" of the cross section is consistent with it being constant with energy. Thus, it could be that the full rise in the total cross section down to the lowest energies is somehow due to QCD effects. This may have very important implications for total interactions at SSC energies that will be dominated by QCD jet production.

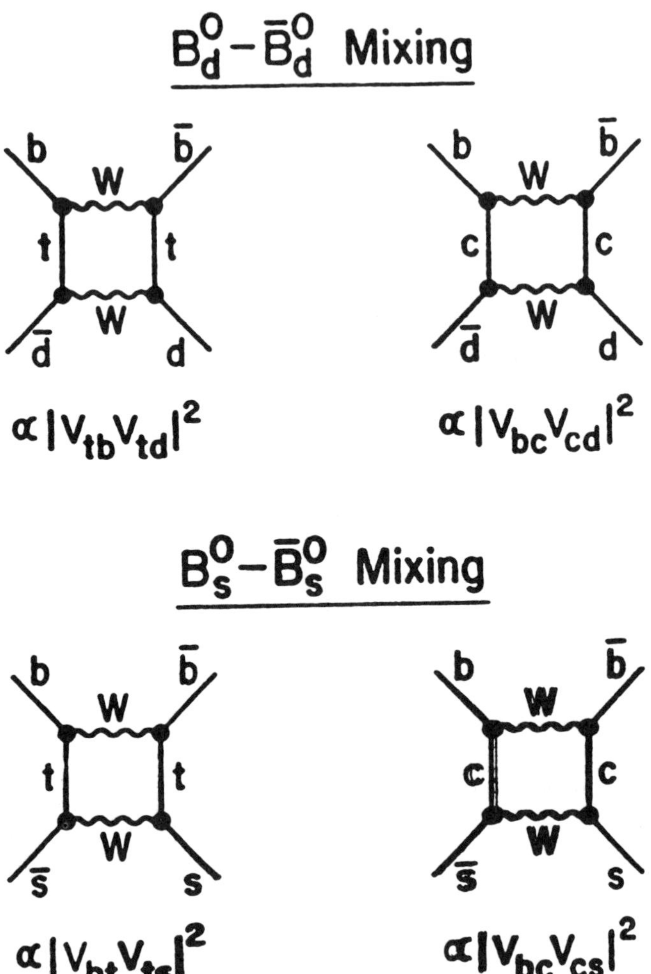

FIGURE 13. The various diagrams that contribute to the mixing of B_d^0 and B_s^0 mesons. Note that $|V_{ts}| \gg |V_{td}|$, $\delta M_s/\Gamma_s \gg \delta M_d/\Gamma_d$, and M_t is large compared to M_c.

TECHNIQUES TO SEARCH FOR NEW HEAVY LEPTONS IN THE DECAYS OF THE *W*

The *W* with a mass of 83 GeV gives a large available mass to create new particles. One possibility is the decay of the *W* into the *t* quark yond *b* particles. The phase factor allows for the detection of a *t* quark with a mass up to 60 GeV (which is beyond that available at Z^o factories). A similar situation holds for the detection of the fourth generation of leptons through the process

$$W \rightarrow L + \nu_L$$

(as first pointed out by Cline and Rubbia in 1983).[12] An intensive search is under way in UA1 for evidence for this decay process and for a new heavy lepton. Eventually, this search might be extended to a mass of 60 GeV (again, beyond SLC and LEP I energies). The present search involves the kinematical signature

$$W \rightarrow L + \nu_L$$
$$ \longrightarrow q + \bar{q} + \nu_L,$$

where the q, $\bar{q}$ "jets" are both on one side of the $\bar{p}p$ line of interaction. This search is being carried out by M. Mohammadi (a Wisconsin graduate student). We will present some very preliminary results of this search here for the purpose of understanding the

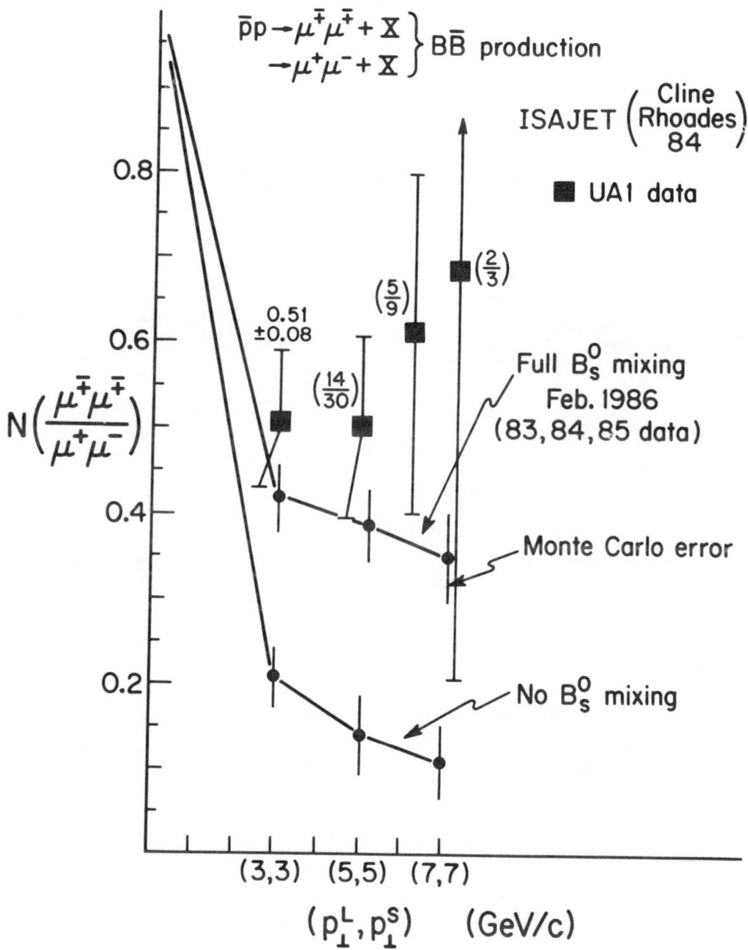

FIGURE 14. Preliminary data from 1983–85 indicating strong evidence for B^0 mixing.

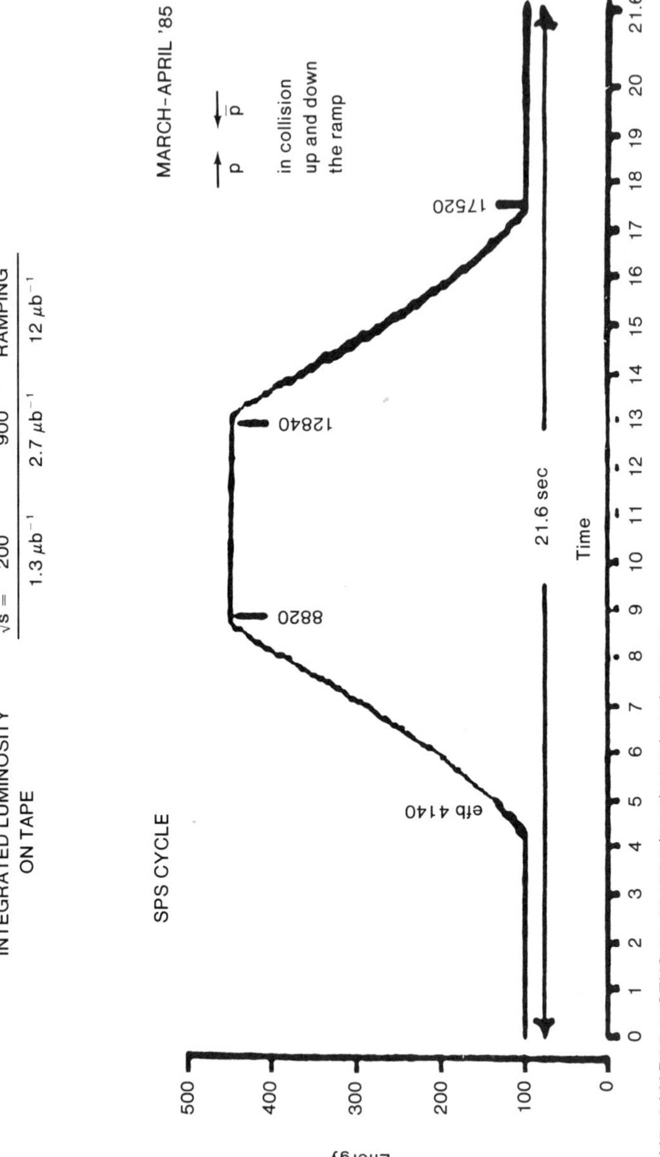

FIGURE 15. The shape of the ramping cycle and the kinds of average luminosity obtained at various center-of-mass energies. This is for the variable energy run (1985) with $\sqrt{s} = 200$–900 GeV.

TABLE 6. Number of Events Analyzed at Each Energy in the Data Sample

● 1985 $\sqrt{s}$ = 0.2–0.9 TeV	230k trig.	$\int L dt \sim$
35%	Flat top (0.9 TeV)	
15%	Flat bottom (0.2 TeV)	
50%	Ramping (0.2–0.9 TeV)	
● Present analysis based on (~all data analyzed):		
0.2 TeV	43k (32k) events	
0.9 TeV	84k (74k) events	
Ramping	106k (91k) events	
● Ramping subdivided in 5 blns:		
260 ± 60 GeV	13k (10.5k) events	
380 ± 60 GeV	14k (11.5k) events	
500 ± 60 GeV	16k (13.5k) events	
620 ± 60 GeV	19k (16.5k) events	
790 ± 110 GeV	45k (39.5k) events	

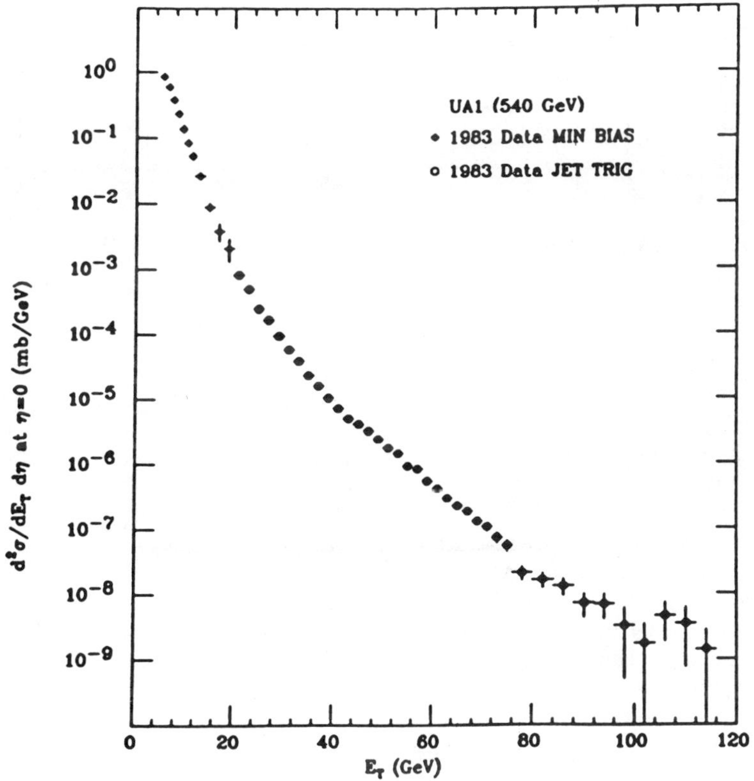

FIGURE 16. The jet transverse energy distribution at 540 GeV c.m. energy.

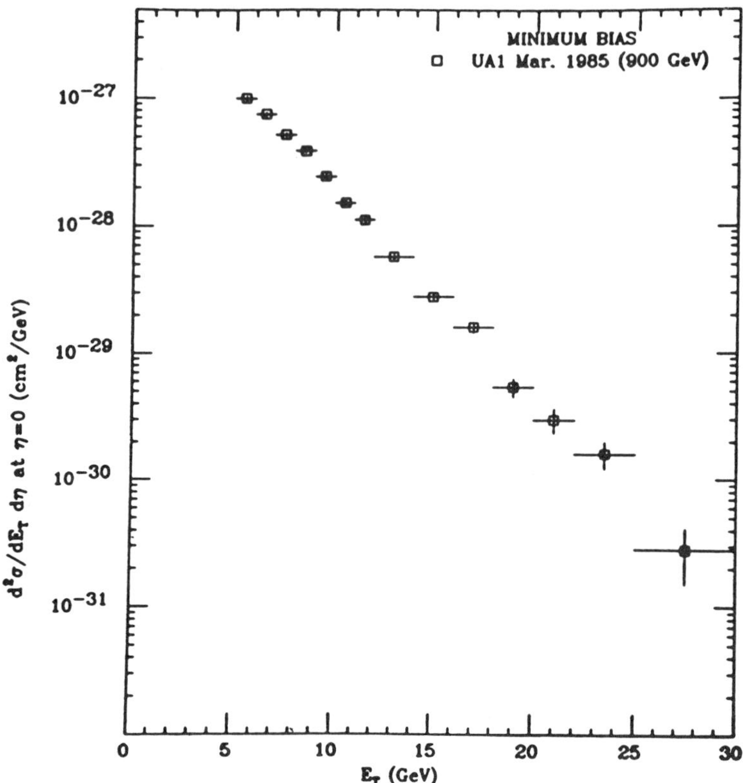

FIGURE 17. The E_T distribution for 90 GeV center-of-mass energy.

implications for SSC experiments. The data analyzed are from the 1983 and 1984 UA1 runs. TABLE 7 shows the breakdown of the different event types as more and more restrictive cuts are applied to this topology of events. The final events contain five candidates. These events form the basis for a search for L. The characteristics of these five events are shown in FIGURE 20, where ETM refers to the missing energy in the event and E_TJET1, E_TJET2 refers to the transverse energy of the highest and lowest energy jet in each event. We now turn to the expectations for the ETM and jets from the heavy lepton decay (FIGURE 21). In this figure, a heavy lepton of mass 30 GeV is studied. The results for $M_L = 40$ GeV or $M_L = 50$ GeV are similar. Note that there are no events with ETM above 40 GeV. Comparing this distribution with FIGURE 20, it is clear that the highest ETM events in the plot cannot be due to a new heavy lepton. We will not discuss these events here further.

We now turn to the backgrounds expected from all possible sources. Full detector simulation and Monte Carlo event generation (using programs like ISAJET) were used in this analysis. TABLE 8 gives a breakdown of these backgrounds. If we ignore the top quark contribution, we obtain 3.3 events, which is in remarkable agreement with the three candidate events. FIGURE 22 shows the distribution for the background events. Thus, the present search for a new lepton has achieved a signal-to-background ratio of 1—a remarkable accomplishment with only 400 nb^{-1} of integrated luminosity. With the 1985–86 data and especially with the ACOL improvement of the $\bar{p}p$ luminosity, it should be possible to extract a signal if the mass of the fourth generation lepton is less than 60 GeV.

TECHNIQUES TO SEARCH FOR A LOW MASS HIGGS BOSON OF THE DECAY OF THE *W*

At a much lower level of branching ratio than the heavy lepton decay of the *W*, it may be possible to search for the Higgs boson by

$$W \rightarrow \text{``}W\text{''} + H,$$

where "*W*" is a virtual *W*. The mass of the "*W*" is approximately $M_W - M_H$. Thus, for

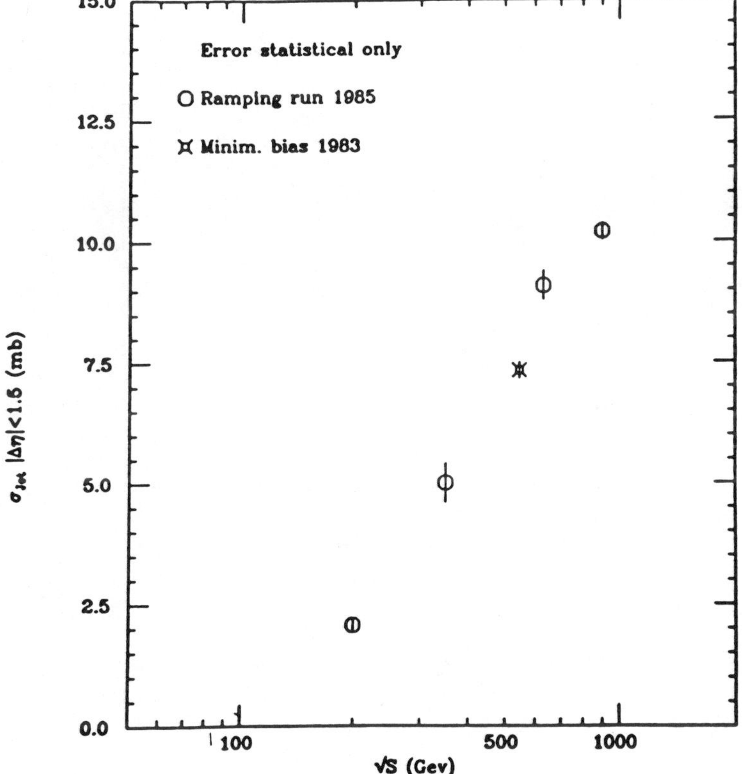

FIGURE 18. The estimated cross section for the production of the minijet events indicating a large rise between 200 and 900 GeV center-of-mass energies.

low mass Higgs bosons, the "W" is nearly on the mass shell (in fact, if $M_H \leq \Gamma_W$, the Higgs boson will be on the mass shell.) An active search is under way in UA1 to study the debris of the events with W's in an attempt to observe the Higgs boson. A very interesting possibility is the detection of the so-called Coleman-Weinberg Higgs boson with a mass of 10.7 GeV. This level is well above the mass range of the upsilons, and thus, the only present viable technique to search for the Higgs boson is at the $\bar{p}p$

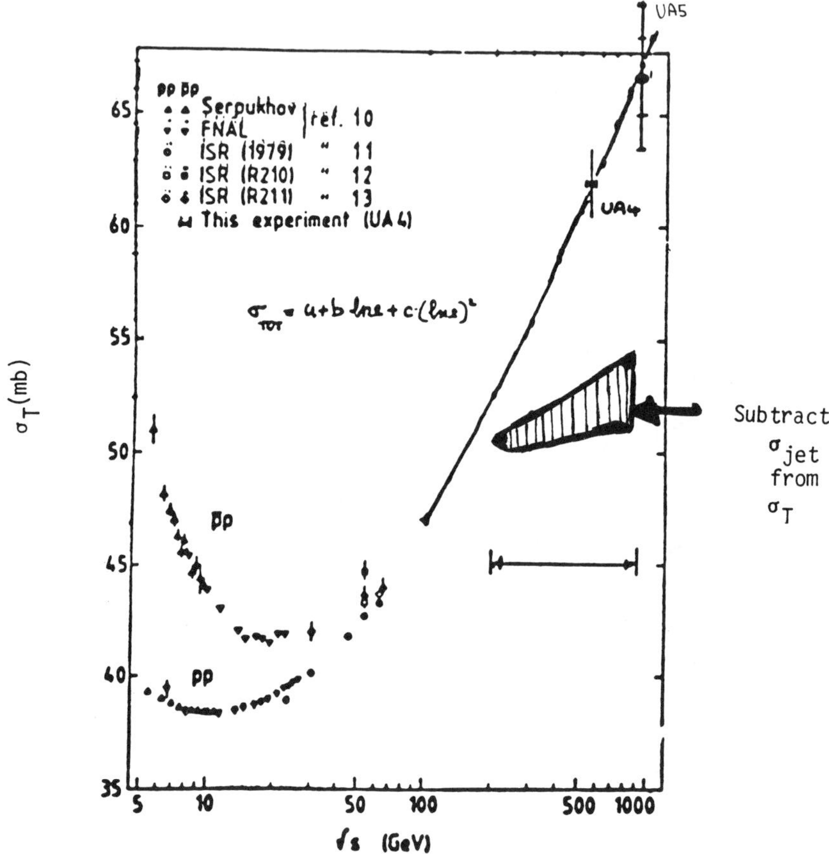

FIGURE 19. The total cross section with the measured minijet cross section subtracted.

collider. It has been pointed out that the best final state to search for is

$$W \longrightarrow \text{"}W\text{"} + H$$
$$ \mathrel{\rightarrow} \ell + \nu \longrightarrow \tau\tau.$$

Although this process has a small branching ratio, it may be very clear if the $\tau\bar{\tau}$ final state is actually identified. The expected branching ratio of the $W \longrightarrow$ "W" $+ H$ is

TABLE 7. Selection of Events

1984 Data	Number of Events: Passing Cuts:	
Initial sample (x-line, split)	56,388	Total
	19,356	x-line
ETM > 15 GeV and $>3\,\sigma$	12,541	Total
ETM not pointing to within 20° of vertical		
No CALJET ($E_T > 12$ GeV) within 20° of vertical	5,011	x-line
At least 2 jets ($E_T > 8$ GeV)	6,996	Total
	2,228	x-line
(% Events Cut)		
Jet 1 $\equiv$ Highest E_T jet ≥ 12 GeV: 3%		
Jet 2 $\equiv$ Second highest E_T jet ≥ 10 GeV: 27%		
No CALJET ($E_T > 8$ GeV, $171 < 2.5$) within 90°: 75%		
No CD jet ($P_T >$ GeV, $171 < 2.3$) within 90°: 52%	425	Total
$\Sigma_{1\text{-}0}\, P_T/E_T$ (JET 1) ≥ 0.1: 24%	125	x-line
Calibration + vertex cut		
Technical cuts, that is,		
No CD jet vertical (7.5 GeV, 15°)		
Halo		
Gondola, bouchon, c-cell problems		
TDC cuts ...	276	Total
(As in Job ETM3NEW)		
Hadronic $\chi^2 > -5$. For JET1, JET2: 85%		
No electron (ELTAG): 24%	23	Total

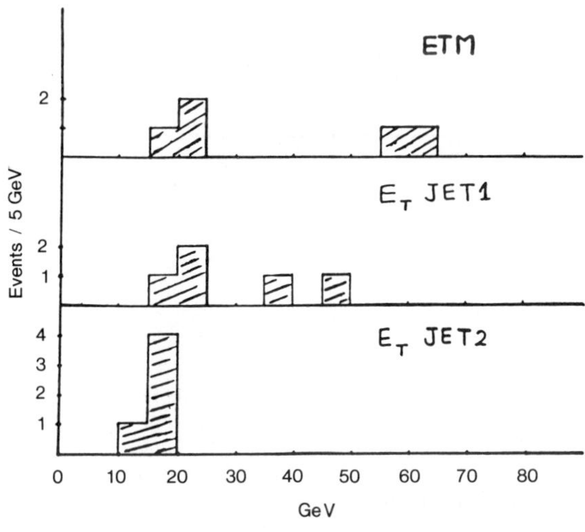

Gondola ϕ prob.	9
Bouchon prob.	5
Leakage	2
Halo	2
Good Events	5

FIGURE 20. The characteristics of the five events that form the basis for a search for L.

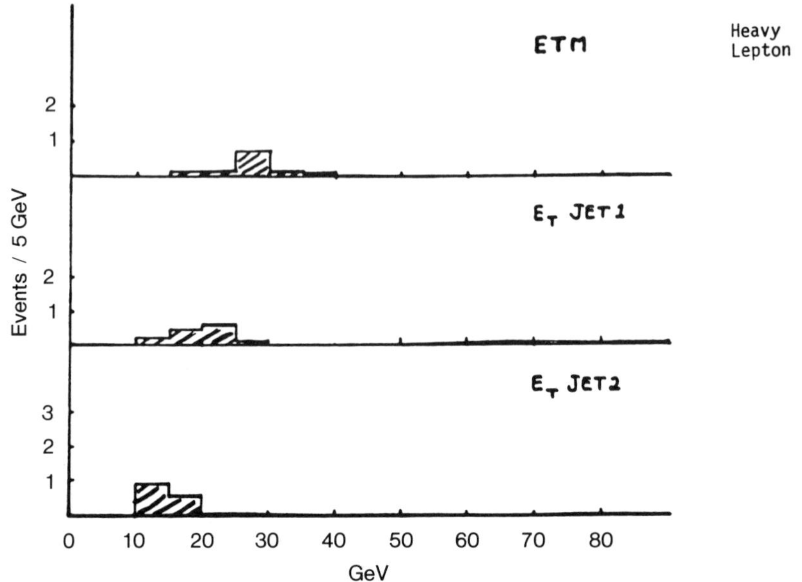

FIGURE 21. The expectations for the ETM and jets from the heavy lepton decay. $W \rightarrow L + \bar{\nu}_L$, $M_L = 30 \text{ GeV}/c^2$: 1.4 events.

TABLE 8. Processes Generated by ISA410 or Event Mixing (Full Detector Simulation and Trigger Simulation)[a]

		$83 + 84$ (381 nb^{-1})
JET-JET Fluctuations		1.2 ± 0.6
$W \rightarrow e\nu$ {Mixed 0.5		0.5
{ISA410 0.4		
$W \rightarrow \tau\nu$ {Mixed 0.03		0.08
$\rightarrow e\nu$ {ISA410 0.08		
$W \rightarrow \mu\nu$ Mixed		0.0
$W \rightarrow \tau\nu$ Mixed		0.0
$\rightarrow \mu\nu$		
$W \rightarrow \tau\nu$ {Mixed 1.3		1.3
$\rightarrow$ hadrons $+ \nu$ {ISA410 1.1		1.3
$Z^o \rightarrow \nu\bar{\nu}$ {Mixed 0.0		0.1
{ISA410 0.1		
$Z^o \rightarrow \tau\bar{\tau}$ ISA410		0.1
$Z^o \rightarrow c\bar{c}, b\bar{b}, t\bar{t}$ ISA410		0.0
$W \rightarrow c\bar{s}$ ISA410		0.0
$W \rightarrow t\bar{b}$ ISA410		0.4
$\bar{p}p \rightarrow c\bar{c}, b\bar{b}$ ISA418		0.0
$\bar{p}p \rightarrow t\bar{t}$ ISA418		0.6
Total		4.28 Events
without top		3.28 Events
top mass $= 48 \text{ Gev}/c^2$		

[a]The processes give the same selection as DATA.

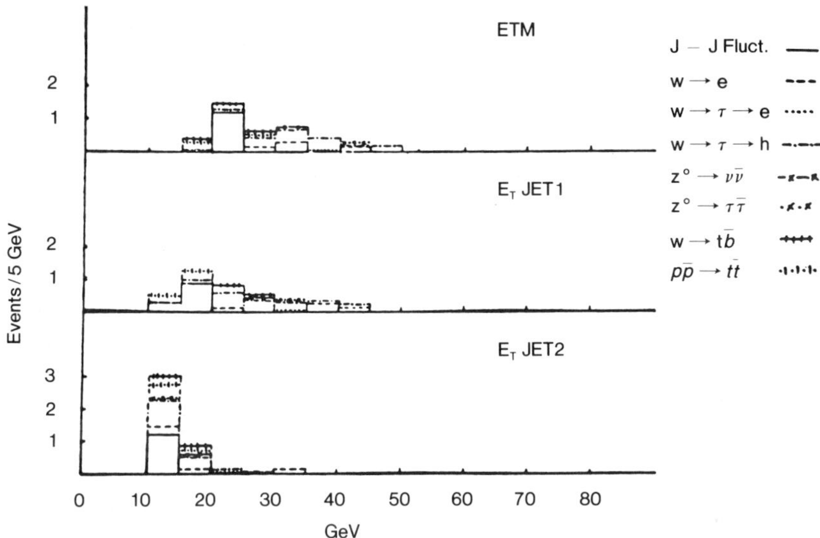

FIGURE 22. The distribution for the 4.3 background events.

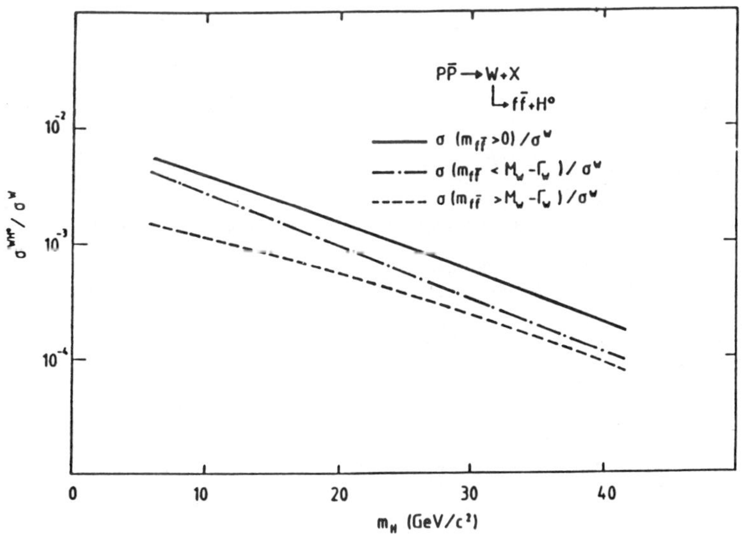

FIGURE 23a. The expected branching ratio for Higgs-W associated production.

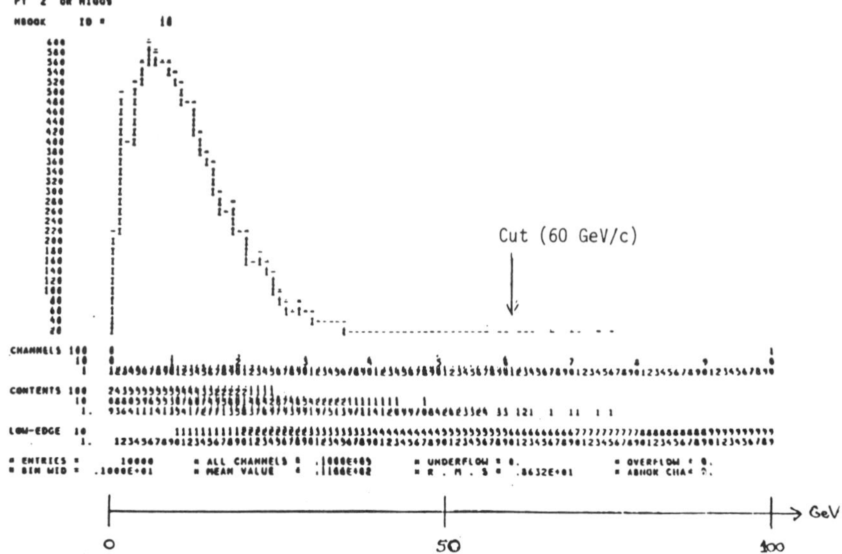

FIGURE 23b. The expected $P_{\perp H}$ distribution for the $M = 7$ GeV case from a recent Monte Carlo calculation.

shown in FIGURE 23a as a function of the Higgs mass. It should be possible to search for a Higgs up to ~10 GeV with the 84/85/86 data and to reach 20 GeV with the ACOL improvement. FIGURE 23b shows the expected $P_{\perp H}$ distribution for the $M = 7$ GeV case from a recent Monte Carlo calculation.

REFERENCES

1. UA1 COLLABORATION. 1983 (August). UA1 upgrade proposal. CERN/SPSC/83-48.
2. UA1 COLLABORATION. 1984 (September). Technical report on the design of a new combined electromagnetic/hadronic calorimeter of rUA1. CERN/SPSC/84-72.
3. UA1 COLLABORATION. 1985. Prospects for a warm liquid TMP calorimeter for UA1. UA1 Internal Notes.
4. The proposal to determine the number of neutrino types by this technique was given independently by: CABBIBO, N. 1983. Rome $\bar{p}p$ Meeting; and CLINE, D. & J. ROHLF. 1983. UA1 TN83-40, where the technique was first applied.
5. ARNISON, G. *et al.* 1986. Phys. Lett. **166B:** 484.
6. See, for example: DESFPANDE, N. G. *et al.* 1985. Phys. Rev. Lett. **54:** 1757.
7. ARNISON, G. *et al.* 1985. Published by Phys. Lett. and CERN EP/85-108 Report.
8. See various reports on UA1 data in 1985: Kyoto Conferences; Oregon DPF Meeting; New Particles '85, Madison, Wisconsin; etc.
9. CLINE, D. 1986. First Aspen Winter Physics Conference. Ann. N.Y. Acad. Sci. **461:** 487–502.
10. CLINE, D., F. HALZEN & J. ROTH. 1973. Phys. Rev. Lett.
11. CERADINI, F. 1985. Study of minimum bias trigger events, etc. CERN EP/85-196.
12. CLINE, D. & C. RUBBIA. 1983. Phys. Lett. **127B:** 277.

First Events and Prospects
at the Fermilab Collider

Presented by MORRIS BINKLEY

CDF Collaboration[a]
Fermi National Accelerator Laboratory
Batavia, Illinois 60510

INTRODUCTION

The Collider Detector at Fermilab (CDF) is a large detector currently being assembled at the Fermilab Tevatron by an international collaboration of physicists from the United States, Italy, and Japan. During September–October 1985, there was an engineering run to test the antiproton source[1] and the CDF data acquisition system. Significant parts of the CDF detector were in place and proton-antiproton collisions were observed at 1.6 TeV center-of-mass energy. Many of the detector components were not installed for this run because of time limitations and because of a temporary physical constraint imposed by the location of an accelerator beam pipe 18 inches above the Tevatron. The constraint will be moved and the full detector will be in place for the next run scheduled for the winter of 1986–87.

DETECTOR

An isometric view of the CDF detector is shown in FIGURE 1 and a cut through one-half of the detector is shown in FIGURE 2. The detector is built around a solenoidal magnet with an axial field (i.e., parallel to the beam direction). Inside the magnet and in the forward direction, there are tracking chambers, and behind these, there are electromagnetic and hadronic calorimeters. In addition, there are muon detectors, trigger counters, and small angle silicon strip detectors.

Small Angle Counters

Scintillation trigger counters and silicon microstrip detectors measure the charged tracks produced at small angles with respect to the beam. The trigger counters form a scintillation hodoscope surrounding the beam pipe that participates in the lowest level triggering. The sixteen trigger counters at each end of the detector cover the angular range from 5.8 mr $< \theta <$ 80 mr and provide precise timing signals. These timing signals are used to provide the event timing for the time to digital converters (TDCs) and to measure the z position of the interaction vertex.

The silicon strip detectors are inside the beam pipe at positions ranging from 6 to 55

---

[a]The members and their institutions are listed in APPENDIX A.

meters from the detector center and they cover an angular range of 0.2 mr $< \theta <$ 6.0 mr. They will be used to measure diffractive and elastic scattering, and in conjunction with the rest of the detector, they will allow a luminosity independent measurement of the total cross section.

Magnet

The superconducting solenoidal magnet[2] is three meters in diameter by five meters long, which produces a 1.5-tesla field coaxial with the beam direction. The iron in the endplug and endwall calorimetry, along with the yoke, form the return path for the magnetic field; only a small part of the flux is returned through the central calorimetry

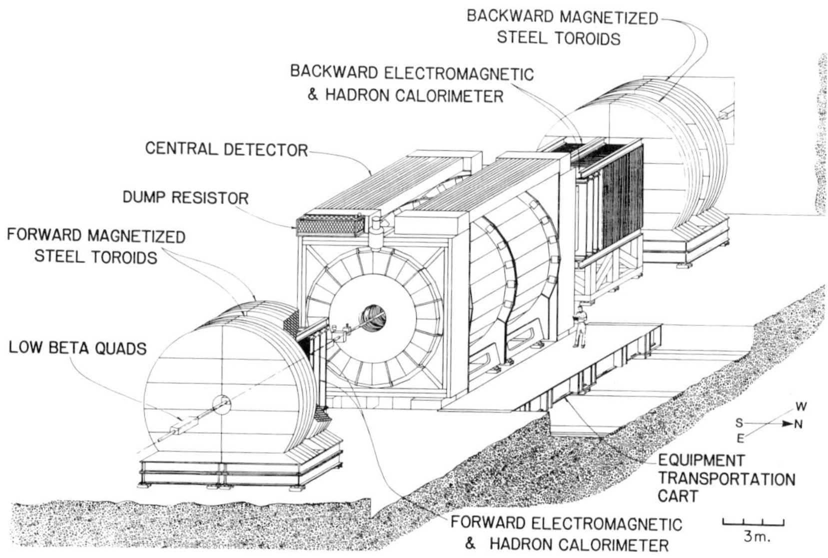

FIGURE 1. An isometric view of the CDF detector.

iron. The magnet was not operated during the first test run because the accelerator constraints kept the endplugs from being installed, but it was tested earlier with the endplugs in place and it performed well.[3] The magnet installation between the endwalls of the central detector is shown in FIGURE 3.

Calorimetry

The calorimetry is characterized by a projective tower structure, good granularity, and good energy resolution; the specifics are summarized in TABLE 1. The central detector and associated calorimetry are shown while in transit from the assembly area

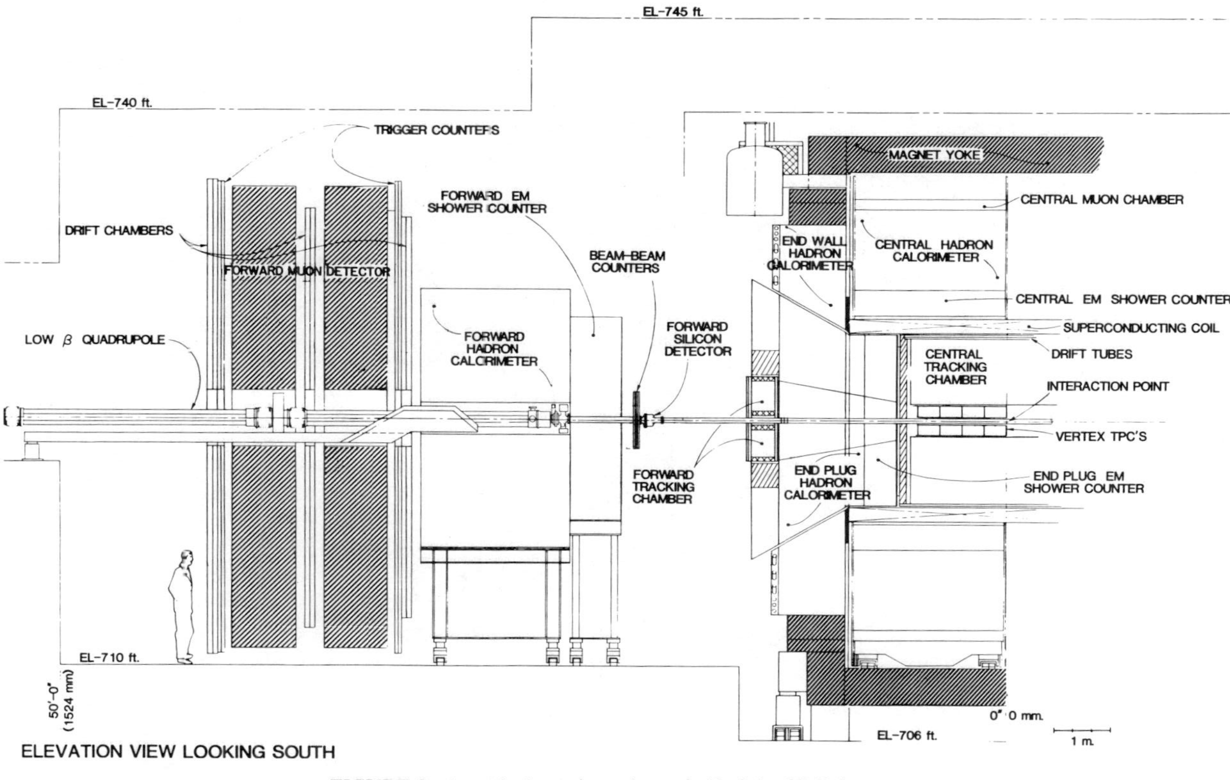

FIGURE 2. A vertical cut through one-half of the CDF detector.

FIGURE 3. Installation of the superconducting solenoid between the central detector endwalls.

TABLE 1. Summary of EM and Hadron Calorimetry

EM Calorimetry Summary			
	Forward	End Plug	Central
Ang. range	2° to 10°	10° to 37°	37° to 90°
Towers in η	18	16	10
Tower size	$\Delta\eta \sim 0.1$	$0.05 < \Delta\eta < 0.1$	$0.087 < \Delta\eta < 0.13$
	$\Delta\phi = 5°$	$\Delta\phi = 5°$	$\Delta\phi = 15°$
Construction	Pb-gas tubes	Pb-gas tubes	Pb-scintillator
R.L. per layer	$0.9X_o$	$0.53X_o$	$0.6X_o$
Layers	15 15	5 24 5	31
R.L. per sect.	$13X_o$ $13X_o$	$2.7X_o$ $13X_o$ $2.7X_o$	$19X_o$
$\Delta E/E$	$\sim27\%/\sqrt{E}$	$\sim24\%/\sqrt{E}$	$\sim14\%/\sqrt{E}$
Position resol.	2–4 mm	1–2 mm	1.5–3 mm

Hadron Calorimetry Summary				
	Forward	End Plug	End Wall	Central
Ang. range	2° to 10°	10° to 30°	30° to 45°	45° to 90°
Tower size	$\Delta\eta = 0.1$	$\Delta\eta = 0.09$	$0.08 < \Delta\eta < 0.12$	$0.1 < \Delta\eta < 0.15$
	$\Delta\phi = 5°$	$\Delta\phi = 5°$	$\Delta\phi = 15°$	$\Delta\phi = 15°$
Construction	2 in. Fe + gas tubes	2 in. Fe + gas tubes	2 in. Fe + scintillator	1 in. Fe + scintillator
Layers	28	20	15	32
$\Delta E/E$	$\sim125\%/\sqrt{E}$	$\sim130\%/\sqrt{E}$	$\sim14\%$ at 50 Gev	$\sim70\%/\sqrt{E}$

to the collision hall in FIGURE 4. The central calorimetry has a scintillator as the sampling medium in order to optimize the energy resolution. In the forward direction ($\theta < 30°$), all of the calorimetry employs gas proportional tubes for sampling in order to minimize the effects of radiation damage. Considerable care has been taken in the design of the calorimetry to provide good gain monitoring systems and to make sure the calorimetry is hermetic. There also are "crack" counters covering the boundaries at

FIGURE 4. The central detector moving out of the assembly area toward the collision hall in preparation for the 1985 test run. The outside of two central calorimetry arches, along with their cabling and RABBIT crates, can be seen. The cable carrier is above the detector.

constant ϕ between the central calorimetry wedges. They measure the position and energy of the electromagnetic showers pointing at these boundaries and eliminate a "hot spot" in the central hadronic calorimetry caused by these showers. The "crack" counters consist of nine radiation lengths of uranium followed by a multiwire gas proportional chamber.

Tracking

There are three systems of tracking chambers. Together, they measure all charged tracks down to approximately two degrees of the beam direction. Future plans include a silicon strip vertex detector surrounding the beam pipe.

Vertex Time Projection Chamber (VTPC)

The innermost tracking system consists of eight VTPC modules that surround the beam pipe. An isometric view of two modules is shown in FIGURE 5. Each module has two 15-cm drift regions separated by a center high voltage screen. The electrons drift from the center screen toward the endcap proportional wire chambers, which are divided into eight octants with 24 wires and pads each. The wires (chords of circles centered on the beam) measure points in the $r - \theta$ plane and are read out using multihit TDCs. The pads measure the position along the wire or the ϕ coordinate and are read out with flash analog to digital converters (FADCs). The active area extends from 6.8 cm to 21.4 cm in radius, and the total length in z of one module is approximately 35 cm. The eight modules with a total of 3072 wires and 3072 pads are sensitive for particles with angles larger than $\theta \sim 3°$. During the test run, only the wires were instrumented. Wire resolutions vary depending on drift distance and track angle,

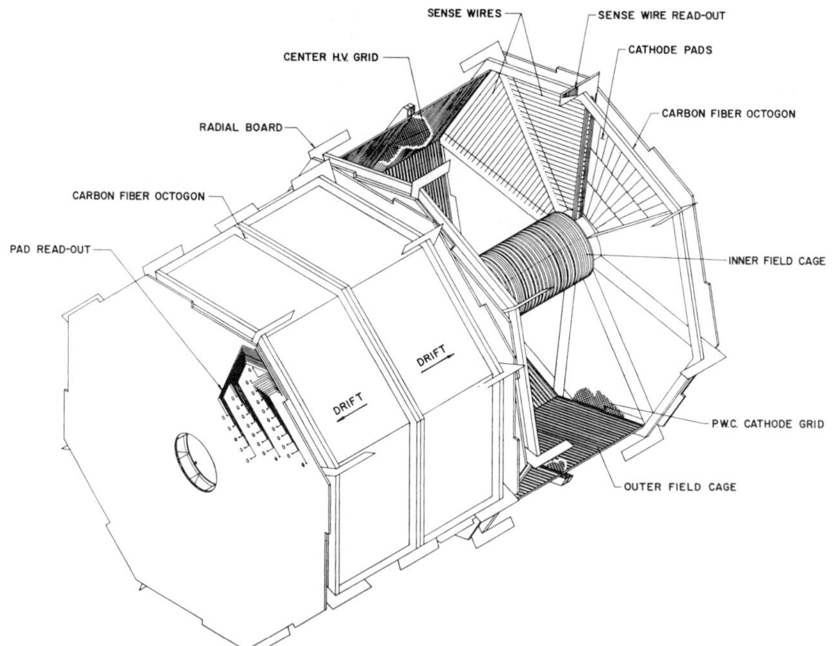

FIGURE 5. An isometric view of two VTPC modules. They are rotated in ϕ by 11.3° with respect to each other.

FIGURE 6. VTPC modules being installed inside the central detector for the fall 1985 run.

but they can be characterized as being of the order of 500 μ in the θ direction. FIGURE 6 shows the installation of these chambers for the fall 1985 test run. The data from these chambers for two good beam-beam interactions are shown in FIGURES 7 and 8.

Central Tracking Chamber (CTC)

The CTC is a "jet type" axial drift chamber surrounding the VTPC that gives good resolution with a cell structure that facilitates pattern recognition. The chamber consists of five axial superlayers interspersed with four small angle stereo superlayers having a stereo angle of $\pm 3°$. There are 12 sense wires in axial cells and 6 sense wires in stereo cells, and the cells are tilted ~45° with respect to the radial direction. This tilt compensates for the ~45° Lorentz angle of the electrons drifting in the magnetic field,

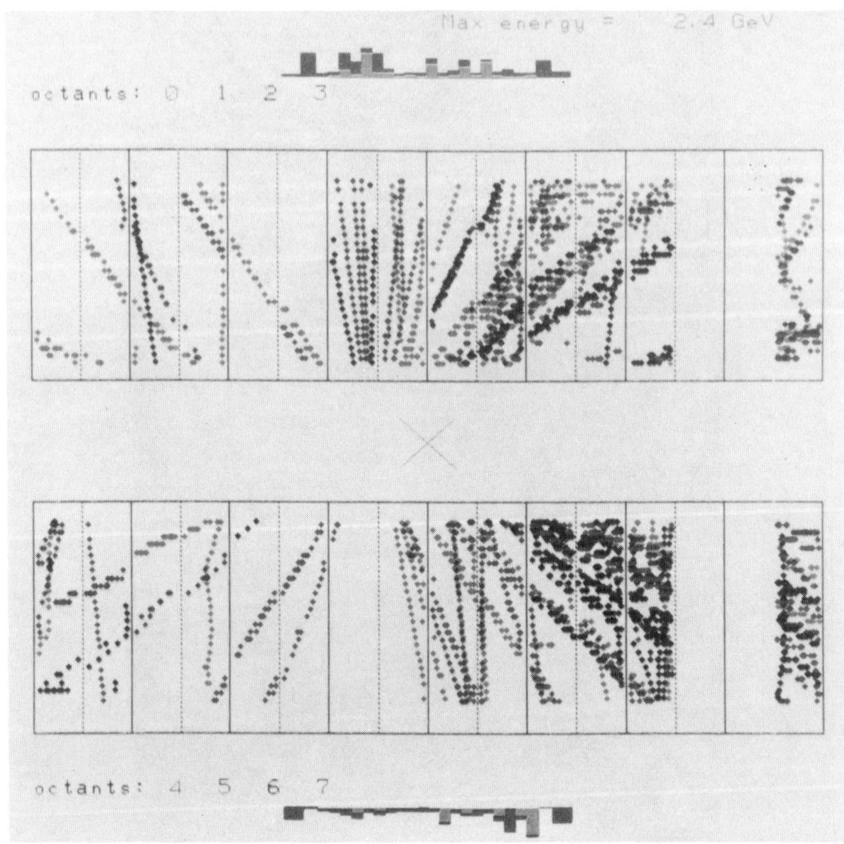

FIGURE 7. Event display of the VTPC wire data for a good beam-beam interaction (fall 1985 test run). The ϕ projection of the calorimetry is also shown. Note that the aspect ratio is distorted.

so the actual drift is perpendicular to the radial direction (i.e., perpendicular to the trajectory of high momentum particles). The sense wire spacing is $\sim$7 mm in the radial direction and the maximum drift distance is $\sim$35 mm. The 6156 sense wires are read out with multihit TDCs. The length of a wire is 3214 mm and the radii of the innermost and outermost sense wires are 309 mm and 1320 mm, respectively. The outermost superlayer covers the region $\theta \geq 40°$ and the innermost superlayer covers $\theta \geq 14°$. The single-hit precision is $\sim$200 μ/wire and the two-hit resolution is <5 mm. The momentum resolution is $\Delta P_t/P_t < 0.002\ P_t$ in GeV for primary tracks that pass through all superlayers.

At the outer edge of the CTC, there are also three layers of axial drift tubes ($\sim$3 m long) that are instrumented for charge division in order to determine a space point. During the next run, two drift tubes will be ganged together for readout purposes; this, though, limits the z resolution to ~6 mm. Upgrading the readout would improve the z resolution to ~3 mm.

Forward Tracking Chamber (FTC)

The FTC is a radial wire chamber that measures the ϕ coordinate of particles from $2° < \theta < 10°$ that pass through the hole in the endplug. The chamber is divided into 72 ϕ cells that are 21 wires deep in z and that have a 2° tilt with respect to the z axis for resolving left-right ambiguities; the cell size is $\Delta\phi = 5°$. The chamber is read out with multihit TDCs and has a single-hit precision of ~125 μ/wire and a double-hit resolution of ~4 mm. A few wires in each ϕ cell will be instrumented for charge division and will be read out with an FADC.

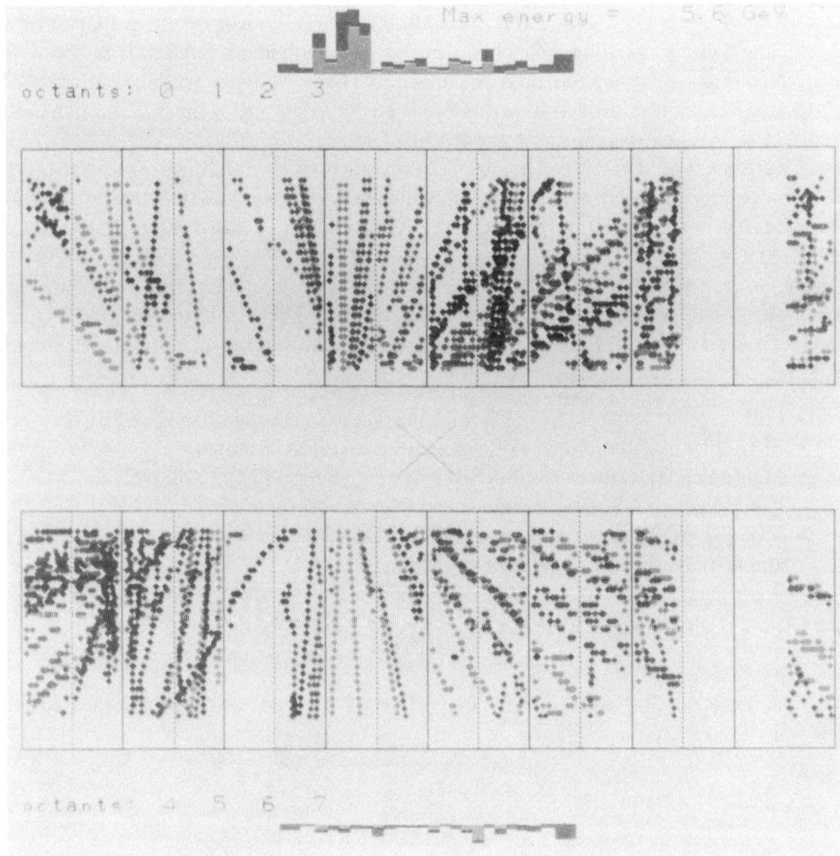

FIGURE 8. Event display of the VTPC wire data for the beam-beam interaction with the most energy in the calorimetry (fall 1985 test run). Note that the aspect ratio is distorted.

Muon Detection System

Behind the central and forward calorimetry are muon detectors. The central muon detector is four planes of large drift tubes operated at high gain in order to enhance the charge division measurement used to determine the third coordinate. They cover an angular range from $50° < \theta < 90°$. In the forward direction, the muon detector consists of two magnetized steel toroids between three sets of drift chambers. The expected momentum resolution is $\Delta P/P \sim 20\%$. The forward muon detector covers the angular region between $2° < \theta < 17°$. Upgrades to the muon detection that would increase the angular coverage are being considered.

DATA ACQUISITION SYSTEM

The general data acquisition system is FASTBUS based and there are two distinct types of front-end electronics that feed this system. One type sits on the detector in RABBIT (Redundant Analog Bus Based Information Transfer) crates that have a backplane with an analog bus. The data in the individual RABBIT modules is converted into a voltage and stored on capacitors. These voltages are multiplexed over the analog bus and digitized by a precision 16 bit ADC (allowing a large dynamic range) that sits in a special slot in the RABBIT crate.

The other type of front-end electronics consists of pulse amplification and shaping on the detector followed by digitizing modules residing in FASTBUS crates (digital bus) located in the counting room. Currently, this scheme is used for multihit TDCs and FADCs.

In order to control the data acquisition system and move the data into the computer, four complicated modules are required. These multiported, intelligent devices are all new, and the fall 1985 run was an important test of them under actual running conditions. They are:

- UPI (Unibus Processor Interface): Interfaces the computer to FASTBUS.
- SI (Segment Interconnect): Networks the FASTBUS crates.
- MX/MEP: Interfaces the RABBIT system to FASTBUS. Controls the ADCs, stores pedestals, buffers the data, and does calibrations.
- SSP (SLAC Scanner Processor): Reads FASTBUS front-end modules, buffers and formats the data, and does calibrations.

TRIGGER

The trigger is divided into three levels. Level 1 was the only one implemented for the fall 1985 test run.

Level 1

The level 1 trigger decision is made between beam crossings (i.e., less than ~ 3.5 μsec) and therefore introduces no deadtime. The goal of level 1 is to reduce the trigger rate to less than ~5000 events/sec in order to limit the deadtime due to level 2. The ingredients of level 1 can include the following:

- Calorimetry—Channels are added and weighted (usually with sin θ and any necessary calibration constants) to form trigger towers. The electromagnetic and hadronic energy are kept separate and there is no clustering. Up to four thresholds are defined, and energy sums are made for towers above the different thresholds. Each sum can then be compared to a sum threshold. The calorimetry was used for the fall 1985 run with the requirement of at least one tower above ~2 GeV.
- Trigger Counters—Hits in both banks of hodoscopes are required with proper timing for a beam-beam interaction. Vetoes can be made on events that have out-of-time hits in hodoscope elements. The trigger counters were used for the fall 1985 run with the requirement that at least one counter in each direction would have a hit within a loose timing window.
- Muon Detectors—Fast signals from both the central and forward muon detectors are available (not implemented for the 1985 test run).
- Central Tracking Detector—At least one stiff track in the central tracking detector can be required (not implemented for the 1985 test run).

Level 2

The level 2 decision typically takes about 10 μsec (sometimes longer for interesting events). During this time, the detector is dead and the data in the front-end electronics are held on sample-and-hold capacitors or in shift registers awaiting the decision on whether the event will be encoded. The hardware is designed to handle ~100 events/sec satisfying level 2. Level 2 was not used in the fall 1985 test run, but it should be ready this year. Some items planned for level 2 are:

- Calorimetry—Clusters can be done for both electromagnetic and electromagnetic plus hadronic energies. Triggers for jets and missing transverse energy are also possible.
- Tracking—High momentum tracks will be available from the central tracking chamber. These can be correlated with energy clusters and muon detector segments to form electron and muon signals.

Level 3

The level 3 trigger is a microprocessor-based system that has access to all the data in the event and can do at least partial event reconstruction. A computer "farm" of approximately 100 microprocessors running in parallel is planned with each microprocessor working on a different event. Each processor will have about one second for each event if the rate of events passing level 2 is ~100 events/sec. Level 3 is needed at full luminosity in order to reduce the rate of events written to tape to a few per second. Tests of level 3 will be made during the next run.

RESULTS OF THE FALL 1985 TEST RUN

The results of the test can be divided into two categories: first, the experience gained with the hardware and software necessary for doing colliding beam physics, and second, the actual events themselves.

Experience with the Detector and Other Components

As expected, the benefits of bringing up and exercising such a complicated system were substantial. Some of the major points can be summarized as follows:

- Data Acquisition System—A system with all the major components was demonstrated to work under actual running conditions. All of the complicated new modules mentioned earlier worked together along with the supporting software.
- Level 1 Trigger—The basic hardware all worked and the calorimetry information seen by level 1 was essentially the same as that encoded by the data acquisition system. Energy levels could be set to ~1 GeV.
- Detector Components—The central and endwall calorimeters, the central muon detectors, the VTPC, and the trigger counters all worked and data were written to tape from these detectors.
- Calorimetry Calibration—The calibration systems for the central and endwall calorimeters were demonstrated to have worked. Energy response was tracked to ~1%.
- Forward Silicon Strip Detectors—Prototypes were demonstrated to work very close to the beam (~6 mm from beam) with low rates in the detector. Electronic noise was also demonstrated not to be a problem.
- Accelerator and Antiproton Source—All the important pieces of the antiproton source worked and the Tevatron was demonstrated to work as a collider. The luminosity was low, but as a result of the run, most of the problems were understood and much higher luminosity is expected for the next run.

Events

Collisions at a center-of-mass energy of 1.6 TeV were seen from three different shots of antiprotons. Typical conditions were $\sim10^{10}$ protons mostly in one bunch and $\sim10^7$ antiprotons in several bunches, thus giving a luminosity in the range of 10^{23}–10^{24} cm^{-2} sec^{-1}. The level 1 trigger required a trigger tower with more than ~2 GeV and a particle in the trigger counter hodoscopes in both directions. This trigger was expected to see 2–5 mb of cross section. Roughly, the expected number of events were observed, and their multiplicity and energy distributions were reasonable.

The result of about three hours of running was ~280 triggers and 23 definite beam-beam interactions inside the VTPC active region, plus a couple of possible events. All 23 good events were found independently by the following three methods:

- Hand scan of the VTPC data: Physicists visually scanned the VTPC displays looking for good events. Two displays for good events are shown in FIGURES 7 and 8. All 23 good events plus ~3 additional candidates were found.

- Computer scan of the VTPC data: All the track segments with more than seven wires in a single VTPC octant were projected onto the beam axis. A 5-cm wide window was moved along the beam axis to find the point of maximum clustering of z intercepts. A good event was defined as having six or more segments pointing at the window with a forward-backward asymmetry $|(N_f - N_b)/(N_f + N_b)| \leq 0.6$. Only the 23 good events were found.
- Computer scan of the trigger counter data: At least three-eighths of the channels in both the proton and antiproton direction were required to have a hit within ~5 nsec of the proper time. All 23 good events were found, along with one possible event that was also found by hand-scanning. By looking in a wider timing window, there was evidence of nine other beam-beam interactions due to satellite beam bunches occurring outside the VTPC active region. This number is consistent with the predicted decrease in luminosity for interactions away from the detector center.

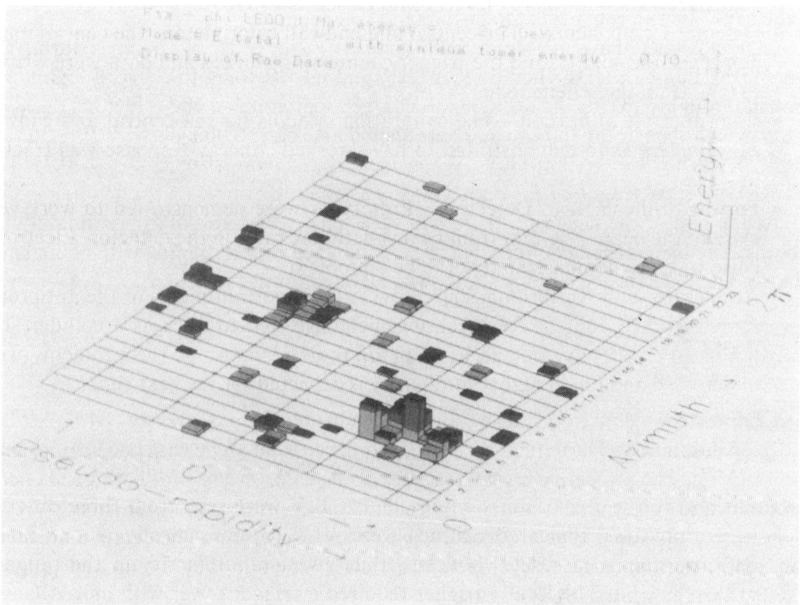

FIGURE 9. Central calorimetry Lego plot for the event shown in FIGURE 8. The electromagnetic contribution is shown with lighter shading.

In the VTPC event displays shown in FIGURES 7 and 8, the upper four octants are superimposed at the top and the lower four octants at the bottom. Note that in these displays, the aspect ratio is distorted by a factor of four (i.e., an apparent angle of 40° with respect to the beam is really ~10°). Also, two half-modules on the right side were dead due to high voltage problems.

During the test run, there was a relatively thick stainless steel beam pipe that led to a considerable number of soft secondaries in the VTPC. Changing to a thin beryllium beam pipe should cause more than an order of magnitude reduction in the number of secondaries seen in the VTPC during the next run.

In addition to the data from the VTPC and trigger counters, some energy was always seen in the calorimetry because of the level 1 trigger requirements. FIGURE 9 shows a Lego plot of the calorimetry for the event with the biggest energy deposition—a cluster of energy greater than 10 GeV. This is the same event as shown in the VTPC display in FIGURE 8.

FUTURE PLANS

The current plans call for the next run to start about December 1, 1986. This will still be a test run for the antiproton source. It will also be a commissioning run for the full CDF detector. The initial luminosity will be low with one-bunch operation. The goal is to achieve three-bunch operation with a luminosity of 10^{29} cm^{-2} sec^{-1} by the end of the run. The run is expected to last about three months.

CDF expects to have an integrated luminosity of the order of 100 nb^{-1} during this run. With 100 nb^{-1}, CDF should see a few hundred Ws going to charged lepton plus neutrino, roughly twenty Zs decaying to charged lepton pairs, and a few Ws decaying to top, which then decay to beauty. There should also be 100,000 jets with $P_t \geq 50$ GeV and 200 jets with $P_t \geq 200$ GeV. When looking for compositeness, $\Lambda = 0.75$ TeV changes the cross section at $P_t = 200$ GeV by a factor of two.

At the end of the next run, there will be a fixed target run of approximately five months. The plan is then to alternate collider and fixed target running with about equal time for each. There are no long shutdowns planned for construction.

CONCLUSIONS

CDF will take data in the near future with a center-of-mass energy of 1.6–2.0 TeV and good luminosity. The detector covers a very large solid angle with good calorimetry and tracking. The projective tower structure of the calorimetry with both good energy resolution and good spatial segmentation should make the detector extremely powerful in answering questions that are inaccessible at lower energies. The detector construction and antiproton source development are on a schedule that should yield physics results from the winter 1986–87 run.

REFERENCES

1. PEOPLES, J. 1983. Fermilab antiproton source. IEEE Trans. Nucl. Sci. **NS-30** (no. 4): 1970; DUGAN, G. 1985. TeV I, energy saver, and anti-proton source. IEEE Trans. Nucl. Sci. **NS-32** (no. 5).
2. WANDS, R. *et al.* 1983. Design of an indirectly cooled 3-m diameter superconducting solenoid with external support cylinder for the Fermilab Collider Detector Facility. IEEE Trans. Magn. **MAG-19**: 1368; GRIMSON, J. *et al.* 1983. A 3-m diameter × 5-m superconducting solenoid for the Collider Detector at Fermilab. Proceedings of the 12th International Conference on High-Energy Accelerators, p. 639. Fermilab. Batavia, Illinois.

3. FAST, R. W. *et al.* 1985. Superconduction solenoid for the Fermilab Collider Detector: installation and testing. Proceedings of the 9th International Conference on Magnet Technology, p. 156. SIN. Zurich.

APPENDIX A

CDF Collaboration

Argonne National Laboratory: R. Diebold, W. Li, L. Nodulman, J. Proudfoot, P. Schoessow, D. Underwood, R. Wagner & A. Wicklund.

Brandeis University: J. Bensinger, C. Blocker, M. Contreras, L. DeMortier, P. Kesten, L. Kirsch, H. Piekarz, L. Spencer & S. Tarem.

University of Chicago: D. Amidei, M. Campbell, H. Frisch, C. Grosso-Pilcher, J. Hauser, T. Liss, G. Redlinger, H. Sanders, M. Shochet, R. Snider, J. Ting & Y. Tsay.

Fermi National Accelerator Laboratory: M. Atac, E. Barsotti, P. Berge, M. Binkley, J. Bofill, A. Brenner, J.T. Carroll, T. Collins, J. Cooper, C. Day, F. Dittus, T. Droege, J. Elias, G.W. Foster, J. Freeman, I. Gaines, J. Grimson, D. Hanssen, J. Huth, H. Jensen, R. Kadel, H. Kautzky, R. Kephart, C. Nelson, C. Newman-Holmes, J. O'Meara, S. Palanque, A. Para, J. Patrick, R. Perchonok, D. Quarrie, S. Segler, D. Theriot, A. Tollestrup, K. Turner, C. van Ingen, R. Vidal, R. Wagner, V. White, R. Yamada, G.P. Yeh & J. Yoh.

INFN—Frascati, Italy: S. Bertolucci, M. Cordelli, M. Curatolo, B. Esposito, P. Giromini, S. Miozzi, S. Miscetti & A. Sansoni.

Harvard University: G. Brandenburg, D. Brown, R. Carey, M. Eaton, A. Feldman, E. Kearns, R. Schwitters, M. Shapiro & R. St. Denis.

University of Illinois: G. Ascoli, S. Bhadra, R. Downing, S. Errede, L. Holloway, I. Karlinger, H. Keutelian, U. Kruse, R. Sard, V. Simaitis, D. Smith & T. Westhusing.

KEK, Japan: Y. Arai, Y. Fukui, S. Mikamo & M. Mishina.

Lawrence Berkeley Laboratory: W. Carithers, W. Chinowsky, R. Ely, M. Franklin, C. Haber, R. Harris, B. Hubbard & J. Siegrist.

University of Pennsylvania: D. Connor, L. Gladney, S. Hahn, N. Lockyer, M. Miller, T. Rohlay, R. VanBerg, J. Walsh & H. Williams.

INFN—University of Pisa, Italy: G. Apollinari, F. Bedeschi, G. Bellettini, N. Bonavita, L. Bosisio, F. Cervelli, R. Del Fabbro, M. Dell'Orso, E. Focardi, P. Giannetti, M. Giorgi, A. Menzione, R. Paoletti, G. Punzi, L. Ristori, A. Scribano, P. Sestini, A. Stefanini, G. Tonelli & F. Zetti.

Purdue University: V. Barnes, A. Byon, K. Chadwick, A. Di Virgilio, A. Garfinkel, S. Kuhlmann, A. Laasanen, M. Schub & J. Simmons.

Rockefeller University: S. Belforte, T. Chapin, G. Chiarelli, N. Giokaris, K. Goulianos, R. Plunkett & S. White.

Rutgers University: A. Beretvas, T. Devlin, U. Joshi, K. Kazlauskis, N. Pearson & T. Watts.

Texas A&M University: J. Buchholz, S. Cihangir, D. DiBitonto, F. Marchetto, P. McIntyre, T. Meyer & R. Webb.

University of Tsukuba, Japan: F. Abe, Y. Hayashide, M. Ito, T. Kamon, S. Kanda, Y. Kikuchi, S. Kim, K. Kondo, M. Masuzawa, T. Mimashi, S. Miyashita, H. Miyata, S. Mori, Y. Morita, T. Ozaki, M. Sekiguchi, M. Shibata, Y. Takaiwa, K. Takikawa, A. Yamashita & K. Yasuoka.

University of Wisconsin: J. Bellinger, D. Carlsmith, D. Cline, R. Handler, J. Jaske, G. Ott, L. Pondrom, J. Rhoades, M. Sheaff, J. Skarha & T. Winch.

Visitors: M. Sivertz—Haverford College, Haverford, Pennsylvania; S. Kobayashi and A. Murakami—Saga University, Japan; Y. Muraki—ICRR, Tokyo University, Japan.

Status of the SLC[a]

KENNETH C. MOFFEIT

Stanford Linear Accelerator Center
Stanford University
Stanford, California 94305

INTRODUCTION

The goals of the SLAC Linear Collider (SLC) are to develop the techniques of linear colliders and to do physics at and slightly above the energy necessary to produce the Z^0. In a linear collider, the electrons and positrons collide once and are then discarded. This is in contrast to a storage ring where the beams are stored for many hours and collide many thousands of times per second. In order to obtain a luminosity large enough to do interesting physics, the beams are focused to a small spot of a few microns and they are only a few millimeters in length. Using the repetition rate of 180 hertz at SLAC, luminosities of 6×10^{30} cm^{-2} sec^{-1} can be reached. A genuine linear collider would require two accelerators directed at one another. The SLC prototype will use only the existing linear accelerator upgraded to reach 50 GeV. The positrons and electrons are accelerated simultaneously to the desired energy, and then they enter separate arcs that transmit them to an intersecting region where they collide head-on. In this way, the SLC machine can reach center-of-mass energies high enough to produce the Z^0 via its coupling to e^+e^- pairs. At the design luminosity of the SLC, Z^0's will be produced in the excess of one million per year.

The outline of this paper will be as follows: A short review of the physics goals will be followed by the status of the SLC and the detectors. Finally, in the last section, the plans for polarized electrons will be given.

PHYSICS AT THE SLC

The physics program of the SLC concentrates on production of Z^0's and their decay. Measurements of the Z^0 mass and width will be among the first and most important physics results from the SLC. Using spectrometers in the extraction lines of the SLC, the Z^0 mass will be determined to an accuracy of around 50 MeV/c^2. Once the Z^0 mass is known to this accuracy, all of the couplings of the standard model of electroweak interactions will be completely determined by three fundamental constants: α, G_F, and M_Z. The Z^0 width is both a test of the standard model and a measure of the particle content of Z^0 decays. For example, a fourth neutrino species contributes 160 MeV/c^2 to the Z^0 width.

The SLC is unique among existing or planned colliding-beam facilities in its

[a]This work was supported by the Department of Energy, Contract No. DE-AC03-76SF00515.

potential to accelerate longitudinally polarized electrons. The polarization sense is reversible from pulse to pulse at the operator's control, and it thereby allows precise tests of the couplings of the fermion through the measurement of the left-right asymmetry:

$$A_{\mathrm{LR}} = \frac{\sigma_\mathrm{L} - \sigma_\mathrm{R}}{\sigma_\mathrm{L} + \sigma_\mathrm{R}}.$$

In the standard model, A_{LR} is uniquely predicted once M_Z is known. In lowest order, for $e^+ e^- \rightarrow f\bar{f}$, independent of the final fermion type,

$$A_{\mathrm{LR}}(M_Z) = \frac{g_\mathrm{L}^{e2} - g_\mathrm{R}^{e2}}{g_\mathrm{L}^{e2} + g_\mathrm{R}^{e2}} = \frac{2a_e v_e}{a_e^2 + v_e^2}.$$

The accuracy of the measurement of A_{LR} depends on the uncertainty in the polarization measurement and statistics. The precision anticipated at the SLC is displayed in FIGURE 1 as a function of the number, N, of observed Z decays. The three curves correspond to different levels of precision of measuring the electron polarization. The scales on the right show the resulting precision of the A_{LR} measurement to those of $\sin^2\theta_W$ and M_Z. The left-right asymmetry will replace the ratio of neutral-to-charged currents for neutrino scattering as the best test of the standard model when the number of Z's is between 10^3 and 10^4. With the more precise measurement of the beam polarization at the 1% level and with 10^6 Z^0's, the accurate measurements of M_Z and A_{LR} at the SLC will make very high precision tests of the standard model.

Searches for Z-decays to the predicted top and Higgs particles will be made, along with searches for new or unexpected phenomena beyond the standard model.

THE STATUS OF THE SLC

FIGURE 2 shows the layout of the systems of the SLC. These include a new injector and booster, two damping rings to provide the small beam emittance, a new positron source, the existing linac structure upgraded for higher energy and better control of the beams, beam transport arcs, a final focus section, and experimental halls and detectors. The SLC works as follows: two bunches of 5×10^{10} electrons separated by 59 ns are accelerated to 1.2 GeV and stored in the north damping ring on opposite sides of the ring for the time between accelerator cycles (5.6 msec at 180 hertz). The two e^- bunches are extracted 59 ns behind the positron bunch from the south damping ring and accelerated. The e^+ bunch and the first e^- bunch are sent around the colliding arcs so as to be focused to less than two microns in size for collison at the interaction region. The second e^- bunch generates new positrons for a later e^+ bunch. A brief status report of these systems is given below.

The electron source and injector are required in order to deliver to the damping ring two bunches of greater than 5×10^{10} electrons that are 59 ns apart at 180 hertz. The test in 1984 showed that the new thermionic gun and the first linac section upgraded with 80 focusing quadrupoles passed these specifications.

The damping rings are used to reduce the beam emittance to values low enough to be forced to the small sizes at the $e^+ e^-$ interaction region. After the electrons are collected into a single bunch and the positrons are returned from the e^+ source, their emittances are still too high for this. The south ring was built a few years ago and the experience showed the need for stronger sextupoles, better and more position monitors, and septum cooling. These improvements are in progress. The north ring was built with these changes and was commissioned in February 1986. Both rings will be running in

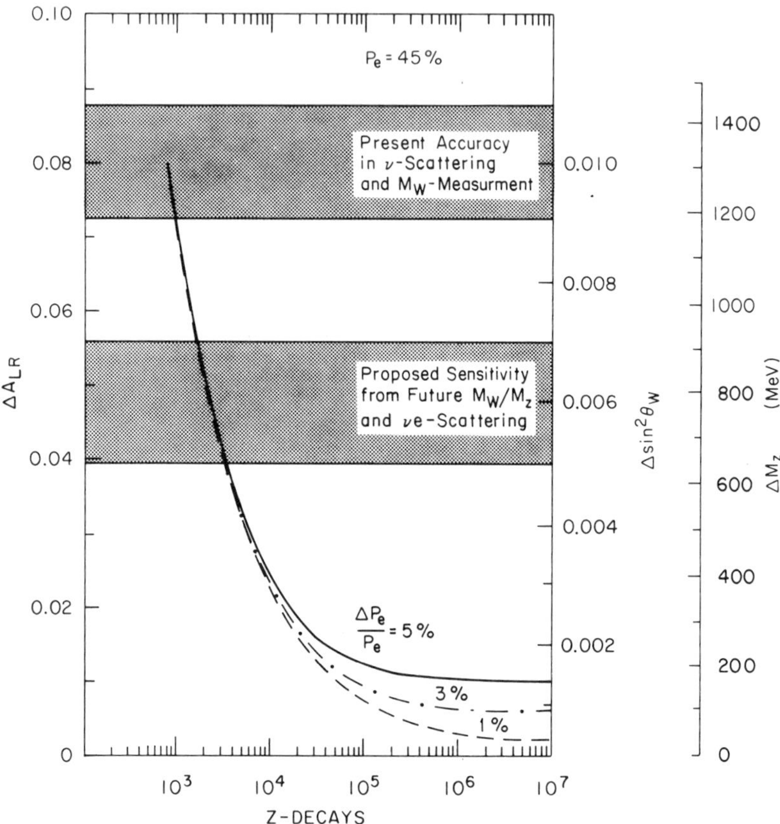

FIGURE 1. The uncertainty in the left-right asymmetry, ΔA_{LR}, as a function of the number, N, of observed Z decays. The three curves correspond to $\Delta P/P = 5\%$, 3%, and 1%. The scales on the right give the corresponding uncertainty in $\sin^2\theta_W$ and M_Z. The shaded regions indicate the present accuracy and proposed sensitivity from M_W/M_Z measurements and ν-scattering experiments.

the summer of 1986. The maximum single bunch beam stored is $4 \times 10^{10}\,e^-$, and this was limited only by the time available for tuning. The electron ring requires a double pulse kicker to handle the two bunches.

The main improvements to the linac are the upgrading for higher energy and the addition of beam focusing and position monitoring every 40 feet. The RF control and

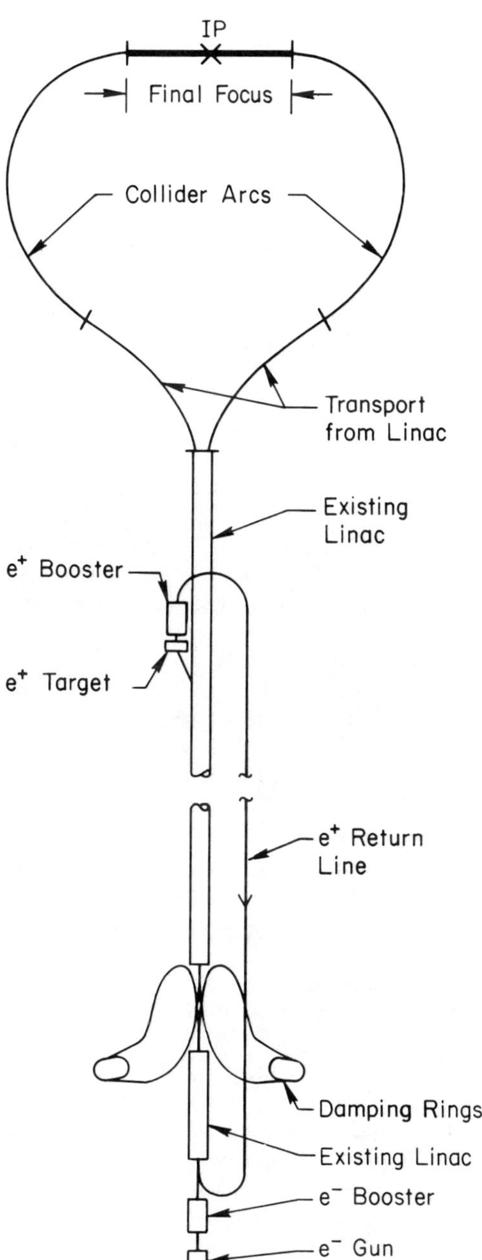

FIGURE 2. Schematic layout of the SLC systems.

monitoring is also being upgraded. The first third of the accelerator is complete and the rest is scheduled to be finished by summer. Replacement of the 38-megawatt klystrons with 67-megawatt ones are in progress. About 75% of the 232 units are made and it is expected that the complete complement will be on the machine by January 1987. Twenty sectors have the new high-powered klystrons installed. Experience was made in the January and February 1986 running period with four of these sectors.

The SLC requires a positron source with a yield large enough to give equal numbers of positrons and electrons at the interaction point. The colliding positrons must have an emittance and bunch length similar to the electron beam. A schematic diagram of the positron source is shown in FIGURE 3. A single bunch of electrons trailing the primary bunch by about 59 ns is extracted from the linac at the two-thirds point. These 33-GeV electrons are transported to the positron vault where they strike a tantalum–10% tungsten target. Positrons are collected in a focusing solenoid system using an iron-shaped solenoidal field of 1 tesla at the target and a pulsed solenoidal field of 5 teslas. The positron beam is accelerated to 200 MeV in a 1.5-m accelerator section of 50 MeV/m and standard accelerator sections. They are then transported back to the beginning of the linac for further acceleration to 1.21 GeV, whereupon which they are sent to the positron damping ring. The extraction and return beam lines are finished and the other components are scheduled to be completed by summer. The goal of the positron group is to have the entire positron system commissioned by the end of the summer of 1986.

Magnets for the collider arcs are fabricated and installation is in progress. The 950 magnets combine bending, quadrupole, and sextupole fields into one element. The very strong focusing uses second-order achromatic optics to transport a $\Delta E/E = \pm 0.5\%$ beam to the final focus. Commissioning to the arcs is planned for late fall 1986.

The final focus system takes the linac e^+ and e^- beams as they emerge from the arcs with elliptical size of 250μm $\times$ 30μm to round spots with sigmas less than $2\ \mu$m at the IR. This all has to be done with $\pm 0.5\%$ momentum spread and without background from off-axis $e^\pm$ and synchrotron radiation. The control of the beams occurs within 500 feet of the IR using a set of 8 bending magnets, 27 quadrupoles, 8 sextupoles, and numerous correctors and profile monitors. The vacuum is to be 2×10^{-9} torr within ± 30 meters of the IR. After the beams collide, they are transported through the final focus to a kicker magnet where they are extracted and directed toward a dump. The installation of these components is in progress and the commissioning is planned for the end of 1986.

Initially, the quadrupole triplet near the interaction region will be conventional magnets. There are plans to replace these with superconducting magnets so that the beams can be focused to a smaller spot size at the IR. The earliest these could be installed is in the summer of 1988.

Mechanical stability of the arc magnets and final focus magnets must be held to tolerances of 100 μm. Automatic alignment mechanisms are used to adjust for thermal effects and earth movements.

The experimental hall has also beam support and assembly areas. To minimize beam downtime, there are two staging areas for two detectors, and either can move onto the beam line. The facilities are nearing completion. Experimental equipment is being moved into the east assembly area. The contractor is expected to be finished in June 1986.

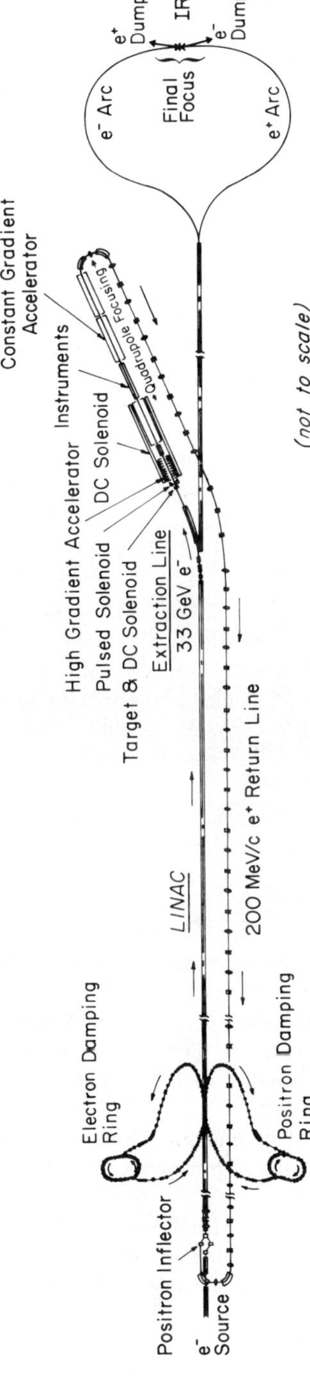

FIGURE 3. Schematic diagram of the SLC positron source system (not to scale). The source from the target to the end of the constant gradient accelerator is about 15 meters.

While the ultimate goal of the SLC calls for a luminosity of $6 \times 10^{30}\,\text{cm}^{-2}\,\text{sec}^{-1}$, a turn on the machine will be unable to deliver that luminosity. For the first couple of years of data-taking, the maximum luminosity will be $6 \times 10^{29}\,\text{cm}^{-2}\,\text{sec}^{-1}$ due to the limitation of the 120 pps operation and the conventional quadrupole triplet in the final focus. In addition, $e^{\pm}$ beams with 7.5×10^{10} particles per bunch will be required to reach the design luminosity. Undoubtedly, it will take time to learn how to achieve high luminosity, so for spring 1987, a luminosity of $6 \times 10^{28}\,\text{cm}^2\,\text{sec}^{-1}$ is a reasonable goal.

STATUS OF THE DETECTORS

Two detectors have been approved for SLC. The first detector for the SLC will be an upgrade of the Mark II that was used at PEP. The upgrade involved replacing the main tracking chamber and associated electronics, the endcap chambers, the time-of-flight counters, and the coil. In addition, new vertex detectors of silicon strips and a precision drift chamber are being made for installation at the SLC. In FIGURE 4, a sectional view of the Mark II/SLC detector is shown. The upgraded Mark II with the exception of the vertex detectors and small angle monitors was completed last summer, and data acquisition occurred at PEP with the new components up until the end of February 1986. The move of the Mark II from PEP to SLC started in March 1986. After checking out this detector with cosmic rays in the staging area at the SLC experimental hall, it is planned that it move on the beam line in February 1987. Data-taking is scheduled to start in March 1987.

The MARK II/SLC collaboration has nine institutions and some 120 physicists. The original Mark II collaboration of SLAC and LBL has been joined by groups from the California Institute of Technology, the University of Colorado, the University of Hawaii, Indiana University, Johns Hopkins University, the University of Michigan, the University of California at Santa Cruz, and two other SLAC groups.

The Mark II central drift chamber consists of 12 layers of cells, each containing 6 sense wires. Alternate layers have their wires parallel to the axis or at $\sim \pm 3.5°$ to the axis to provide stereo information. The inner radius is 19 cm, the outer radius is 148 cm, and the active length is 2.3 m. The magnetic field of the new coil will be at 0.5 tesla. Readout of the data is by FASTBUS. Both TDC and FADC data are accumulated. The dE/dx information will provide e/π separation for particles with energies less than 10 GeV. FIGURE 5 shows an event taken at PEP during the checkout. Preliminary analysis shows that the resolution from tracking averages is about 175 μm, which corresponds to $\sigma(p)/p = 0.1p\,(\text{GeV}/c)$.

The SLD collaboration consists of institutions from the United States (California Institute of Technology, Boston, California State Northridge, Cincinnati, Colorado, Columbia, Illinois, MIT, Northeastern, San Francisco State, SLAC, Santa Barbara, Santa Cruz, Tennessee, Vanderbilt, Washington, Wisconsin), Canada (British Columbia, Triumf, Victoria), and Europe (Frascati, Ferrara, Padona, Perugia, Pisa, Rutherford). This large detector goes beyond the Mark II/SLC in having hadron calorimetry and full particle identification over nearly 4π solid angle. FIGURE 6 shows a cross section of a quadrant of the SLD.

The schedule for building the detector shows completion in 1989. The research and development projects will conclude by the end of 1986. The construction of the magnet is under way, with delivery of the iron core and the 0.6T conventional coil in spring 1987.

Tracking in the SLD consists of a CCD vertex detector and central and endcap drift chambers. The resolution from the 200 CCD cylindrical vertex detector is expected to be 5 μm with a double-track separation of 40 μm. The central tracker is 2 m

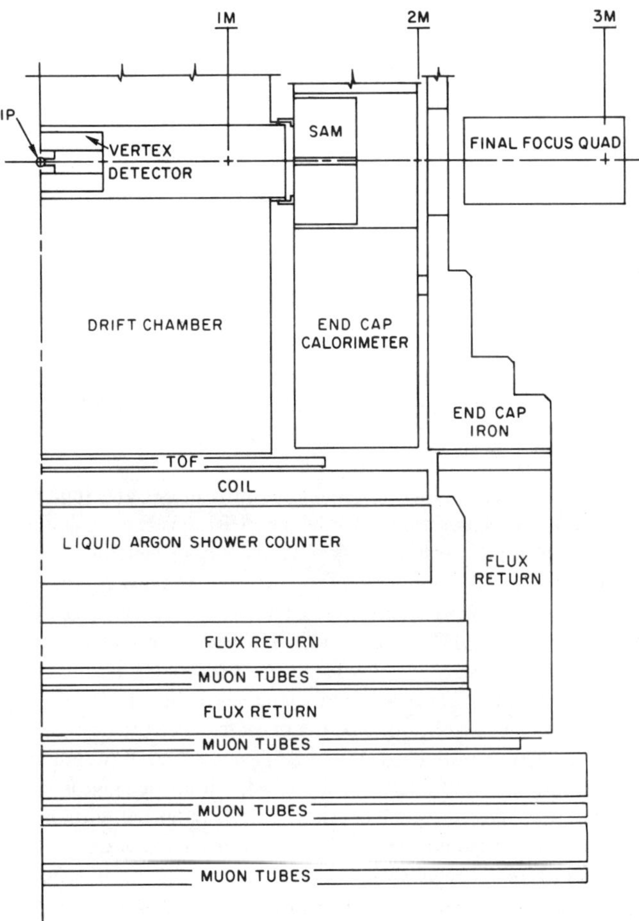

FIGURE 4. A sectional view of a quadrant of the Mark II/SLC detector.

long and has inner and outer radii of 20 cm and 100 cm, respectively. The resolution for a single track is expected to be 100 μm with a double-track separation of 2 mm. Tests with a prototype using the cell configuration and gas of the design chamber have shown resolutions to be considerably better.

Cherenkov ring imaging detectors (CRID) will provide the particle separation in

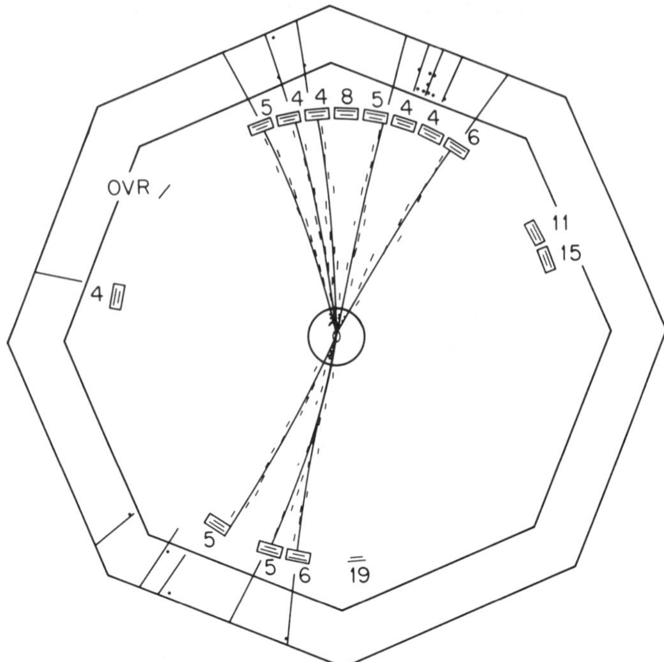

FIGURE 5. Drift chamber tracks in a multihadronic event from the Mark II/SLC detector. The short lines represent the six hit sense wires in a cell.

the SLD. Liquid freon and gaseous freon provide the radiators, and the Cherenkov rings are detected by photon detection in a quartz box filled with a carrier gas that is doped with tetrakis-dimenthyl-amino-ethylene (TMAE). The photoelectrons are drifted toward a detector of sense wires at the ends of the CRID volume. Prototype tests look promising with performances at the expected levels.

Calorimetry in SLD is achieved with liquid argon and lead as a radiator. The electromagnetic shower is measured in the first 22 radiation lengths in towers of 33 milliradians, which is followed by 2.2 collision lengths in 66-milliradian towers for the hadronic section. Instrumented warm iron calorimetry outside the aluminum coil completes the hadron calorimeter and muon tracking. The resolutions expected are $\sigma(E) = 8\% \sqrt{E}$ (electromagnetic) and about $55\% \sqrt{E}$ (hadronic).

POLARIZATION AT THE SLC

The SLC has the capability of accelerating polarized electrons and transporting these through the damping ring to the interaction region with small depolarization of the beam. A collaboration of physicists from Indiana, LBL, SLAC, Rome, Genoa, and Wisconsin was recently approved to provide a polarization capability at the SLC.

The electrons are produced longitudinally polarized by irradiating a GaAs crystal with circularly polarized light. A source capable of reliably producing beams of the required intensity and pulse structure for SLC has been built and is presently being installed at the SLC. Polarizations of 45% are expected with this source. To preserve the polarization while the electrons are in the damping ring, the spin direction must be transverse to the plane of the damping ring. In addition, the spin direction must be reestablished after leaving the damping ring so that it points in the desired direction for acceleration to the end of the linac and for transportation around the electron arc to the interaction point where longitudinal polarization is required. Three spin rotating solenoids of 6.34 T-m each are needed to rotate the spin. These superconducting solenoids are located at positions in transfer lines to and from the damping ring where the spin direction has been processed 90° by the guide field. Compton and Møller Polarimeters will be used to monitor and measure the polarization.

Full polarization capability is expected to be ready in the summer of 1988. The

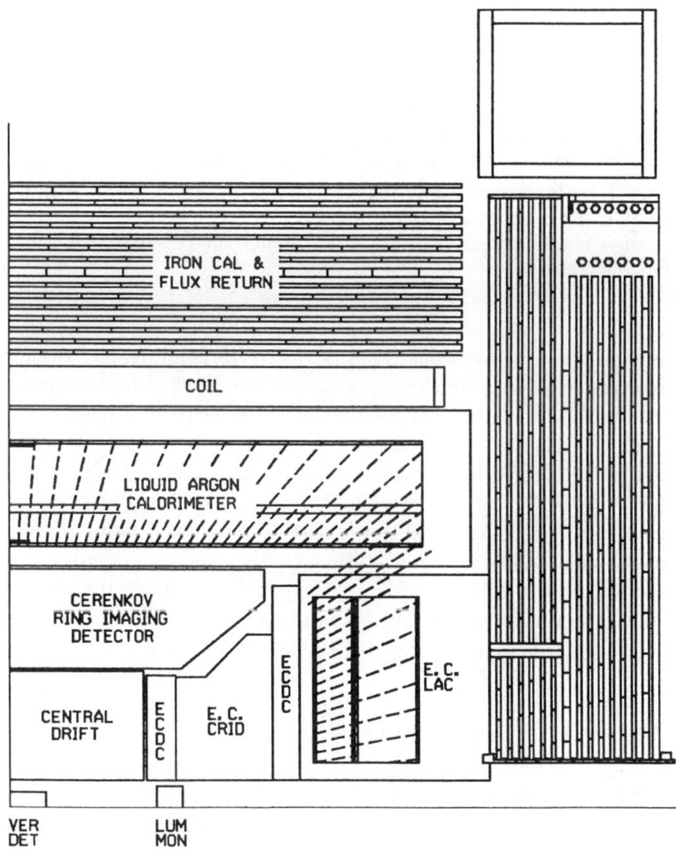

FIGURE 6. A cross section of a quadrant of the SLD.

polarized electron source will be available for use at the beginning of the SLC program. It is also expected that implementation of systems of the polarization will occur in the summer of 1987.

CONCLUSIONS

Colliding beams at the Z^0 mass will occur in the next year at the SLC. The progress on the accelerator and the other systems of the SLC is proceeding according to schedule. The Mark II/SLC detector is on its way from checkout at PEP to the SLC experimental hall. The physics experiments will start in 1987 to study Z^0 properties and to look for new physics. Longitudinally polarized electrons will be available in 1988 and will allow more precise tests of the standard model. The SLD detector to fully exploit the potential physics is under construction and is expected to be ready in 1989.

ACKNOWLEDGMENTS

A great number of people are working on the SLC. I would like to thank some of those for their assistance in preparing this report: R. Stiening, R. Erickson, John Seeman, S. Ecklund, and G. Fisher on the SLC status and schedules, and M. Breidenbach, D. Hitlin, and G. Hanson on providing information on the detectors. In addition, the report on the SLC program at the 1985 Aspen Winter Physics Conference by C. Prescott was very helpful.

Hadroproduction of Heavy Flavors[a]

IAN D. LEEDOM[b]

Fermilab
Fermi National Accelerator Laboratory
Batavia, Illinois 60510

The subject of this review, as one can gather from the title, is the current state of knowledge (or ignorance) of heavy quark production, particularly charm, by hadron beams. The only report of B mesons in a hadron experiment is that of the WA75 collaboration.[1] We shall come back to that experiment because of its interesting implications for B lifetimes, but the rest of the paper will deal exclusively with charm production.

Charm hadroproduction, like all Gaul, can be divided into three parts: total cross section, p_T, and x_F dependence. The only other question that arises concerns the species that have been observed. Of the mesons, the charged and neutral D and D^* have certainly been seen.[2-4] The F has probably been found,[5] but the evidence is sketchy and somewhat contradictory. The Λ_c and the A^+ baryons appear secure,[6-8] while the T^0 is on much shakier ground.[9]

CROSS SECTIONS

Quoted charm cross sections are large, but there exist great uncertainties as to how large. The reason for this is due primarily to: uncertainties in nuclear cross sections and the extrapolation to hydrogen; large and difficult acceptance and efficiency calculations; and poorly known exclusive branching ratios. As D production is the best known, we begin with it. The published values are shown in TABLE 1.

Most hadron-nucleon charm cross sections have been derived from results on heavy nuclei. Such cross sections are normally calculated by assuming that

$$\sigma_A = A^\alpha \sigma_p, \tag{1}$$

with α a constant. If heavy quarks are formed in hard scattering processes, then one expects that the exponent should be equal to unity. This would be in agreement with J/Ψ production. The E613 collaboration has measured prompt ν fluxes, which presumably come from charm decays, on three different target materials: beryllium, copper, and tungsten.[10] The best fit gives $\alpha = 0.75$. The inferred cross section of 28.6 μb/nucleon is twice as large as that found by the NA27 collaboration using the hydrogen bubble chamber LEBC.[11] In fact, the beam dump values shown in TABLE 1 have been calculated using $\alpha = 1$ so that they agree with the LEBC data. The smaller

[a]This work was supported by the U.S. Department of Energy.
[b]New address: Department of Physics, Northeastern University, 360 Huntington Avenue, Boston, Massachusetts 02115.

TABLE 1. Measured $D\bar{D}$ Cross Sections for $x_F > 0$

Experiment	Beam/Target	$\sqrt{s}$ (GeV)	σ (μb/nucleon)
CHARM[a]	p/Cu	27	9.4 ± 3.1
E595[a]	π^-/Fe	24	$17.5^{+5.5}_{-3.9}$
	p/Fe	27	$10.7 \pm 1.1 \pm 1.8$
E613[a]	p/W	27	7.8 ± 1.7
NA16	π^-/p	26	$12.2^{+7.6}_{-3.8}$
	p/p	26	$15.6^{+8.2}_{-4.6}$
NA18[a]	p/Freon	25	25 ± 11
NA27	π^-/p	26	7.9 ± 1.3
	p/p	27	11.2 ± 2.3
R407 408[b]	p/p	53	$(100–500) \pm 60\%$
R415[b]	p/p	62	$(120–350) \pm 40\%$
R416[b]	p/p	62	$(100–150) \pm 50\%$

[a] Assuming a nuclear dependence of A^1.
[b] Results are highly model dependent.

value of α found by E613 would give a consequent increase in these cross sections by factors of 1.7, 2.7, and 3.7 for beryllium, iron, and tungsten, respectively.

The question of the relation of p-p to p-nucleus cross sections was investigated by Barton et al.[12] for the production of strange and nonstrange species. FIGURE 1 shows their findings for α as a function of x_F. They showed that cross sections on nuclei extrapolate to larger proton cross sections than are measured in hydrogen and that α is a function of x independent of beam or product type. However, a recent CERN experiment[13] seems to indicate that α may depend on the produced species. What is certain is that the simple assumption embodied in equation 1 is naive and that charm production on nuclei is a much more complicated process than was anticipated. One is led to conclude that the only reliable cross sections are those found using proton targets.

This only leaves the ISR and LEBC results. The latter are small statistics experiments (NA16 and NA27), but with very clean, reliable data samples and cross sections that are inferred from topological rather than exclusive decay channels. The ISR experiments typically have correction factors $\geq 10^6$ and depend critically on exclusive branching ratios into the observed final states. These cross sections have been calculated using values of the D branching ratios (which have changed by factors of 2–4, depending on the mode) over the past eight years.[14] The ISR results cannot, therefore, be considered to give reliable cross sections. The only known cross sections for D production are the LEBC measurements for pions at $\sqrt{s} = 26$ GeV and for protons at $\sqrt{s} = 27$ GeV.

The ACCMOR collaboration (NA11) has published evidence for F production in π^-Be reactions.[5] The signal sought is the $\phi\pi$ decay mode as is seen in e^+e^- reactions. Data from both ϕ and electron triggers were used. F and Cabbibo-suppressed D decays are seen at approximately the same rate. The data reduction takes 5×10^6 triggers to yield 14 events in the final sample. Of these 14, 5 are F candidates, 3 of these being ambiguous with a Λ_c interpretation. The calculated lifetime is $\tau_F = 3.2^{+3.0}_{-1.2} \times 10^{-13}$. The Fermilab experiment E623 also looked for F production in the $\phi\pi$ final state.[15] While the decay of the charged D is easily seen (FIGURE 2), there is no enhancement in the F region. Additionally, NA27 has set a limit on F production in πp interactions by searching for decays in which the outgoing K is of the same sign as the decaying particle. This would correspond to either Cabbibo suppressed D or allowed F decay. No signal is seen at the level of 750 nb.[16] It should be noted that this limit assumes an F lifetime greater than 2×10^{-13} s. A shorter F lifetime would imply a less restrictive limit. More information is needed before the F can claim to have been seen in hadronic reactions.

Unlike the F, there is no doubt that the Λ_c is produced hadronically. The first reports came from the ISR,[6,7] and it has since been seen both in SPS experiments (e.g., NA27) with π and proton beams and in a neutron on carbon experiment at $\sqrt{s}$ of 9 GeV.[17] The ISR cross sections for Λ_c production suffer from extreme model dependence due to the restrictive triggers used and the limited apparatus acceptance. All of the experiments are plagued with very poorly known branching ratios. In most cases, the exclusive decay modes are known to no better than 50%. Therefore, it is best to say that the Λ_c cross section is unknown at any energy.

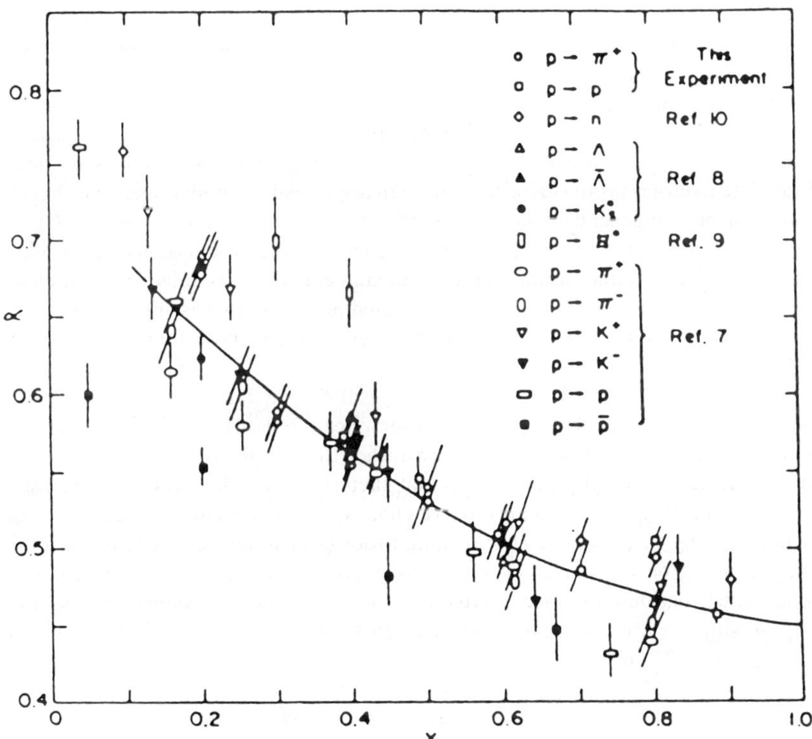

FIGURE 1. The nuclear dependence parameter, α, as a function of x_F, as given by Barton *et al.* in reference 12.

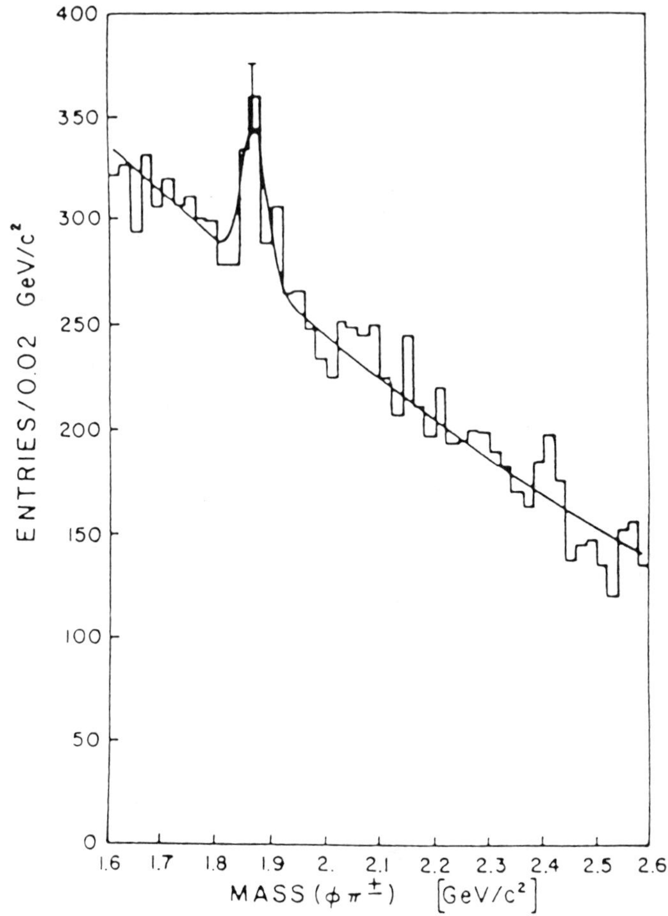

FIGURE 2. The observation of the decay $D^\pm \rightarrow \phi\pi^\pm$ as reported by the E623 collaboration (reference 15). No evidence is seen for the decay of the $F^\pm$ into this same final state.

The A^+ baryon (quark content: csu) has been reported by the WA42 collaboration.[8,9] The experiment used a 135-GeV Σ^- beam on a Be target, but because the spectrometer is only sensitive to $x_F > 0.6$, the calculated cross section is highly dependent on the apparatus acceptance to the $\Lambda K^- \pi^+ \pi^+$ final state. The same group also claims evidence for the T^0 (quark content: css), but with much lower significance. Both observations require confirmation.

To summarize our knowledge of charm cross sections:

$$\sigma_{pp}(D\bar{D}) = 11.2 \pm 2.3 \; \mu b, \qquad x_F > 0, \; \sqrt{s} = 27 \text{ GeV},$$

$$\sigma_{\pi p}(D\bar{D}) = 7.9 \pm 1.3 \; \mu b, \qquad x_F > 0, \; \sqrt{s} = 26 \text{ GeV},$$

$$\sigma_{\pi p}(F) \leq (750 \pm 250) \text{ nb}, \qquad x_F > 0, \; \tau_F > 2.0 \times 10^{-13} \text{ s}.$$

The unfortunate side of the coin is what we do not know:

- — any of the above cross sections at other energies;
- — the Λ_c cross section at any energy at all;
- — how to convert p-nucleus to p-nucleon cross sections.

The last point is one of the most telling. Until we can understand the A dependence of heavy target data, a wealth of information on production cross sections will be essentially inaccessible. This problem and that of the poorly known charmed baryon branching fractions remain the most pressing today.

p_T DEPENDENCE

The one area in which there seems to be general agreement is that of the p_T dependence of charm hadroproduction. The standard parameterization for this is

$$d\sigma/dp_T^2 \propto \exp\left(-bp_T^2\right). \tag{2}$$

TABLE 2 shows the results from several representative experiments. The most striking feature is that the average p_T and slope are independent of the incident beam, beam energy, target, and produced species. From this, we can gather that

$$\langle p_T \rangle \simeq 800 \text{ MeV} \qquad \text{and} \qquad \langle b \rangle \simeq 1.1 \text{ (GeV}^2)^{-1}.$$

The average p_T is then about twice that seen for π and K production.

x_F DEPENDENCE

Although there is a general consensus on the transverse momentum dependence, none exists for the longitudinal momentum. Much of the interest in this subject revolves around the question of forward charm production in which one of the outgoing charm particles would be expected to carry a valence quark from the beam. The reader is directed to the review of Kernan and Van Dalen,[18] and the references therein, for a full description of the various theories of charm production.

TABLE 2. p_T Dependence of Charm Production[a]

Experiment/Reaction	$\langle p_T \rangle$ (MeV)	b (GeV^2)$^{-1}$
E595/π^-Fe $\rightarrow \mu$	700 ± 150	—
/pFe $\rightarrow \mu$	920 ± 140	—
NA11/π^-Be $\rightarrow D$	—	1.1 ± 0.5
NA16/$\pi^- p \rightarrow D$	850 ± 120	1.1 ± 0.3
/$p\bar{p} \rightarrow D$	750 ± 120	1.1 ± 0.3
NA18/π^-Freon $\rightarrow D$	780 ± 140	~ 1.1
WA42/Σ^-Be $\rightarrow A^+$	—	$1.1^{+0.7}_{-0.4}$

[a]See equation 2.

The importance of the longitudinal momentum stems as well from its impact on total cross sections for those experiments that cannot cover charm momenta down to x_F of zero. If we adopt the parameterization that

$$d\sigma/dx_F \propto (1 - x_F)^n, \tag{3}$$

then TABLE 3 contains measured values of n for a variety of experiments. As usual, the bulk of these are from heavy metal targets and one is left to guess what may be the effect of a nuclear target on the momentum spectrum of the outgoing charm. (This may be counted as another outstanding reason for us to solve the problem of nuclear production.) One point, though, is apparent from TABLE 3. All those experiments whose sensitivity extends to $x_F = 0$ measure a softer charm spectrum than those with a range of x_F limited to larger values. Such differences can manifest themselves as whopping big changes in the inferred total cross sections because $\sigma \propto 1/(n + 1)$. The NA11 collaboration, for example, measured $n = 0.8 \pm 0.4$ when the apparatus acceptance was limited to $x_F > 0.2$.[19] Their improved spectrometer extended their acceptance down to zero, and the slope became $n = 2.9$. The moral of this story seems to be that the reliable measurements can only be those that encompass central production.

A possible explanation is that one is viewing some sort of leading particle effects. These were first invoked in an attempt to reconcile the differing values for total charm cross sections that came from the dilepton (~ 100 μb) versus discrete channel experiments at the ISR (~ 1 mb).[20-22] Evidence for such leading particles was seen by the CFRS collaboration who measured $\mu^\pm$ production off an iron dump with π^- and p beams.[23,24] In the π^- part, they measured an excess of μ^- events as compared to μ^+, which could be interpreted as a leading $D^0 - D^-$ component (quark content: $c\bar{u}$ and $\bar{c}d$, respectively). No such effect was seen in the proton running. In that case, Λ_c would be the leading particle as it carries two of the valence quarks of the incident proton. Even

TABLE 3. Feynman x Dependence of Charm Production Parameterized by Equation 3

Experiment/Reaction	$\sqrt{s}$ (GeV)	n	$x_{F\text{min}}$
(single component fits)			
BIS-2/$nC \rightarrow \Lambda_c$	9	1.5 ± 0.5	0.5
E595/pFe $\rightarrow \mu$	26	2.8 ± 0.8	0
NA11/π^-Be $\rightarrow D$	19	0.8 ± 0.4 (now ~ 2.9)	0.2
NA16/$pp \rightarrow D$	26	2.8 ± 0.8	<0
WA42/Σ^-Be $\rightarrow \Lambda^+$	16	1.7 ± 0.7	0.6
(two component fits)			
E595/π^-Fe $\rightarrow \mu$	23	(80%) $5.9^{+3.2}_{-1.6}$ (central) (20%) $0.9^{+0.9}_{-0.6}$ (forward)	
NA27/$\pi^-p \rightarrow D$	26	(80%) $7.5^{+2.5}_{-1.7}$ (central) (20%) $0.7^{+1.0}_{-0.7}$ (forward)	

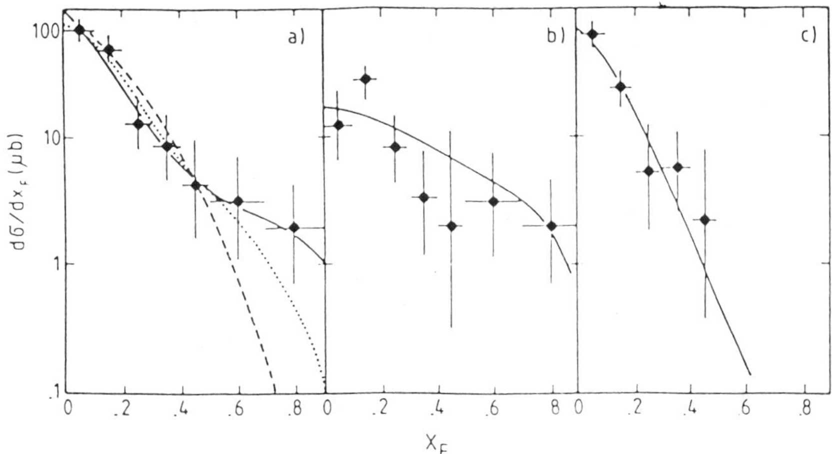

FIGURE 3. The x_F distributions for D mesons from NA27 (reference 14). (a) All D's: The continuous line is a two-component fit, whereas the dotted and dashed lines are single-component fits to two different production models. See the text for details. (b) Leading D's (i.e., D^-, D^0, D^{*-}) only: The solid line is from the two-component fit of (a). (c) Nonleading D's $(D^+, \overline{D}^0, D^{*+})$: The solid line comes from the two-component fit of (a).

more striking evidence for leading particles comes from NA27.[25] Their running was also broken into π^- and proton parts. Examination of the π-produced D sample gave a bad fit to a single $(1 - x_F)^n$ distribution. A much better fit comes from a two-component model with 80% of the production being central with $n = 7.5$ and 20% being forward with a much harder spectrum characterized by $n - 0.7$. These distributions and the associated fits are shown in FIGURE 3. When the same sample was divided into "leading" D's, that is, D^- and D^0, and central ones, D^+ and $\overline{D}^0$, then the fits to the two samples yielded

$$D^-, D^0 \qquad n = 2.0^{+0.6}_{-0.5} \qquad \text{and} \qquad \langle x_F \rangle = 0.232,$$

$$D^+, \overline{D}^0 \qquad n = 7.0^{+1.4}_{-1.2} \qquad \text{and} \qquad \langle x_F \rangle = 0.107.$$

Again, there was no evidence for leading particles in the proton sample. The case for leading charmed mesons seems well established, while leading baryons have yet to be confirmed.

A FEW OTHER TOPICS

D* Production

One of the most interesting recent results was that of the UA1 collaboration's report of the observation of charmed mesons in jets at $\sqrt{s} = 540$ GeV. The signal was isolated by demanding that the mass of a $K\pi_1\pi_2$ combination minus the $K\pi_1$ mass be equal to 145 MeV. This is the standard trick that uses the small Q value of the $D^* \rightarrow$

$D\pi$ decay. Their published result[26] found that the fragmentation of the charmed quark was softer than that seen in e^+e^- collisions and that $N(D^{*\pm})/\text{jet} = 0.65 \pm 0.2$ (stat) $\pm$ 0.33 (sys). This would have seemed to imply that there is, on average, one charm per jet at $S\bar{P}PS$ energies—a truly heroic charm production cross section. As has been remarked at this conference earlier, a reanalysis of the data has yielded $N(D^{*\pm})/\text{jet} = 0.08 \pm 0.02$ (stat) ± 0.04 (sys), which is more in keeping with conventional theoretical calculations.[27]

Both NA11 and NA27 (π^- data) have published D^* results as well. The two give compatible results. Both find D^* production to be central with $n = 4.3^{+1.8}_{-1.5}$ from the LEBC result (see FIGURE 4), which goes down to $x_F = 0$.[14] NA11 finds the smaller value of $n = 3.2 \pm 1.5$, but they are limited to $x_F > 0.2$, which may help explain the discrepancy. In any event, the two are equal within errors. The other question of interest is that of the D^*/D ratio, that is, the measure of indirect to direct D production. The findings are as follows:

$$D^*/D = 0.9^{+3.1}_{-0.6} \qquad \text{(NA11)},$$

$$D^*/D^0 = 1.00 \pm 0.25 \pm 0.12 \qquad \text{(NA27)},$$

$$D^*/D^\pm = 0.48 \pm 0.22 \pm 0.15 \qquad \text{(NA27)}.$$

As noted in reference 14, because of the central nature of D^* production, it cannot account for the forward component seen in the π data.

B *Production*

As alluded to earlier, only one experiment has presented evidence for hadronically produced B mesons: WA75. The apparatus used an emulsion for vertex detection coupled with a muon spectrometer. There was no other momentum analysis. The event in question is notable because the identification is done on topological grounds. FIGURE

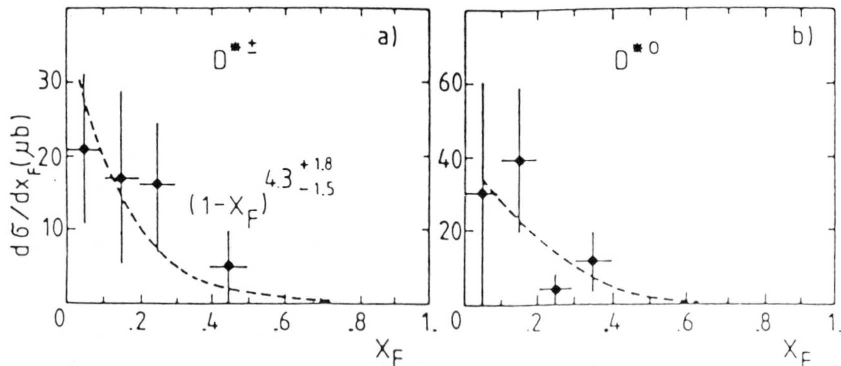

FIGURE 4. Fit to the x_F distributions of D^* mesons from NA27 (reference 14). The production is seen to be central. The fit could only be done for charged D^*'s as seen in (a), but the value found for n is seen to be compatible with the neutral D^*'s as seen in (b).

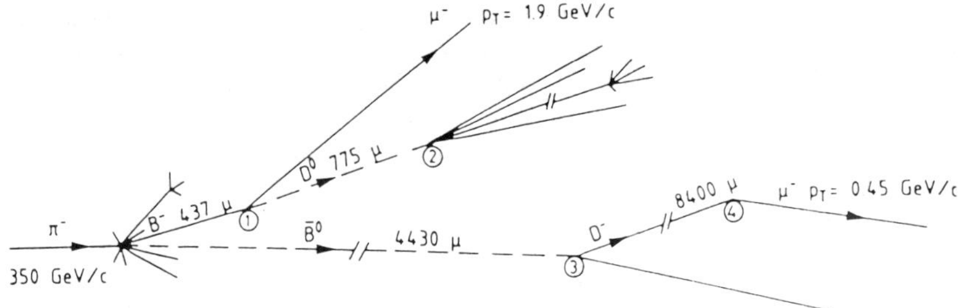

FIGURE 5. The first observation of B meson production in a hadron-induced reaction. The identification depends on the large transverse momenta of the two muons in the event and the topology of the decay of the daughter D at vertex 2. Further details can be found in the text and reference 1.

5 shows the event with one muon coming from a charged decay into a muon and a neutral 4-prong that is found downstream of this semileptonic decay vertex. This is interpreted as a B^-. The other muon comes about from the cascade decay of $\bar{B}^0 \rightarrow D^- + X$, with the second muon coming from the subsequent decay of the D^-. Identification of the D^- in this case is done on the basis of the p_T of the muon, which is too large to have come from a K or π decay. The cross section is ~ 10 nb. The most interesting feature of the event is the short B lifetime that can be inferred. Whereas the world average (based solely on e^+e^- results) is $\tau_B = (11.1 \pm 0.16) \times 10^{-13}$ s,[28] the average found from these two candidates is $\langle \tau_B \rangle = 3^{+2}_{-1} \times 10^{-13}$ s. This is a very intriguing result. It emphasizes the necessity for fixed target results on B mesons. The only advantage that a fixed target has over colliders is the ability to "sit on top" of the primary and secondary vertices, and it behooves us to make use of it.

CONCLUSIONS

After ten years of hadroproduction experiments, we seem to know some things well. These include: the production parameters of D's and D^*'s; the p_T dependence of charm production; and that there are leading particle effects in meson beams. What we do not know well are: σ_C as a function of energy; the production ratio of $F^\pm$ to D; the charmed baryon sector; and finally, how nuclear target data relate to production on protons.

What is needed for charm production is clean, reliable data at Tevatron energies with production on hydrogen and good vertex detection. To this goal, I submit, less than humbly, Fermilab experiment E743, an experiment that utilizes the Fermilab MPS for tracking and LEBC for vertex detection. The combination of chamber and spectrometer is virtually the same as the LEBC-EHS combination that has provided the most reliable production results to date.

[**Note added in proof:** Things have changed considerably since this talk was given. The ACCMOR collaboration has announced new results on F production (where it is seen in a kaon beam and not in a pion one). E743 has published its first finding on the inclusive D cross section at $\sqrt{s} = 39$ GeV. UA1 has published results on B cross sections and B-$\bar{B}$ mixing, and the CERN experiment WA78 has also announced a determination of α, the exponent of the A dependence of charm production, that is consistent with the value given by E613. While it is unfortunate that one is out-of-date before one is in print, it is a striking commentary on the vitality of the subject. A more up-to-date review can be found in "Hadroproduction Characteristics of Charm and Beauty" by Stephen Reucroft in the *Proceedings of the Sixth International Conference on Physics in Collision* (Chicago, Illinois, 1986).]

ACKNOWLEDGMENTS

I would like to thank the organizing committee and the Aspen Center for Physics for their hospitality during my stay here. As always, my E743 collaborators were invaluable in their efforts and assistance to me. In addition, I would like to thank the Aspen Valley Hospital for putting me back together in time to give this talk.

REFERENCES

1. ALBANESE, J. P. *et al.* 1985. Phys. Lett. **158B:** 186.
2. DRIJARD, D. *et al.* 1979. Phys. Lett. **81B:** 250.
3. AGUILAR-BENITEZ, M. *et al.* 1983. Phys. Lett. **122B:** 312.
4. BAILEY, R. *et al.* 1985. Z. Phys. **28C:** 357.
5. BAILEY, R. *et al.* 1984. Phys. Lett. **139B:** 320.
6. GIBONI, K. L. *et al.* 1979. Phys. Lett. **85B:** 437.
7. DRIJARD, D. *et al.* 1979. Phys. Lett. **85B:** 452.
8. BIAGI, S. F. *et al.* 1985. Phys. Lett. **150B:** 230.
9. BIAGI, S. F. *et al.* 1985. Z. Phys. **28C:** 175.
10. DUFFY, M. E. *et al.* 1985. Phys. Rev. Lett. **55:** 1816.
11. TOUBOUL, M. C. 1985. EHS Internal Note no. 85-14. The E613 results were communicated by: BALL, R. C. Univ. of Michigan.
12. BARTON, D. S. *et al.* 1983. Phys. Rev. **D27:** 2580.
13. ABREU, M. G. *et al.* 1984. CERN/EP 84-51.
14. CASO, C. 1985. CERN/EP 85-180.
15. GEORGIOPOULOS, C. H. *et al.* 1985. Phys. Lett. **152B:** 428.
16. AGUILAR-BENITEZ, M. *et al.* 1985. Phys. Lett. **156B:** 444.
17. ALEEV, A. N. *et al.* 1984. Z. Phys. **23C:** 333.
18. KERNAN, A. & G. VAN DALEN. 1984. Phys. Rep. **106:** 297.
19. BAILEY, R. *et al.* 1983. Phys. Lett. **132B:** 237.
20. CLARK, A. G. *et al.* 1978. Phys. Lett. **78B:** 339.
21. BAUM, L. *et al.* 1977. Phys. Lett. **68B:** 279.
22. LOCKMAN, W. *et al.* 1979. Phys. Lett. **85B:** 443.
23. BODEK, A. *et al.* 1982. Phys. Lett. **113B:** 77.
24. BODEK, A. Proc. XIV International Symposium on Multiparticle Dynamics Conf., Lake Tahoe.
25. AGUILAR-BENITEZ, M. Submitted to Phys. Lett.
26. ARNISON, G. *et al.* 1984. Phys. Lett. **147B:** 222.
27. ELLIS, K. Private communication (Fermilab).
28. BERKELMAN, K. 1987. These proceedings.

Production of Hadrons and Leptons at High p_t and Pairs at High Mass

CHARLES N. BROWN

E605 Collaboration
Fermi National Accelerator Laboratory
Batavia, Illinois 60510

INTRODUCTION

The study of particle production at high transverse momentum (p_t) and pair production at high mass in energetic collisons between nucleons has been a fruitful area of research for over a decade.[1-3] The approach goes back to Rutherford[4] and is based on the notion that scattering at high momentum transfer should reveal the internal structure of the scatterers. Nowadays, these processes are analyzed in terms of the parton model and Quantum Chromo-Dynamics (QCD). Previous experiments by this collaboration have confirmed the parton-scattering model of large-p_t and high-mass production[2] and have discovered the Υ particles[3] and the fifth quark. The present experiment extends these measurements to higher beam energy and improves resolution, particle identification, and luminosity. Besides addressing QCD issues, the data also allow limits to be set on the mass and lifetime of the axion.

APPARATUS

Experiment 605 at Fermilab uses a large spectrometer that is designed to accurately distinguish pions, kaons, protons, electrons, and muons. It can make precise measurements of their trajectories in the region near 90° in the center-of-momentum frame of the colliding nucleons. Measurements extend from a few GeV/c transverse momentum up to values near the kinematic limit. The large "SM12" magnet (see FIGURE 1) deflects high-p_t charged particles around the beam dump and into the sensitive volume of the detectors, while neutral particles produced in the target are absorbed in the dump or in the magnet walls and shielding. Charged-particle trajectories are measured at three detector stations by proportional and drift chambers, and consistency of the trajectory upstream and downstream of the "SM3" magnet verifies that the particle was produced in the target. Scintillation-counter hodoscopes at the three stations furnish a crude measurement of the trajectory for triggering purposes. Particle identification is provided by the ring-imaging Cherenkov counter,[5] the electron and hadron calorimeters, and the muon proportional tube arrays located behind the calorimeters and additional shielding. The large p_t kick of SM12 (8 GeV/c at full excitation) combined with the good resolution of the drift chambers (200 μm rms) yields a mass resolution better than 0.2% rms at $m = 10$ GeV/c^2.

RATE LIMITATIONS AND CHRONOLOGY OF RUNS

Luminosity was limited by counting rates in the detectors. The main problem was the production of neutrals in the walls of SM12, which could then illuminate the detectors (see FIGURE 2). During our first run, in the spring of 1982, this background limited our luminosity to 3×10^{34} cm^{-2} s^{-1}. The next run took place in the winter of 1984; this was the first run of the Tevatron, but it was at a beam energy of 400 GeV . In preparation for this run, we lined the inner walls of SM12 with carefully designed Pb and W absorbers. We were thus able to operate at 10^{35} cm^{-2} s^{-1}. Finally, in 1985, we erected a 4′-thick Pb wall at the exit of SM12 and concentrated on dimuon measurements at a luminosity of 3×10^{36} cm^{-2} s^{-1}. Data were taken using a variety of targets in order to study nuclear effects. TABLE 1 gives the integrated luminosities for

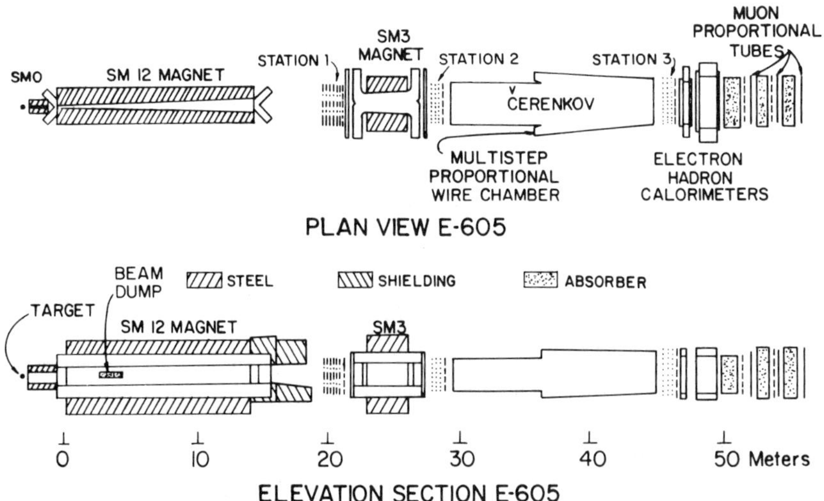

FIGURE 1. Plan and elevation of the apparatus showing the Pb absorber and the Station 0 chamber that were added for the 1985 run. Dotted line: drift chamber; dash-dotted line: proportional chamber; dashed line: counter bank.

each run and target. To date, only the 1982 data have been fully analyzed,[6,7] but preliminary results from the 1984 and 1985 runs will also be presented.

NUCLEAR EFFECTS

The differential cross section versus p_t has been observed to depend in a complicated way on the atomic weight (A) of the target nucleus. The dependence might be expected to be exponential in A, with exponent $\alpha = 2/3$ (if the nucleons on the surface of the nucleus shadow nucleons in the interior) or with $\alpha = 1$ (if there is no shadowing). The former case would be expected (and has been observed) at low p_t (<1

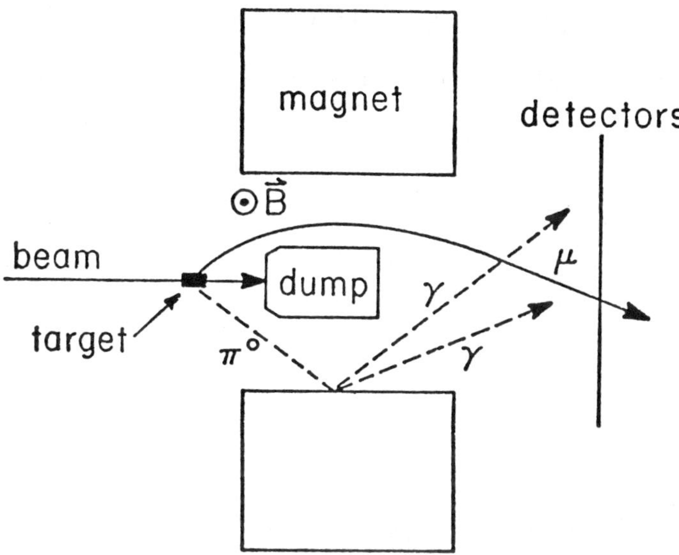

FIGURE 2. Schematic diagram of the apparatus illustrating the photon background problem.

GeV/c), where interaction cross sections are large and where there is thus a significant absorption of the incident nucleon as it penetrates a nucleus. However, at high p_t (>1 GeV/c), one is sensitive to processes involving hard scattering of partons and small interaction cross sections, so α should approach 1.

The observation[1] of $\alpha > 1$ at $p_t > 2$ GeV/c suggests that collective behavior of nucleons in the target nucleus is being observed. Multiple hard scattering of partons is capable of explaining these results, at least at a qualitative level. FIGURE 3 shows our single-hadron data (from the 1982 run)[7] and data from the Chicago-Princeton[1] and Columbia–Fermilab–Stony Brook collaborations,[8] along with predictions of the constituent multiple scattering (CMS) model of Lev and Petersson.[9] The model is seen to reproduce the trend of the data, though there is some disagreement in detail.

A further prediction of the CMS model is independence of α on mass for symmetric hadron-pair production because symmetric hadron pairs tend to arise from single hard scatters. This prediction is borne out by the data, as shown in FIGURE 4. For fixed mass, α should rise as the net p_t of the pair increases. Our acceptance for net $p_t \neq 0$ is small in the (approximately vertical) production plane; however, we can define p_{out} as the

TABLE 1. Data Recorded for the Integrated Luminosity/Nucleon (10^{39} cm^{-2})

	Target						
	Be	Cu	W	LH$_2$	LD$_2$	$\sqrt{s}$ (GeV)	Aperture
1982	0.2	0.3	0.3			27.4	open
1984	1.5	2.2	3.5	0.7	2.6	27.4	open
1984	12.9		2.2	0.8		38.8	open
1985		2800			38.8		closed

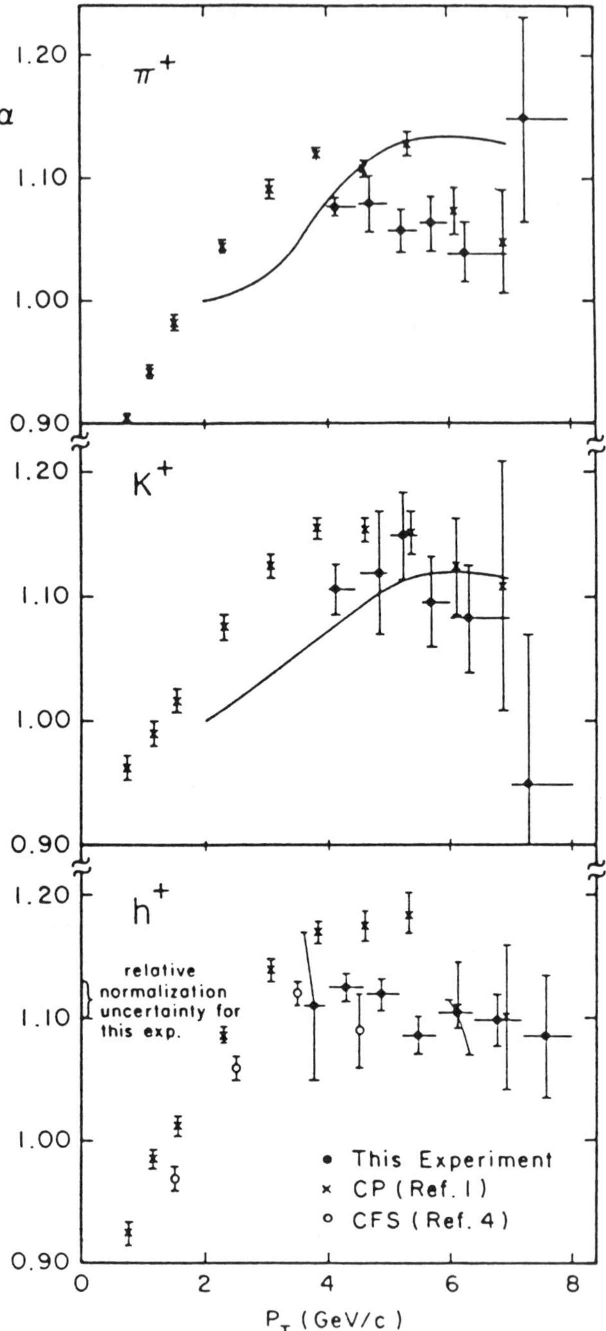

FIGURE 3. The exponent α of the A-dependence of positive-hadron production cross sections at 400 GeV versus p_t; the curves indicate the predictions of the constituent multiple scattering model for π^+ and K^+ production.

momentum component of one hadron perpendicular to the plane defined by the beam direction and the momentum vector of the other hadron. According to the CMS picture, α should rise with increasing p_{out} because multiple scatters are required to give a momentum component out of the plane. FIGURE 5 bears out this prediction.

CROSS SECTIONS VERSUS p_t AND θ

Given some information about how the scattered partons fragment into the observed particles, the magnitude and shape of hadron production cross sections at large p_t versus p_t and scattering angle θ are calculable from QCD. We have used the Lund Monte Carlo[10] to perform such a calculation, using the PYTHIA (4.2) and JETSET (6.2) routines to model parton scattering and fragmentation. FIGURE 6 shows the inclusive invariant cross section versus p_t to produce positive hadrons, as compared

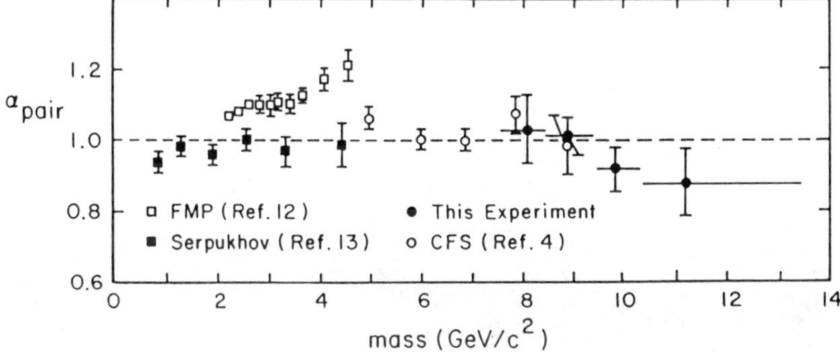

FIGURE 4. The exponent α of the A-dependence of the hadron-pair production cross section at 400 GeV versus pair mass.

with the Lund Monte Carlo prediction. The shapes are seen to agree well; however, the magnitude of the Lund prediction is too low by a factor of 2.6.

The dependence of the cross section on θ has not been measured by any previous experiment in this range of p_t and θ. FIGURE 7 presents the power α of the atomic-weight dependence versus cos θ, as measured in our 1982 run using Be, Cu, and W targets. The data are consistent with a constant value of α in our range of cos θ. FIGURES 8 and 9 present the invariant cross sections versus cos θ for positive and negative hadrons from the three targets. We also show an extrapolation to $A = 2$ (labeled "deuterium"), as compared with QCD predictions from the Lund Monte Carlo and from a calculation to leading-log order by J. F. Owens.[11] The Lund prediction has been multiplied by 2.6 to facilitate the comparison, but the leading-log prediction agrees in magnitude with the experimental results.

For positive hadrons, both the data and the QCD predictions show an enhancement at angles forward of 90°, but the data show a stronger enhancement than either

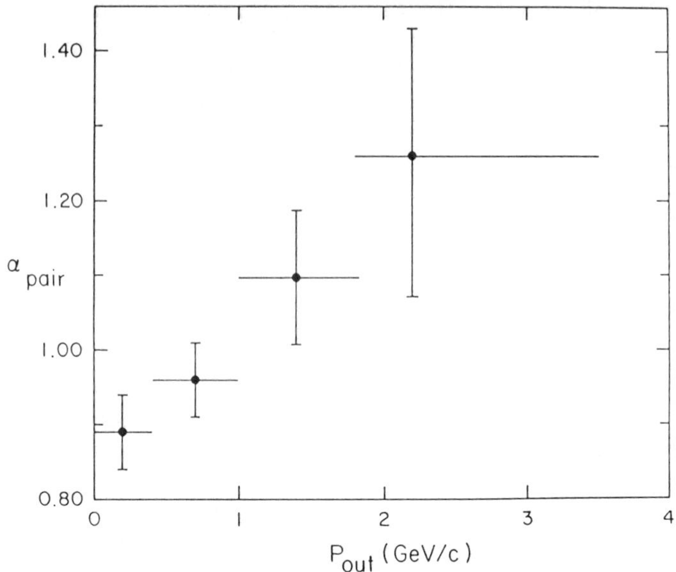

FIGURE 5. The exponent α of the A-dependence of the hadron-pair production cross section at 400 GeV versus p_{out}.

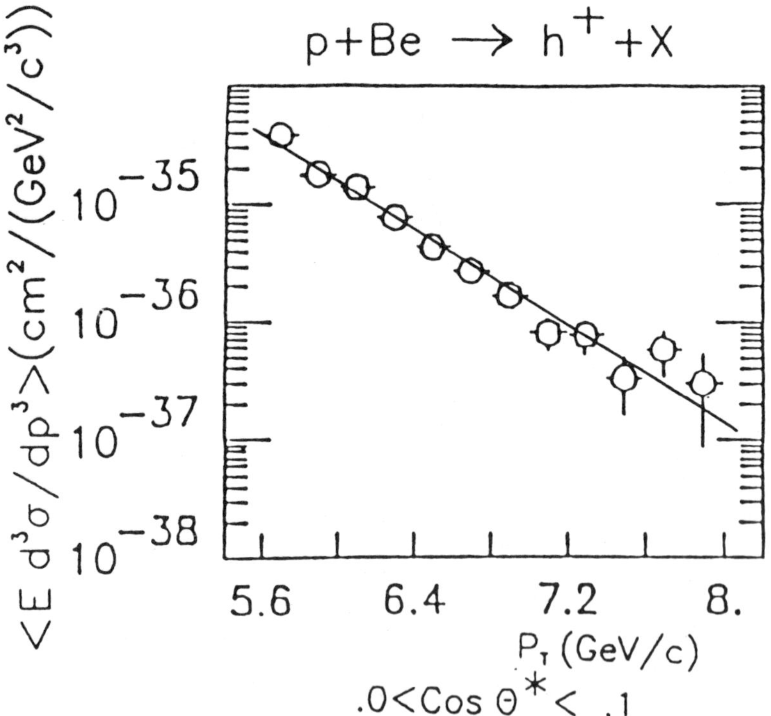

FIGURE 6. Inclusive invariant cross section versus p_t for positive hadron production off of Be at 400 GeV; the curve represents the Lund Monte Carlo prediction multiplied by 2.6.

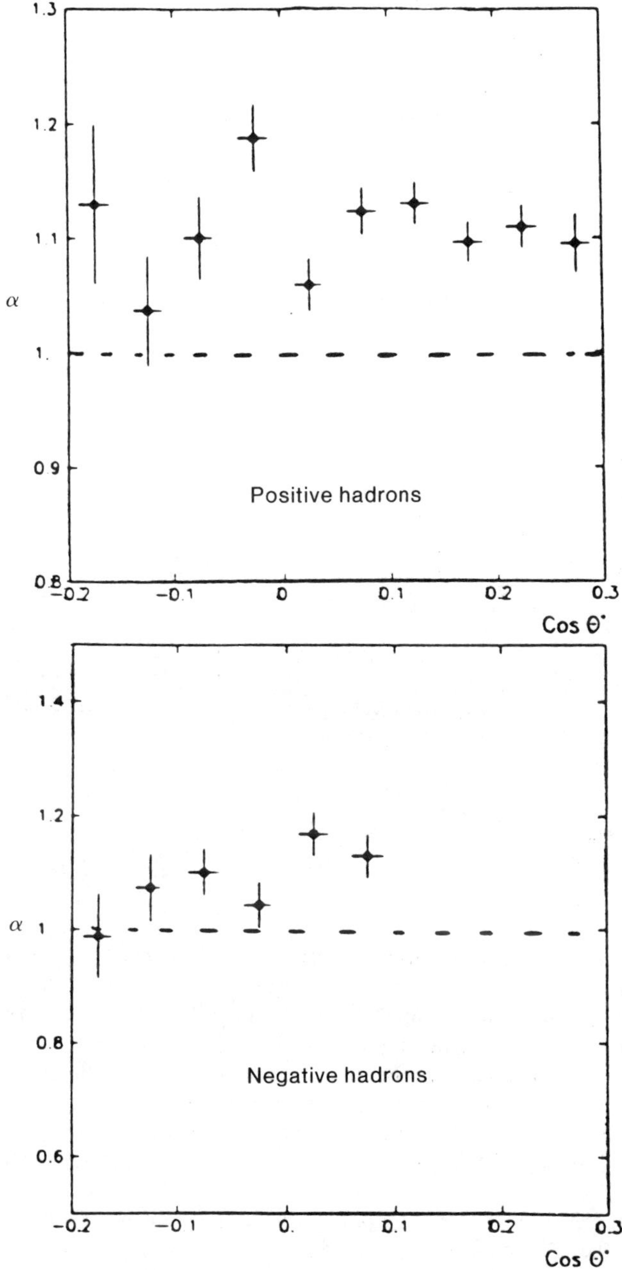

FIGURE 7. The power α of the A-dependence of positive- and negative-hadron production cross sections versus $\cos \theta$.

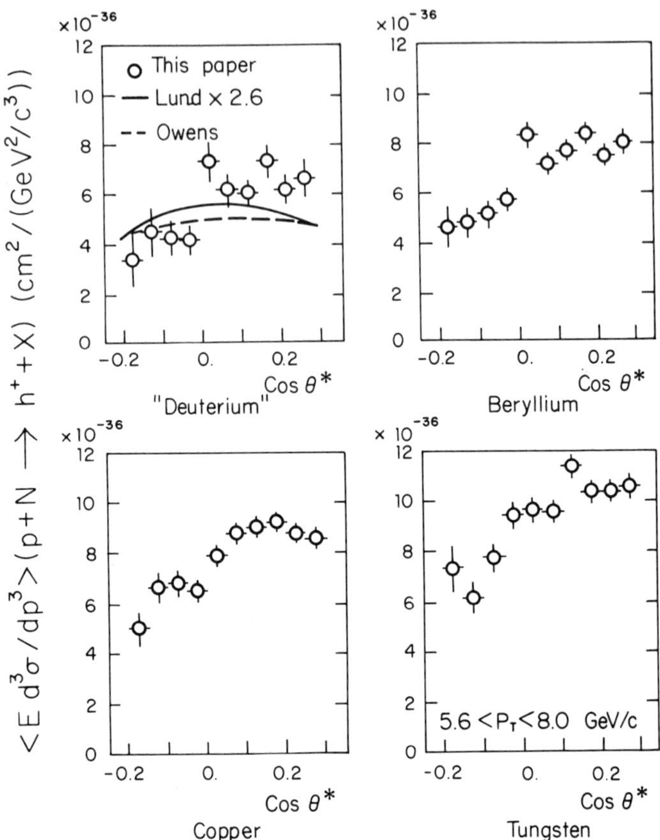

FIGURE 8. The invariant cross section versus cos θ for positive hadron production at 400 GeV off Be, Cu, and W targets, and its extrapolation to $A = 2$ ("deuterium").

calculation. This forward enhancement arises in the QCD models due to the presence of neutrons in the target, as follows: At these high values of $x_t = 2p_t/\sqrt{s}$, the cross section is dominated by scattering of valence quarks off of each other and off of gluons; also u quarks tend to fragment into positive hadrons, whereas d quarks tend to fragment into negative hadrons. Because the u-quark structure function is harder in the proton and the d-quark structure function is harder in the neutron, then in proton-neutron collisions, positive hadrons tend to be produced in the proton direction (forwards) and negative hadrons tend to be produced in the neutron direction (backwards).

Note that the results heretofore presented are all based on the 1982 run. Once the analysis of the 1984 data is complete, we will have better statistics by an order of magnitude, as well as better acceptance and better control over systematic errors. Some preliminary results from the 1984 data are presented below.

PARTICLE RATIOS

FIGURE 10 presents preliminary measurements of particle production ratios at 400 GeV using the LH_2 target, along with results from the Chicago-Princeton collaboration and Lund Monte Carlo predictions. The two data sets are in agreement where they overlap, and both agree with the Lund prediction for π^+/π^-; however, the Lund Monte Carlo is seen to underestimate the production of kaons and to overestimate the baryon production. These predictions are sensitive to two parameters in the fragmentation model: the ratio of s to u quark production and the ratio of diquark to single quark production. These parameters were determined from e^+e^--annihilation data[12] to be 0.3 and 0.1, respectively; our data prefer the values of 0.5 and 0.05, respectively. Similar problems have been noted by two ISR experiments (at lower values of x_t).[13] Note that both the data and the Monte Carlo runs suffer limited statistics so far.

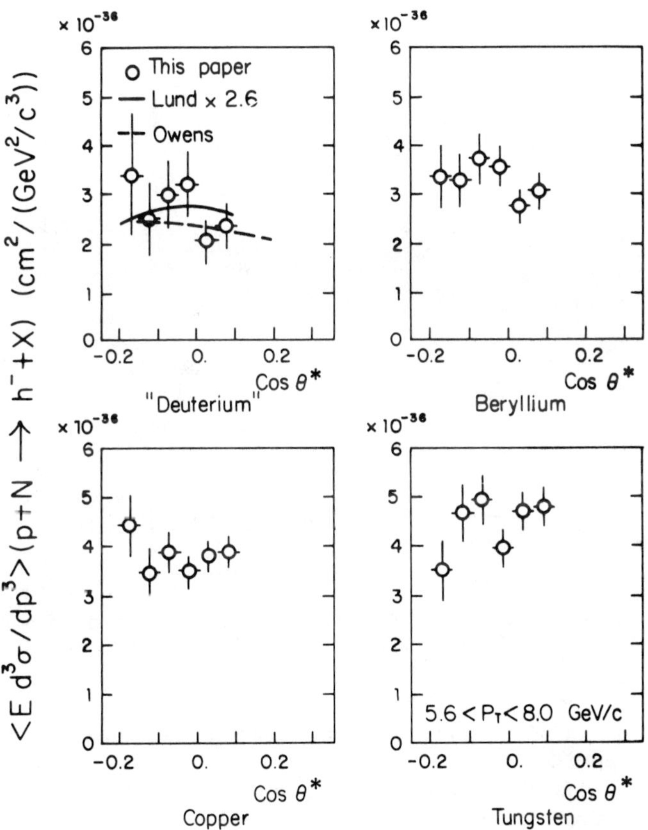

FIGURE 9. The invariant cross section versus cos θ for negative hadron production at 400 GeV off Be, Cu, and W targets, and its extrapolation to $A = 2$ ("deuterium").

800-GeV DATA

Open-Aperture Run

FIGURE 11 presents preliminary results on the yields versus p_t of the positive hadrons and hadron pairs at 800 GeV off of Be. The results shown represent the first

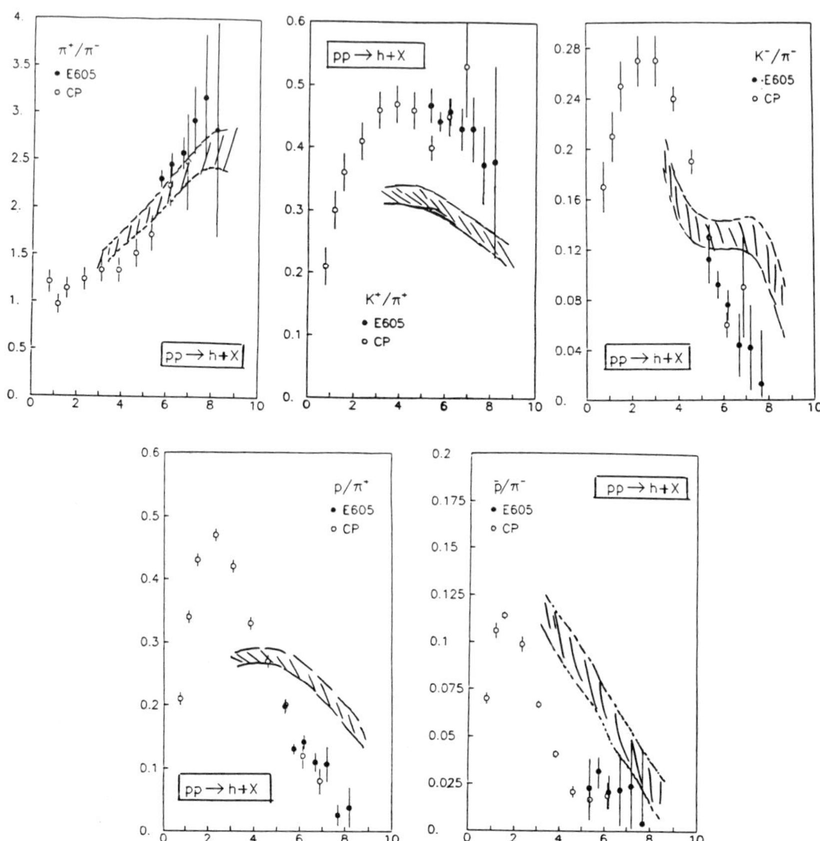

FIGURE 10. Invariant cross section ratios for pions, kaons, and protons produced in 400 GeV *p-p* collisions: (a) π^+/π^-, (b) K^+/π^+, (c) K^-/π^-, (d) p/π^+, and (e) $\bar{p}/\pi^-$. The shaded bands represent the Lund Monte Carlo predictions.

3% of data taken, and they serve to indicate the quality of the results expected once the analysis is complete.

FIGURE 12 shows the dimuon and dielectron yields versus mass. The Υ states are clearly resolved, and the yields in the two modes are equal to within 20%. No large opposite-sign μe signal is seen; the yield is less than 1% of the dimuon and dielectron yields.

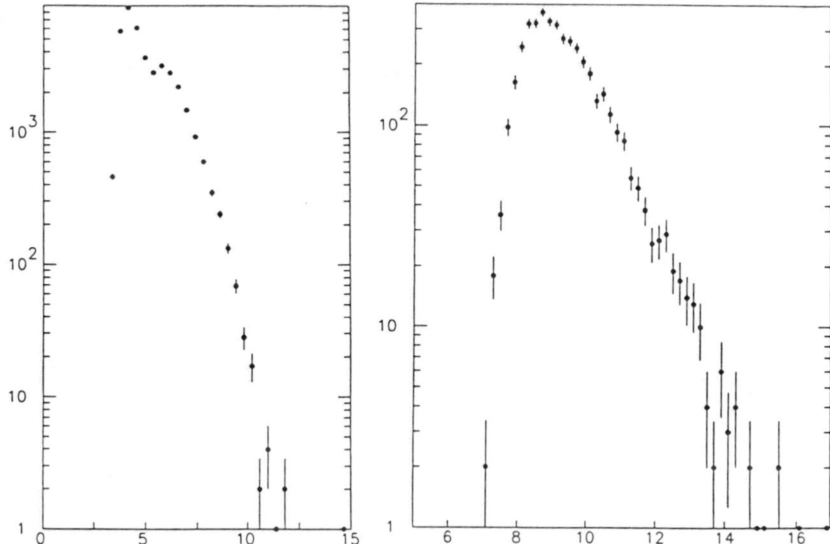

FIGURE 11. Yield of (a) positive hadrons versus p_t and (b) opposite-sign hadron pairs versus mass at 800 GeV off Be for the first 3% of the data taken.

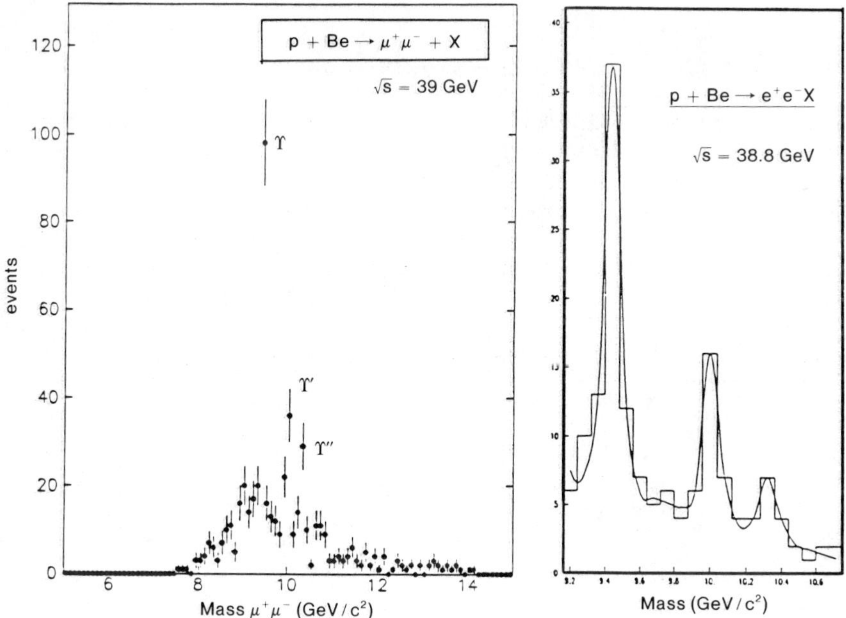

FIGURE 12. Yield of opposite-sign (a) muon pairs and (b) electron-positron pairs versus mass at 800 GeV.

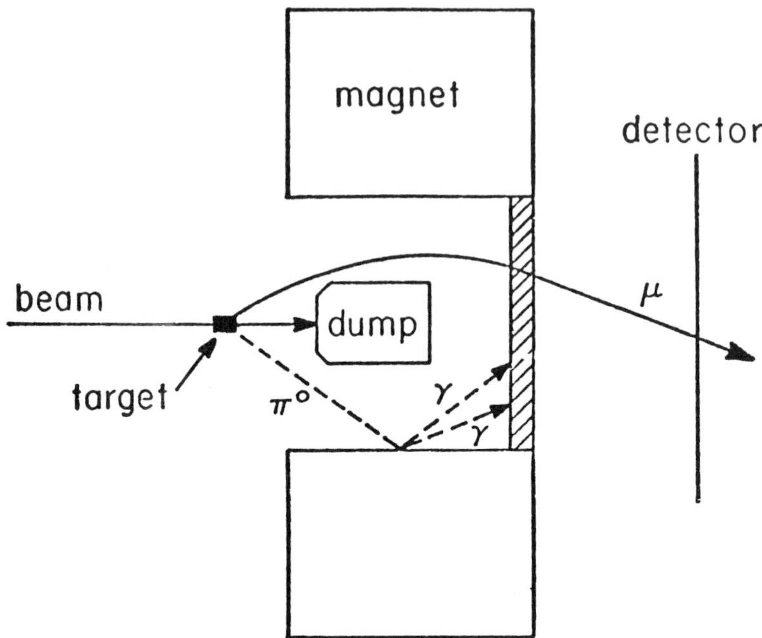

FIGURE 13. Schematic diagram of the apparatus with the Pb absorber installed.

High-Luminosity Dimuon Run

In 1985, we added a 4'-thick Pb absorber at the exit of SM12 in order to eliminate our photon background (see FIGURE 13). We were thus able to increase our luminosity by a factor of 30. We retain good mass resolution (0.2% rms at the Υ) in the presence of the absorber (because SM3 provides a good measurement of momentum), and we trace the trajectory back through the field of SM12 in order to determine the production angle, thus allowing the track to have a kink due to multiple scattering in the absorber.

TABLE 2. Expected Contributions to Closed-Aperture Υ Mass Resolution

Contribution	$\sigma(\mathrm{MeV}/c^2)$
Target size	6.3
dE/dx fluctuations	3.0
Multiple scattering	
in target	9.4
in lead absorber	7.8
in detectors	6.5
in helium	3.7
Chamber resolution	7.0
Other	4.6
Total	18.0

TABLE 3. Rates per 2×10^{12} Protons on Target (Closed Aperture, 800 GeV)

Station 0	3×10^8
Station 1	10^8
Station 3	3×10^7
Muon hodoscope left-right coincidences	10^6
Trigger matrix coincidences	2×10^3
Events satisfying trigger processor	2×10^2
Good high-mass $\mu^+\mu^-$ events found off-line	2

To improve the momentum determination, we added a station of proportional drift tubes ("Station 0") just downstream of SM12, which we successfully operated at rates in excess of 100 MHz (2 MHz on the hottest wire). Using Station 0, SM3 provides 0.1% momentum measurement. TABLE 2 gives expected contributions to the mass resolution at the Υ.

TABLE 3 summarizes rates at various stages in the apparatus and data acquisition. In order for data acquisition not to be the limiting bottleneck, we used a two-stage trigger. The first stage consisted of a trigger matrix that looked for triple coincidences of hodoscopes pointing back to the target in the bend plane. The second stage was a fast parallel-pipelined trigger processor[14] that found wire-chamber tracks in the bend plane pointing back to the target and firing the muon proportional tubes and that required two opposite-sign tracks with a mass exceeding a threshold. The trigger processor not only reduced the trigger rate by an order of magnitude, but it gained another order of magnitude in required off-line computing for those events written to tape because chamber hits inconsistent with processor tracks could be ignored in the off-line trackfinding. (Due to the high beam intensity, there were typically 10 to 20 accidental

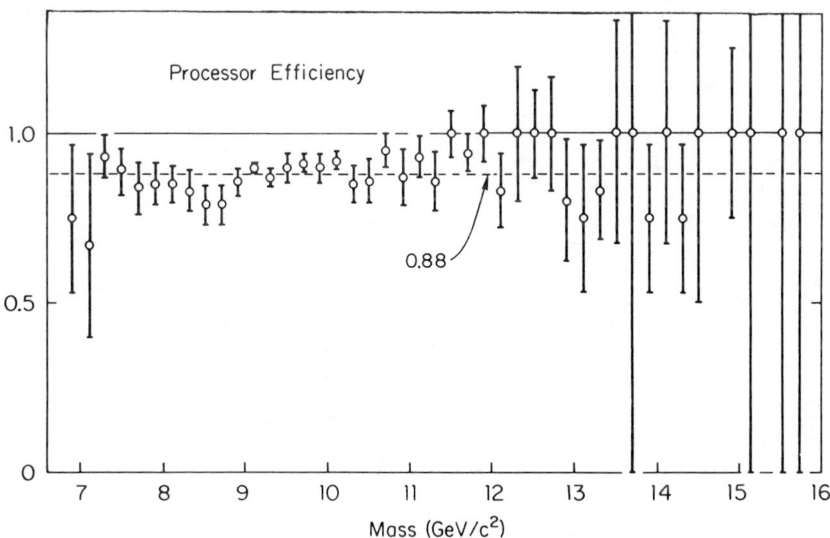

FIGURE 14. Trigger processor efficiency versus mass.

hits per chamber plane per event.) FIGURE 14 shows trigger processor efficiency; the inefficiency is consistent with that expected from chamber inefficiency and deadtime.

FIGURE 15 shows the event yield in the Υ region with half of the data analyzed. The mass resolution has not yet been optimized, but the three Υ states below the B-meson decay threshold stand out clearly and with little background. TABLE 4 gives prelimi-

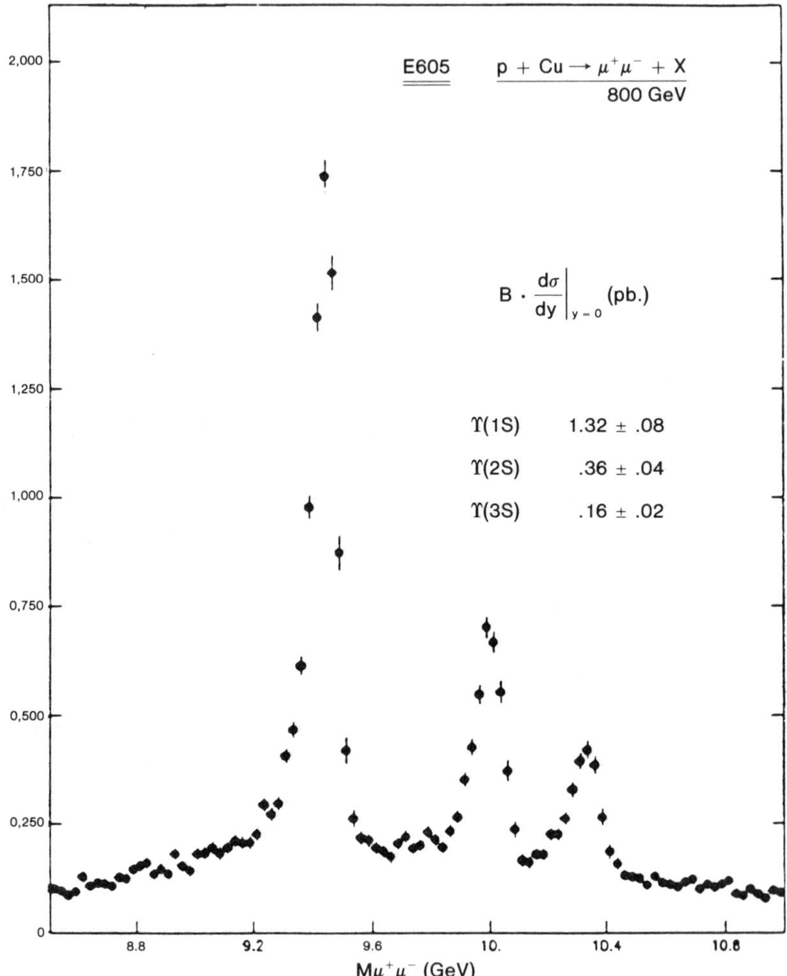

FIGURE 15. Yield of opposite-sign muon pairs versus mass in the Υ region at 800 GeV off Cu for one-half of the closed-aperture data.

nary measurements of the cross sections times the branching ratios for the production of Υ's. Using values measured at e^+e^- colliders for leptonic branching ratios,[15] we can extract differential cross section ratios for the three states (shown in TABLE 5). These have been predicted by Baier and Ruckl,[16] assuming that Υ-production in 800-GeV pN

TABLE 4. Υ Cross Sections (Preliminary)

$p + \text{Cu} \rightarrow \mu^+\mu^- + X$ at 800 GeV	
State	$B\dfrac{d\sigma}{dy}\Big|_{y=0}$ (pb)
$\Upsilon\,(1S)$	1.32 ± 0.08
$\Upsilon\,(2S)$	0.36 ± 0.04
$\Upsilon\,(3S)$	0.16 ± 0.02

collisions is dominated by gluon-gluon fusion into P states, which subsequently decay into the observed S states. Their predictions are also indicated in TABLE 5 and are in reasonable agreement with our result. Barger, Keung, and Phillips[17] and Childress *et al.*[18] used models based on local duality to calculate the sum of cross sections times branching ratios for all three Υ states. In these models, the cross sections are strongly sensitive to the shape of the gluon structure function. The Υ data pin down the gluon structure function to within one power of $(1-x)$: Barger *et al.* find the exponent to be 5 or 6, while Childress *et al.* (who include $q\bar{q}$ annihilation in their analysis) find 7.6 ± 1.

FIGURE 16 gives the dimuon yield over the mass range of 8–16 GeV/c^2. No new resonances are in evidence. FIGURE 17 shows 95% confidence-level upper limits for the production of new resonances (expressed as a ratio to the Drell-Yan cross section). These limits should come down a factor of two when the mass resolution has been optimized and the remainder of the data has been analyzed. We also show the expected levels of signal for the Higgs[19] and the technipion,[20] assuming they have masses in our accessible range. The Higgs production cross section increases with the number of quark generations (assumed here to be four), and a plausible "k-factor" enhancement of two has also been put in.

Search for "Axions"

Recently, interest in axions[21] has been renewed due to the observation of monoenergetic electrons and positrons produced in heavy-ion collisions.[22] This suggests the production and decay of a particle with mass = 1.8 MeV/c^2. Our relatively short (5.5 m) Cu beam dump, located within the field of SM12, followed by a spectrometer with good electron identification turns out to give excellent sensitivity to such a particle, which is produced in a π^0-initiated electromagnetic shower at the upstream end of the Cu beam dump and decays downstream of the dump.

We took a sample of data triggered only on energy deposition in the calorimeter

TABLE 5. Υ $B(d\sigma/dy)|_{y=0}$ Ratios (Preliminary)

$p + \text{Cu} \rightarrow \mu^+\mu^- + X$ at 800 GeV		
Ratio	Observed	Predicted (reference 16)
Υ'/Υ	$45 \pm 8\%$	$\lesssim 30\%$
Υ''/Υ	$16 \pm 3\%$	$\lesssim 15\%$

during the 1984 800-GeV run, and for 4×10^{13} protons on target, we find 74 e^+e^- pairs that reconstruct to a vertex at the downstream face of the dump. These pairs are all consistent with zero mass. FIGURE 18 shows the distributions of the exit angles of the pairs in the vertical and horizontal planes at the downstream face of the dump. During the same data runs, we also recorded a prescaled sample of the copious flux of muons emerging from the downstream face of the dump. These muons were produced by meson decay in the initial hadron shower and were then deflected in the vertical plane due to the 3.1-GeV/c magnetic kick over the length of the dump. The angular

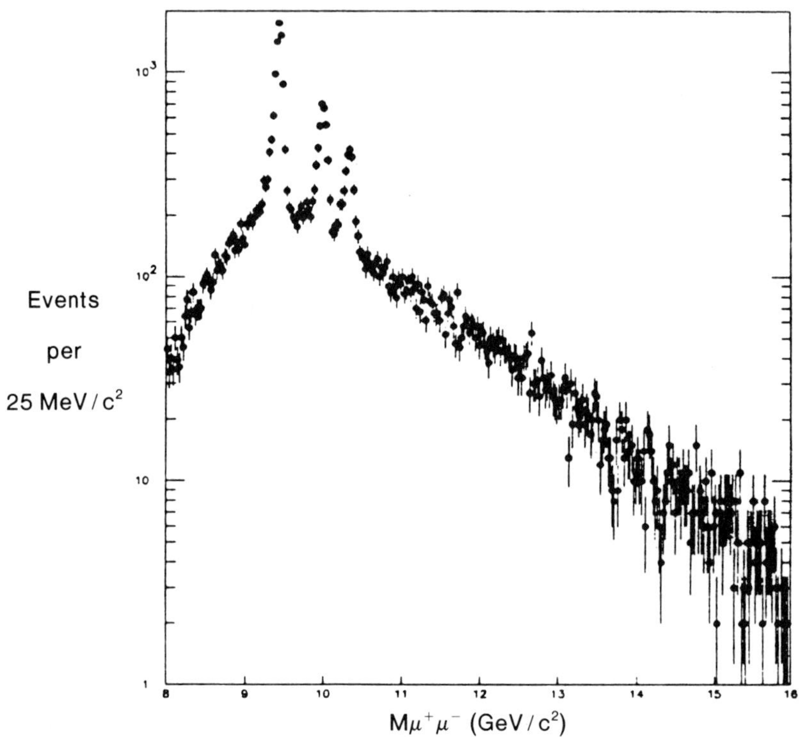

FIGURE 16. Yield of opposite-sign dimuons versus mass at 800 GeV off Cu for one-half of the closed-aperture data.

distributions of the muons were identical to the distributions of the e^+e^- pairs in FIGURE 18. Axions, traversing the dump as neutral particles, would be expected to have narrow angular distributions in both x and y. The e^+e^- angular distributions are consistent with muon bremsstrahlung in the last radiation length of the beam dump and, at most, one pair is consistent with the decay of a neutral particle produced at the upstream end of the dump.

Using a phenomenological fit to the flux of π^0's in thick targets[23] and an axion production formula due to Tsai[24] (assuming pseudoscalar coupling of the axion to

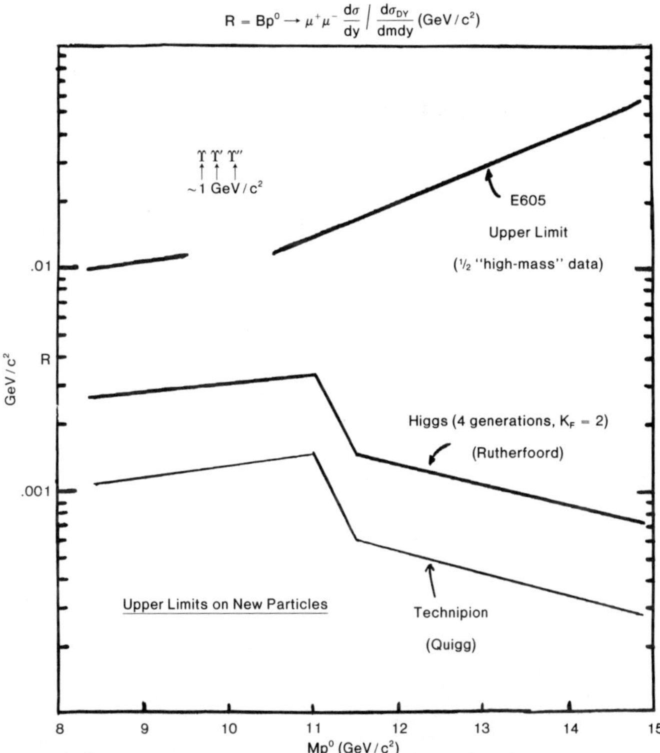

FIGURE 17. The 95% confidence-level upper limits versus the mass for the dimuon resonance production cross sections times the branching ratios, normalized to the Drell-Yan cross section, and compared to predicted cross sections for the Higgs and the technipion.

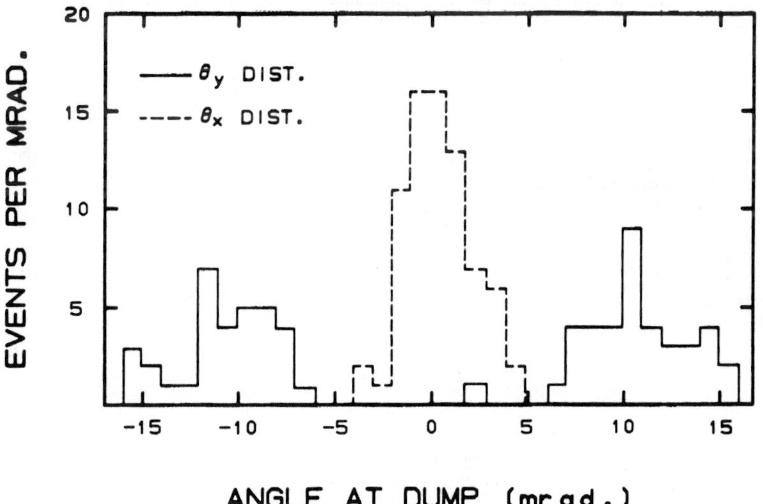

FIGURE 18. Exit angle distribution at the downstream dump face for the observed electron-positron pairs.

e^+e^-), we can compute 90% confidence-level limits on the mass and lifetime of the axion (shown in FIGURE 19). We also show the limits versus mass and lifetime derived from the anomalous magnetic moment of the electron,[25] which would receive a large contribution from axion loops if the axion-electron coupling becomes large. (These limits are not much different for other possible axion couplings.) Together, the two measurements exclude a large region in mass-lifetime space, and a 1.8-MeV axion

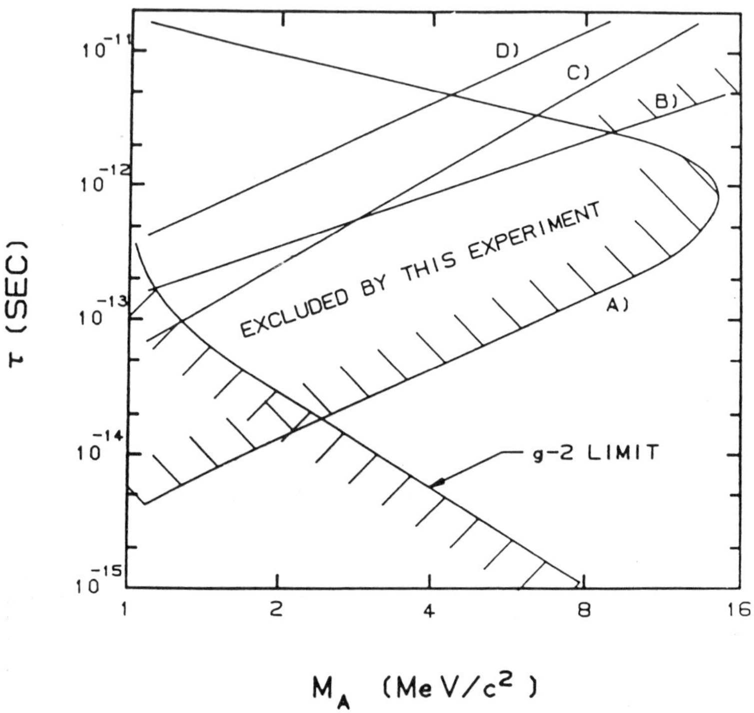

FIGURE 19. Upper limit on mass and lifetime of an axionlike particle from (a) this experiment, and previously published limits from (b) KEK, Konaka *et al.*, (c) FNAL E-613, and (d) SLAC E-56. We also show the lower limit from *g*-2 measurements.

appears to be ruled out (unless it interacts strongly and is absorbed in the beam dump).

REFERENCES

1. ANTREASYAN, D. *et al.* 1979. Phys. Rev. **D19:** 764.
2. JÖSTLEIN, H. *et al.* 1979. Phys. Rev. **D20:** 53; ITO, A. S. *et al.* 1981. Phys. Rev. **D23:** 604; SMITH, S. R., *et al.* 1981. Phys. Rev. Lett. **46:** 1607.
3. UENO, K. *et al.* 1979. Phys. Rev. Lett. **42:** 486; GARELICK, D. A. *et al.* 1978. Phys. Rev. **D18:** 945.

4. RUTHERFORD, E. 1911. Philos. Mag. **6**(xxi): 669.
5. GLASS, H. *et al.* 1985. IEEE Trans. Nucl. Sci. **NS-32**: 692.
6. CRITTENDEN, J. A. *et al.* 1986. Inclusive hadronic production cross sections measured in proton-nucleus collisions at $\sqrt{s}$ = 27.4 GeV. To appear in Phys. Rev. D; SAKAI, Y. 1984. Ph.D. thesis. Kyoto University; GLASS, H. 1985. Ph.D. thesis. SUNY at Stony Brook; HSIUNG, Y. 1985. Ph.D. thesis. Columbia University; CRITTENDEN, J. 1985. Ph.D. thesis. Columbia University.
7. HSIUNG, Y. B. *et al.* 1985. Phys. Rev. Lett. **55**: 457.
8. MCCARTHY, R. L. *et al.* 1978. Phys. Rev. Lett. **40**: 213.
9. LEV, M. & B. PETERSSON. 1983. Z. Phys. **C21**: 155.
10. SJÖSTRAND, T. Private communication; 1982. Comput. Phys. Commun. **27**: 243.
11. OWENS, J. F. Private communication; see also: DUKE, D. W. & J. F. OWENS. 1984. Phys. Rev. **D30**: 49.
12. BARTEL, W. *et al.* 1983. Z. Phys. **C20**: 187.
13. BREAKSTONE, A. *et al.* 1984. Phys. Lett. **135B**: 510; 1985. CERN/EP 85-30. Submitted to Z. Phys. C; ÅKESSON, T. *et al.* 1984. Nucl. Phys. **B246**: 408.
14. HSIUNG, Y. B. *et al.* 1986. Use of a parallel pipelined event processor in a massive-dimuon experiment. To appear in Nucl. Instrum. Methods.
15. PARTICLE DATA GROUP. 1984. Review of particle properties. Rev. Mod. Phys. **56** (no. 2): part II.
16. BAIER, R. & R. RÜCKL. 1983. Z. Phys. **C19**: 251.
17. BARGER, V., W. Y. KEUNG & R. J. N. PHILLIPS. 1980. Z. Phys. **C6**: 169.
18. CHILDRESS, S. *et al.* 1983. Production dynamics of the Υ and the gluon structure function. Unpublished.
19. RUTHERFOORD, J. P. Private communication.
20. QUIGG, C. 1985. FERMILAB-PUB-85/145-T.
21. BROWN, C. N. *et al.* 1986. A sensitive search for the axion. Submitted to Phys. Rev. Lett.
22. SCHWEPPE, J. *et al.* 1983. Phys. Rev. Lett. **51**: 2261; CLEMENTE M. *et al.* 1984. Phys. Lett. **137B**: 41; COWAN, T. *et al.* 1985. Phys. Rev. Lett. **54**: 1761; 1986. Phys. Rev. Lett. **56**: 444.
23. MALENSEK, A. J. Fermilab preprint nos. FN-341, FN-341-A.
24. TSAI, Y. S. 1986. SLAC-PUB-3926 (April).
25. BRODSKY, S. J. *et al.* 1986. Santa Barbara preprint no. NSF-ITP-86-17.

A Precise Determination of the Electroweak Mixing Angle from Semileptonic Neutrino Scattering

Presented by LIVIO LANCERI[a]

CHARM Collaboration[b]
CERN CH-1211
Geneva 23, Switzerland

INTRODUCTION

After the discovery of the weak neutral current,[1] great efforts were undertaken to test the prediction of the Glashow-Salam-Weinberg model that stated that the coupling in all neutral current phenomena depends on one single parameter, $\sin^2 \theta_w$. Indeed, a unique value of this parameter can explain the couplings measured in many different processes, including leptonic and semileptonic neutrino scattering, asymmetries in electron-nucleon, electron-positron, and muon interactions, parity violating effects in atomic transitions, and the masses of the W and Z bosons.[2]

This prediction is based on the Born approximation of the theory.[3] In the corrections to the Born terms, different processes pick up different correction terms. Measurements of $\sin^2 \theta_w$ made with sufficient precision and extracted from the data using the simple zero-order approximations should show differences for different reactions. The differences can be calculated, and a comparison with experiments therefore provides a test of the gauge nature of the theory. The present experiment reaches a precision of $\Delta \sin^2 \theta_w = \pm 0.005$ in neutrino scattering, which is matched in sensitivity to future measurements of the W and Z masses to $\pm 0.1\%$ at the CERN $p\bar{p}$ collider.[4]

EXPERIMENTAL METHOD

For isoscalar targets, Llewellyn-Smith[5] showed that the contributions of u and d quarks to the neutral current (NC) and charged current (CC) semileptonic cross sections were related by assuming isospin invariance alone. The following equation for the differential cross sections applies:

$$d^2\sigma_{NC}^{\nu(\bar{\nu})} /dxdy = [(1/2) - \sin^2 \theta_w + (5/9) \sin^4 \theta_w] \, d^2\sigma_{CC}^{\nu(\bar{\nu})} /dxdy$$

$$+ [(5/9) \sin^4 \theta_w] \, d^2\sigma_{CC}^{\bar{\nu}(\nu)} /dxdy. \quad (1)$$

---

[a]L. Lanceri's permanent address is: Istituto Nazionale Fisica Nucleare, Sezione di Trieste, Via Valerio, 2, I-34100 Trieste, Italy.
[b]Appendix 1 gives a list of the members and their institutions.

Integrating equation 1 over x and y and dividing by $\sigma_{CC}^{\nu(\bar{\nu})}$, one obtains a relation containing only ratios of total cross sections:

$$R^\nu = \sigma_{NC}^\nu / \sigma_{CC}^\nu = (\tfrac{1}{2}) - \sin^2 \theta_w + (\tfrac{5}{9}) \sin^4 \theta_w \, [1 + r]. \qquad (2)$$

From a measurement of the ratio of neutral current and charged current total cross sections for neutrinos (R^ν), $\sin^2 \theta_w$ can be derived. A measurement of the ratio of CC cross sections for antineutrino and neutrino scattering, r, is needed to determine the small correction term, $(\tfrac{5}{9}) \sin^4 \theta \times r$.

We report here results of a new experiment using data taken in 1984 to measure R^ν with high precision.[6] An upgraded version of the narrow-band beam was used at a central momentum of 160 GeV/c with nearly a factor two higher flux than that used previously. The analysis will be described in some detail in the following paragraphs. Emphasis is given on the corrections we applied to extract the physical ratio R^ν from the visible one.

The Experimental Setup

Data were taken in the 160 GeV/c narrow-band beam (NBB) developed by Grant and Maugain[7] as a high flux version of the CERN NBB optics. It satisfies the conditions for a precision measurement of R^ν,[8] namely:

- A sufficiently high average neutrino energy that is needed for efficient pattern recognition in the CHARM detector.
- Low background contributions that are measurable with high accuracy.
- The possibility to measure the relative $\bar{\nu}/\nu$ flux normalization accurately.
- A calculable neutrino energy spectrum.
- High event rate.

The CHARM neutrino detector is a fine-grain calorimeter followed by an iron spectrometer with a toroidal magnetic field. It is surrounded by a magnetized iron frame. The calorimeter has a sampling step corresponding to one radiation length or 0.22 absorption length, with scintillators, proportional drift tubes, and streamer tubes as detecting elements. It is described in detail elsewhere.[9] The fiducial mass in this experiment was 87 tons.

A special feature of the CHARM detector is the low detection threshold that makes it possible to measure showers down to an energy as low as 2 GeV. The overall trigger efficiency for NC showers with an energy above 2 GeV has been determined to be 99.93%.

Care was taken to avoid problems caused by two interactions within the conversion time needed by the electronics of the detector elements. After each trigger, a TDC recorded the time between the trigger and the following interaction in the detector. Events followed by another interaction within 1.2 μsec—indicating the presence of the second interaction in the gate of the proportional drift tubes—were rejected. This procedure induced an additional effective deadtime of $\sim 2\%$, which does not affect the measurement of R^ν.

Event Classification and Corrections

Interactions in the CHARM detector are classified by an automatic pattern-recognition program on an event-by-event basis. Selection criteria are applied in attempts to identify optimally the physical processes and to minimize the corrections needed to relate the visible cross sections to the physical cross sections.

Neutrino interactions are defined as events that have no entering charged tracks. A neutrino event is called a charged current interaction if it contains a muon originating from the event vertex. The muon has to be visible (outside the hadronic shower) over a range corresponding to an energy loss of at least 0.67 GeV. Its total range has to exceed 1 GeV energy loss. All other neutrino event candidates are classified as neutral current interactions. Only those events that have their vertex inside the fiducial volume and that have a shower energy of at least 2 GeV are analyzed. In order to obtain the final CC and NC event numbers from the automatic classification, a number of corrections have to be applied, which will now be discussed.

The classification efficiency of the automatic pattern recognition is very high. The program recognized and flagged event topologies for which the automatic procedure could fail. A visual inspection of these flagged events (~4,000), together with a double scan of 20,000 randomly selected events, allowed us to correct the errors introduced by the classification code to $\Delta R^{\nu}/R^{\nu} \leq 0.1\%$. Off-line representations of typical NC and CC events in the CHARM detector are shown in FIGURES 1 and 2.

CC events in which the primary muon cannot be identified are automatically classified as NC. Some of these lost CC events have muons with an energy less than 1 GeV. Another contribution to the loss of CC events is caused by muons of more than 1 GeV that either leave the sides of the detector before their energy loss is sufficiently large or that are obscured by the hadronic shower. A correction is required for these losses. The precision that can be reached in measuring R^{ν} depends in an essential way

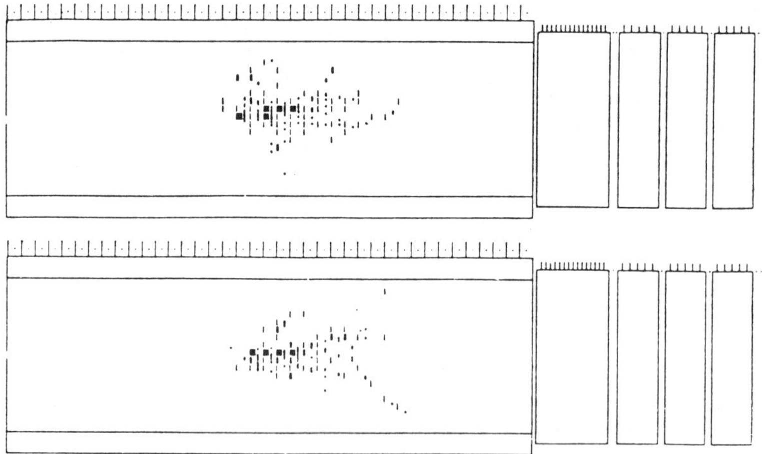

FIGURE 1. A neutral current event recorded in the CHARM detector.

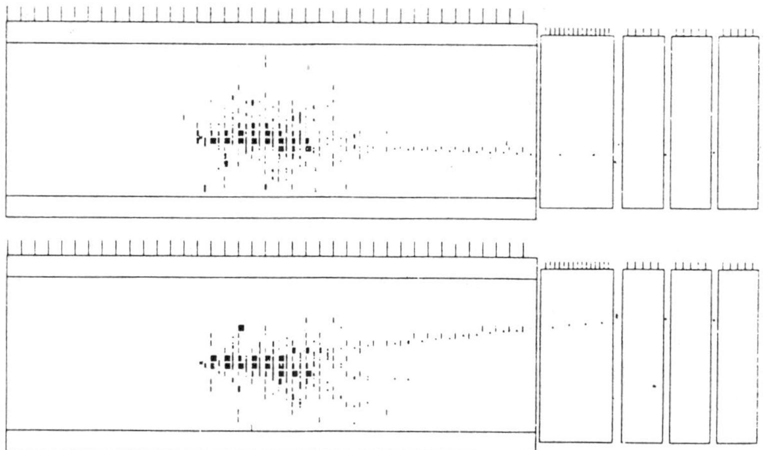

FIGURE 2. A charged current event recorded in the CHARM detector.

on the reliability of this correction. We therefore determined the correction by two independent methods. In one method, the efficiency of recognizing primary muons is obtained by overlaying simulated muons over real showers obtained by removing the original muon from CC events. In the other method, the effective shower length distribution is calibrated by measuring the length of the hidden part of primary muon tracks in CC events and then using this distribution in a Monte Carlo program to determine the muon recognition efficiency. The total correction induced by all sources of unidentified muons is 3.5% of the CC event rate. The two methods give consistent results within 2% of the correction value. The systematic error is dominated by the determination of the shower length (± 0.5 calorimeter samplings) and is estimated to be $\pm 2.8\%$ of the correction, thus contributing an error of $\Delta R^{\nu}/R^{\nu} \simeq \pm 0.4\%$.

A small fraction of NC events contain a track that fulfills the requirements of a primary muon of a CC event; these events are classified as CC events. This background is caused by decays of pions or kaons in the shower or by noninteracting hadrons—the so-called punch-through tracks. The correction is calculated in bins of the shower energy as the fraction of charged current events in which, after removing the primary muon, another track is found that satisfies the muon recognition criteria (a small contribution, present in CC, but not in NC events, is due to prompt muons from charm decays and can be subtracted with sufficient precision). The size of this correction (integrated for shower energy above 2 GeV) is $\simeq 5.5\%$ of the NC rate, determined with a fractional error of $\pm 2.6\%$, thus giving a contribution of $\Delta R^{\nu}/R^{\nu} \simeq \pm 0.2\%$.

Both NC and CC interactions of electron-neutrinos originating from K_{e3}-decays in the beam are classified as NC events. The contribution of these events amounts to $\sim 7\%$ of the muon-neutrino–induced NC candidates. Measurement of the K/π ratio with an accuracy of $\sim \pm 3\%$ determines this correction with a fractional error of $\pm 4\%$, which corresponds to $\Delta R^{\nu}/R^{\nu} \simeq \pm 0.3\%$.

Events induced by the neutrino flux from decays of pions and kaons before the decay tunnel were measured when the entrance of the decay tunnel was blocked by an absorber and subtracted. The uncertainty introduced by this so-called wideband (WB) background corresponds to $\Delta R^{\nu}/R^{\nu} \simeq \pm 0.3\%$ in the neutrino exposure.

The trigger rate induced by cosmic ray events is $\simeq 4$ kHz. A fast filter program reduced the cosmic-ray-induced background to the level of a few Hz, while retaining 99.96% of all NC events with a shower energy above 2 GeV. The remaining background contribution in the live-time of the experiment was found to be $\simeq (1.42 \pm 0.02)\%$ for NC events (negligible for CC events) and was subtracted.

In TABLE 1, these corrections to the data are summarized for shower energies above 4 GeV. The same analysis was repeated with shower energy cuts at 2, 4, 9, 16, and 25 GeV; because the corrected value of $\sin^2 \theta_w$ that we obtain is independent of this cut, we quote (in the following) the results with a 4-GeV cut, which minimizes our uncertainties.

Taking these corrections into account, we obtain for events with hadronic energy

TABLE 1. Corrections to the Data[a]

	NC	CC	$\Delta R^{\nu}/R^{\nu}$
Raw event numbers	$39,239 \pm 198$	$108,472 \pm 329$	0.6%
Corrections:			
Trigger and filter efficiency	7 ± 4	0 ± 0	
Scan correction	40 ± 39	60 ± 44	$\simeq 0.5\%$
WB background	-1998 ± 87	-4308 ± 119	
Cosmic background	-312 ± 8	-3.1 ± 0.8	
Difference in energy cut	—	0 ± 129	0.1%
Muon recognition losses	-3738 ± 105	3735 ± 105	0.4%
π/K decays in the shower	1892 ± 50	-1835 ± 50	0.2%
K_{e3} decays	-2300 ± 88	-139 ± 8	0.3%
Corrected event numbers	$32,831 \pm 283$	$105,982 \pm 408$	
Total systematic error			0.8%
Total error			1.0%

[a]Neutrino exposure, shower energy > 4 GeV.

greater than 4 GeV:

$$R^{\nu} = 0.3098 \pm 0.0031. \tag{3}$$

For the ratio of the total antineutrino and neutrino charged current cross sections, we obtain $r = 0.439 \pm 0.011$. Correcting for the different spectra in neutrino and antineutrino exposures gives

$$r = 0.456 \pm 0.012. \tag{4}$$

The uncertainty on r is dominated by the event-statistics (1.4%), the WB background subtraction (0.8%), and the relative normalization error of the $\bar{\nu}$ to ν flux (2.0%).

Using equation 2 and neglecting all corrections to this equation due to the presence of other than only u and d quarks, the raw value of $\sin^2 \theta_w$ is obtained:

$$\sin^2 \theta_w = 0.235 \pm 0.005 \text{ (assuming } \rho = 1). \tag{5}$$

Quark-Parton Model Corrections

The result (equation 5), deduced in the approximations of equation 2, must be corrected for the following effects in a quark-parton model framework:

- The final state invariant mass threshold depends on the produced quark flavor.
- The strange quark sea and the charm quark sea in the nucleons must be taken into account: their distributions were determined from dimuon production data in deep inelastic neutrino and muon scattering, respectively.
- Charged current interactions that change a light quark into a charm quark are kinematically suppressed by the mass of the charm quark, m_c. A fixed mass, $m_c = 1.5$ GeV, was chosen, and the threshold effects were computed using the slow rescaling procedure.[10]

In these calculations, we used Kobayashi-Maskawa mixing matrix elements obtained[11] assuming three families of quarks and requiring unitarity of the matrix. These corrections gave the following net effect on the electroweak mixing angle:

$$\Delta \sin^2 \theta_w = +0.010. \tag{6}$$

The main contributions to the theoretical error introduced by these corrections can be classified as follows:

- The momentum-weighted content of s and c quarks in the nucleon, and the threshold effects in c-production.
- Uncertainties in the Kobayashi-Maskawa (K.M.) mixing matrix.
- Higher twist terms.

With the present experimental knowledge of the c and s quarks content in the nucleon, along with the knowledge of the c-quark mass and the K.M. matrix elements with the unitarity conditions mentioned above, we estimate a theoretical uncertainty of $\simeq \pm 0.005$ in the determination of $\sin^2 \theta_w$. An additional uncertainty due to higher twist terms is estimated to be smaller than 0.005.[5] These uncertainties are inherent in the hadronic nature of the target and do not occur in neutrino-electron scattering.[12]

Radiative Corrections

The effect of QED and electroweak radiative corrections on $\sin^2 \theta_w$ was estimated following Bardin *et al.*[13] in the on-shell renormalization scheme of Sirlin and Marciano.[14] With the definition of $\sin^2 \theta_w = 1 - M_w^2 / M_z^2$, we find:

$$\Delta \sin^2 \theta_w = -0.009. \tag{7}$$

We estimate a precision of ± 0.002 for this computation. Applying both quark-parton model (equation 6) and radiative (equation 7) corrections, we obtain the best result:

$$\sin^2 \theta_w = 0.236 \pm 0.005 \text{ (experimental error)}. \tag{8}$$

Our result can be compared with the most recent results from the UA1 and UA2 collaborations derived from measurements of M_w and M_z that give:[15]

$$\sin^2 \theta_w = 0.194 \pm 0.031 \qquad \text{(UA1)},$$

$$\sin^2 \theta_w = 0.229 \pm 0.030 \qquad \text{(UA2)}.$$

CONCLUSIONS

The CHARM collaboration has performed an experiment aiming at a precision determination of $\sin^2 \theta_w$ using semileptonic neutrino scattering. The analysis shows consistency with previously published results.[16] An experimental precision of $\Delta \sin^2 \theta_w \simeq 0.005$ has been obtained. Therefore, the experiment matches the precision of future direct measurements of the $W^\pm$ and Z^0 masses to $\pm 0.1\%$ at the upgraded CERN Sp$\bar{\text{p}}$S collider and at LEP.

REFERENCES

1. HASERT, F. J. *et al.* 1973. Phys. Lett. **B46:** 138.
2. See, for example: PANMAN, J. 1984. XIth International Conf. on Neutrino Physics and Astrophysics, Dortmund, p. 741; and references therein.
3. GLASHOW, S. L. 1961. Nucl. Phys. **22:** 579; SALAM, A. & J. WARD. 1964. Phys. Lett. **13:** 168; WEINBERG, S. 1964. Phys. Rev. Lett. **D5:** 1264.
4. MARCIANO, W. 1983. Proc. International Lepton and Photon Symposium, Cornell; MARCIANO, W. J. & A. SIRLIN. 1984. Phys. Rev. **D29:** 945.
5. LLEWELLYN-SMITH, C. H. 1983. Contribution to the SPS Fixed Target Workshop, CERN; Nucl. Phys. **B228:** 205–215.
6. CHARM PROPOSAL. 1984. CERN/SPSC/84-1.
7. GRANT, A. & J. M. MAUGAIN. 1983. CERN/EF/BEAM 83-2.
8. PANMAN, J. 1983. Contribution to the SPS Fixed Target Workshop. CERN yellow report no. 83-02, vol. 2, p. 146.
9. DIDDENS, A. N. *et al.* CHARM COLLABORATION. 1980. Nucl. Instrum. Methods **178:** 27–48; BOSIO, C. *et al.* 1978. Nucl. Instrum. Methods **157:** 35–46; JONKER, M. *et al.* CHARM COLLABORATION. 1982. Nucl. Instrum. Methods **200:** 183–193.
10. BARNETT, R. M. 1976. Phys. Rev. **D14:** 70; GEORGI, H. & H. D. POLITZER. 1976. Phys. Rev. **D14:** 1829.
11. KLEINKNECHT, K. & B. RENK. 1983. Phys. Lett. **130B:** 459.
12. CHARM II PROPOSAL. 1983. CERN/SPSC/83-24 and 83-37.
13. BARDIN, D. YU. & O. M. FEDORENKO. 1979. Sov. Yad. Phys. **30:** 811; JINR preprint no. E2-12085.
14. SIRLIN, A. & W. J. MARCIANO. 1981. Nucl. Phys. **B189:** 442–460.
15. DiLELLA, L. 1985. Proc. International Symposium on Lepton and Photon Interactions at High Energies, Kyoto, Japan, p. 280–311.
16. JONKER, M. *et al.* CHARM COLLABORATION. 1981. Phys. Lett. **B99:** 265.

APPENDIX 1

CHARM Collaboration

NIKHEF, Amsterdam, The Netherlands: F. Bergsma.
CERN, Geneva, Switzerland: J. V. Allaby, U. Amaldi, G. Barbiellini, M. Baubillier, A. Capone, W. Flegel, F. Grancagnolo, L. Lanceri, M. Metcalf, C. Nieuwenhuis, J. Panman, R. Pain, R. Plunkett, C. Santoni & K. Winter.
II. Institut für Experimentalphysik, Universität Hamburg: I. Abt, J. Aspiazu, A. Büngener, F. W. Büsser, H. Daumann, P. D. Gall, T. Hebbeker, F. Niebergall, P. Schütt & P. Stähelin.
Institute for Theoretical and Experimental Physics, Moscow, USSR: P. Gorbunov, E. Grigoriev, V. Khovansky & A. Rozanov.
Istituto Nazionale di Fisica Nucleare, Rome, Italy: A. Baroncelli, L. Barone, B. Borgia, C. Bosio, M. Diemoz, C. Dionisi, U. Dore, F. Ferroni, E. Longo, P. Loverre, L. Luminari, P. Monacelli, S. Morganti, F. de Notaristefani, L. Tortora & V. Valente.

A Precision Measurement of $\sin^2\theta_W$ from Semileptonic Neutrino Scattering

JÖRG WOTSCHACK

CERN-Dortmund-Heidelberg-Saclay-Warsaw Collaboration
CERN CH-1211
Geneva 23, Switzerland

INTRODUCTION

There is considerable interest in measuring the electroweak mixing parameter $\sin^2\theta_W$ of the Glashow-Salam-Weinberg theory[1] as precisely as possible: first, its value may be predicted by models of Grand Unification;[2] second, precise measurements of $\sin^2\theta_W$ from different processes would test the validity of electroweak radiative corrections.[3,4] Different methods have been used to determine $\sin^2\theta_W$ over a large range of Q^2 values. FIGURE 1 gives a compilation of $\sin^2\theta_W$ with remarkable agreement between the results. At present, it is most precisely determined in semileptonic neutrino-nucleon scattering from the ratio of neutral current (NC) to charged current (CC) cross sections,[5–9] and in proton-antiproton collisions from the W boson mass.[10,11]

Here, results are reported from an experiment of the first type. In order to achieve a higher precision in R_ν = NC/CC than in previous experiments, the upgraded CDHS detector was exposed to an optimized version of the CERN narrow-band beam (NBB). This provided a significant increase in statistics and a reduction of the uncertainties related to the neutrino flux. Most of the experiment was performed with neutrinos because the antineutrino NC/CC ratio is less sensitive to $\sin^2\theta_W$.

NEUTRINO BEAM AND DETECTOR

The use of a NBB was chosen in order to ensure clean neutrino (or antineutrino) samples with a calculable fraction of electron neutrinos. The CERN NBB (FIGURE 2) was modified such as to double the neutrino flux per proton on target with respect to previous experiments. This was achieved with 160-GeV/c hadrons produced by 450-GeV/c protons (instead of 200-GeV/c and 400-GeV/c, respectively) and by increasing the acceptance of the beam line.

The largest systematic error in the previous experiment (1.2% error on R_ν) was caused by the so-called wideband beam (WBB) background. This background is due to hadrons of unknown sign and momentum decaying upstream of the decay tunnel. To overcome this uncertainty, a movable 1.5-m-long iron dump was installed just before the decay tunnel (see FIGURE 2b). Data taken with the dump inserted into the beam line enabled the measurement of the WBB background. A precise normalization of "dump in" and "dump out" data taking was made using two beam current transformers that were located upstream of the dump. In addition, an iron absorber was installed just at the end of the 292-m-long evacuated (<0.015 torr) decay tunnel in order to stop

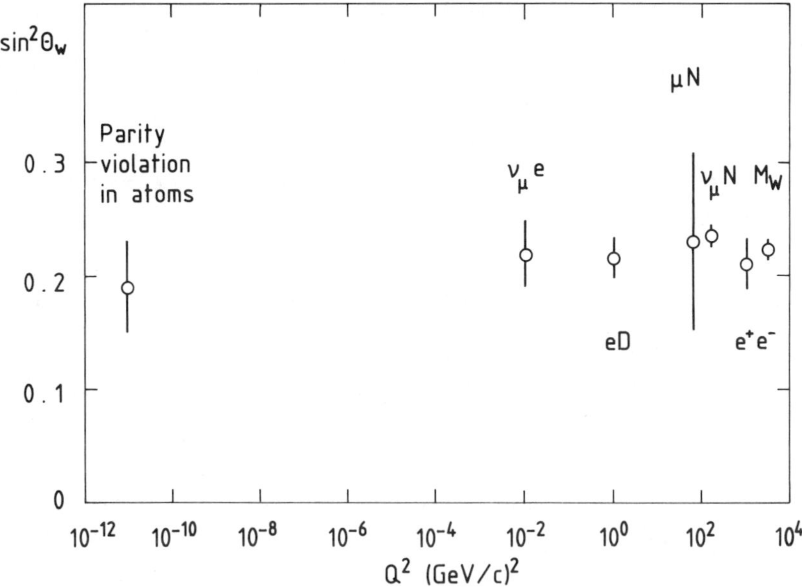

FIGURE 1. A diagram of $\sin^2\theta_W$ as a function of Q^2 as measured in different processes.

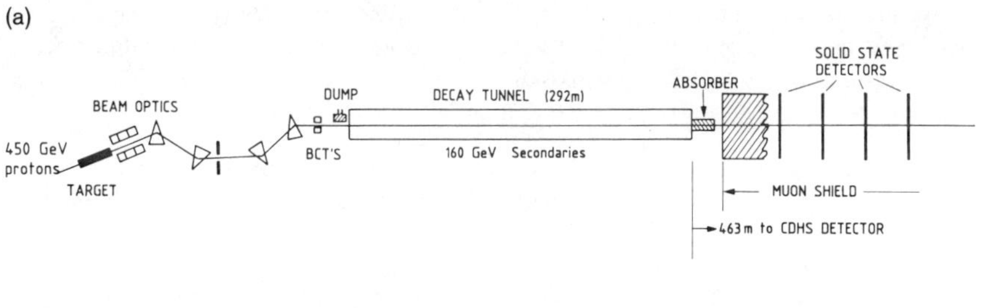

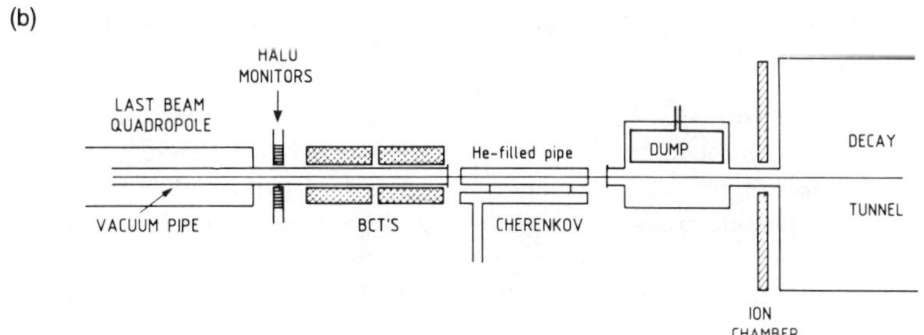

FIGURE 2. Layout of the 1984 NBB at CERN: (a) Overall view; (b) details of the sector before the decay tunnel.

immediately the remaining hadrons. As inferred from an earlier proton beam dump experiment,[12] the WBB background from this source is negligible.

The beam was operated from March to August 1984. A total of 6×10^{18} protons of 450-GeV/c momentum was delivered on target by the CERN SPS. Ninety percent of the data were taken with neutrinos, and 10% with antineutrinos.

For the present experiment, an upgraded version of the CDHS detector[13] was used (see FIGURE 3). It consisted of 21 modules of 1.875 m radius iron plates that were toroidally magnetized in such a way that the final state muon of CC ν_μ events was focused towards the detector axis. The first ten modules had a total iron thickness of 50 cm each; the others had 75 cm each. Scintillator planes inserted between the iron plates served to measure the energy of the hadron shower and to detect penetrating muons. In the first ten modules, hadron showers and muon tracks were sampled every 2.5 cm of iron with 15-cm-wide scintillator sheets. Five successive sheets were viewed by a single phototube. The groups of scintillators were alternated in horizontal and vertical direction, thus allowing the determination of the transverse position of hadron showers to a precision of ± 5 cm. The remaining eleven modules had scintillator sheets of 45 cm width in the horizontal direction only. The sampling thickness was 5 cm of iron in module numbers 11 to 15, and 15 cm in module numbers 16 to 21. Drift chambers in between the modules, made of triplets of wire planes, served to detect and reconstruct the muon tracks.

DATA REDUCTION AND ANALYSIS

The detector was triggered on a local energy deposition (shower trigger) or, independently, on a particle penetrating at least three modules (muon trigger). To avoid any bias between NC and CC events, the event samples were defined from calorimetric and topological information only, with no reference to the muon reconstruction. For every event, the following quantities were measured:

    (i) the longitudinal position of the vertex, defined by the first scintillator in the shower with a pulse height greater than 1.2 times that of a minimum ionizing particle;

    (ii) the lateral vertex position;

    (iii) the event length L, defined as the longitudinal distance between the vertex and the last scintillator hit in the event;

    (iv) the shower energy, E_{sho}, from the pulse height recorded in the first 1.5 m of iron after the vertex.

Events were retained if they had a shower trigger and if their vertex lay inside a cylinder of 1.3 m radius around the detector axis, excluding a 40×40 cm^2 area around the central hole, and ranging from the middle of module no. 3 to the middle of module no. 10. In order to avoid electronic problems at high trigger rates, a time lapse of more than 10 μs with respect to the preceding trigger was required.

The separation of events into NC and CC candidates was made using the event length L as shown in FIGURE 4. The cutoff length L_{cut} is defined as $L_{\text{cut}} = 75 + 38 \cdot \ln E_{\text{sho}}$ [GeV] (in centimeters of iron), which is in such a way as to minimize systematic uncertainties from the spillover of NC events into the CC sample.

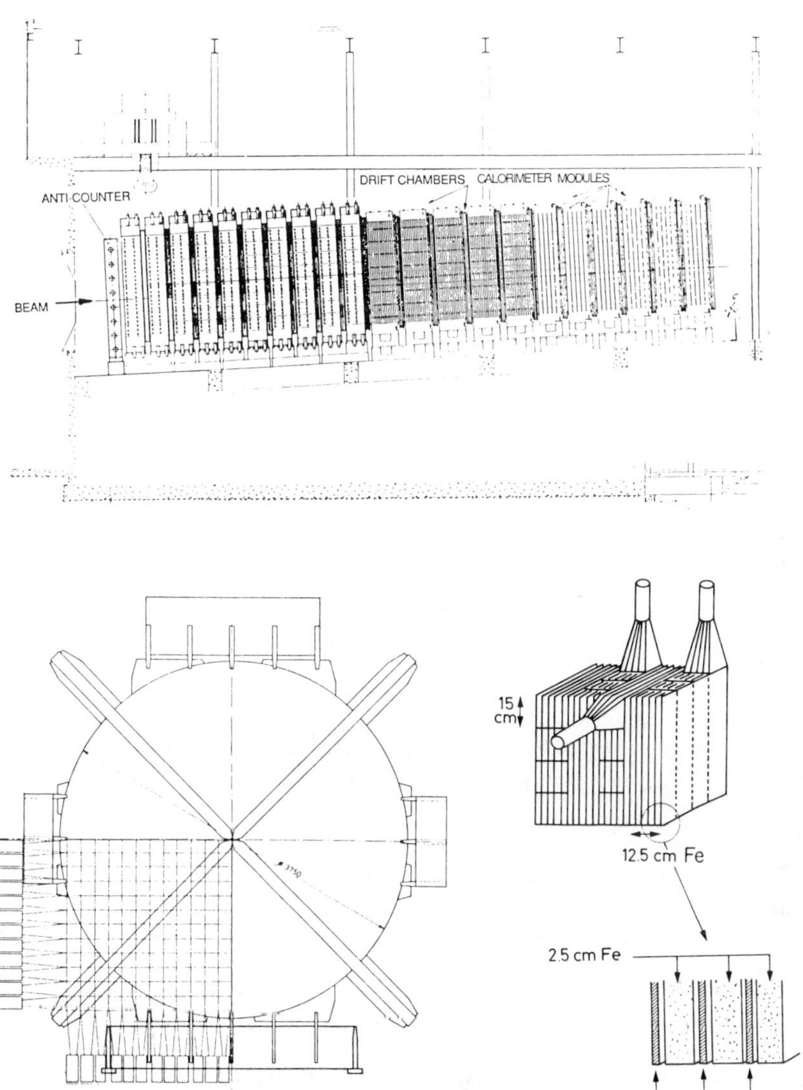

FIGURE 3. Layout of the CDHS neutrino detector (see text) showing a side view, a front view, and a close-up of the scintillator arrangement.

A cut on the hadron energy, $E_h > 10$ GeV, was applied. For NC events, the measured shower energy E_{sho} is identical to E_h. For CC events, E_{sho} also contains the pulse height deposited by the muon in the first 1.5 m of iron. This pulse height, equivalent to typically 3 GeV of hadronic energy, is measured from isolated muon tracks with a precision of $\pm 5\%$. The resulting uncertainty in the E_h cut for CC events

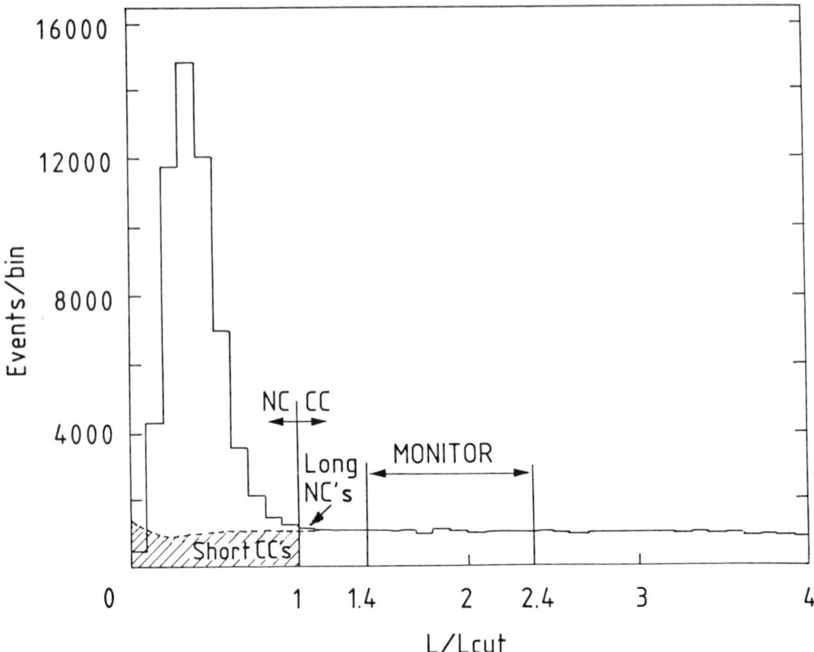

FIGURE 4. Distribution of the event length L, in units of the cutoff L_{cut} (see text), for neutrino events with $E_h > 10$ GeV. The background from cosmic rays and WBB has been subtracted. The dashed line shows the Monte Carlo prediction of the muon length in CC events normalized to the monitor region.

induces a systematic error of $\pm 0.3\%$ on R_ν. The efficiency of the shower trigger for $E_{sho} > 10$ GeV was measured to be better than 99.9% by using events taken with the muon trigger.

Starting from the raw number of NC and CC candidates, various corrections must be made to obtain the genuine NC and CC event numbers. TABLE 1 lists the event numbers, the corrections, and the corresponding systematic errors.

The first correction concerns cosmic ray events, most of which concentrate at short event lengths. They were subtracted statistically using data taken in between beam spills. The WBB background is subtracted using the data taken in "dump in" running conditions.

The "Long Shower" correction takes care of events that are longer than L_{cut}, although they have no primary muon or a primary muon shorter than L_{cut}. Most of these events are due to secondary muons from hadron decay; their rate was determined by Monte Carlo calculations. Including also a small contribution from long hadron showers or event lengthening by noise, this correction amounts to $(+0.5 \pm 0.2)\%$ of the NC/CC ratio.

The largest correction is the subtraction of the so-called short CC events from the NC candidates and their subsequent addition to the CC events. These are mostly CC ν_μ events with a muon range shorter than L_{cut}. Owing to the magnetic field that bends the muons towards the detector axis and also to the small radius of the fiducial volume, only 7% of the short CC events have a muon leaving the detector at the side. The number of short CC events is obtained by a Monte Carlo simulation of the muon length distribution that is normalized to a monitor region defined as $1.4 \le L/L_{\text{cut}} < 2.4$ (see FIGURE 4). This simulation requires the understanding of the y distribution ($y = 1 - E_\mu/E_\nu$) and of the length measurement. No simulation of the hadronic shower or of the muon reconstruction is needed. The normalization of the short CC events, $\langle y \rangle = 0.94$, to the monitor region, $\langle y \rangle = 0.85$, is much less sensitive to physics assumptions than would be the normalization to the whole CC sample. The errors induced by the y distribution uncertainties amount to 0.7% of the correction, and they are primarily from the longitudinal structure function. Because of the normalization to the monitor region, the correction is largely insensitive to errors in the momentum, the divergence, the π/K composition of the parent beam, the muon energy loss, and the hadron energy calibration. It is, however, sensitive to a bias in the measurement of the longitudinal vertex position or of the end of the event. The vertex position is affected by backscattered particles. By varying the minimum energy required in the first scintillator of the shower, the true vertex position was determined to ± 1 cm on average. The end of the muon track is affected by scintillator noise or inefficiency. It was verified that the end of the event distribution after the last drift chamber hit matched the Monte Carlo prediction to better than 1 cm. Altogether, the bias on the event length measurement was estimated to be less than ± 1.5 cm, corresponding to an error of 1% on the correction. In summary, the short CC correction amounts to $(-22.5 \pm 0.35)\%$ of the NC/CC ratio.

The event numbers must finally be corrected for CC events produced by electron-neutrinos from K_{e3} decay. These events are included in the NC sample because the final state electron is hidden in the hadron shower. The absolute rate of CC ν_e events is determined from a Monte Carlo simulation that is normalized to the number of fully reconstructed CC ν_μ events from $K_{\mu2}$ decay. The CC ν_e events are subtracted from the NC sample, and those with $E_h > 10$ GeV are added to the CC sample.

After all these corrections, the NC to CC cross section ratio for neutrino interactions in iron, with $E_h > 10$ GeV, is

$$R_\nu = 0.3059 \pm 0.0025 \text{ (stat)} \pm 0.0020 \text{ (syst)}.$$

TABLE 1. Event Numbers and Corrections of NC and CC Events[a]

	NC	CC	Change of NC/CC and Systematic Error (%)
Candidates	60,936	137,853	± 0.3
Cosmic rays	-1120	-9	-2 ± 0.1
WBB background	-2920	-5187	-1.2 ± 0.1
Long Shower	$+159$	-158	$+0.5 \pm 0.2$
Short CC	-9642	$+9526$	-22.5 ± 0.35
K_{e3} correction	-3150	$+2667$	-8.4 ± 0.2
Corrected event numbers	44,263	144,692	± 0.65

[a]For $E_h > 10$ GeV.

DETERMINATION OF $\text{SIN}^2\theta_W$

Up to small corrections, $\sin^2\theta_W$ can be extracted from the following relation:[4]

$$R_\nu = \tfrac{1}{2} - \sin^2\theta_W + \tfrac{5}{9}\sin^4\theta_W\,(1 + r),$$

where r is the ratio of CC $\bar{\nu}$ to CC ν cross sections integrated over the same neutrino spectrum with the same E_h cutoff. It was measured to be $r = 0.39 \pm 0.01$.

The above formula is valid for an isoscalar target in a world of u, $\bar{u}$, d, $\bar{d}$ quarks and zero quark-mixing angles. Small corrections must be applied to take into account the deviation from isoscalarity of the target, the presence of s and c quarks in the nucleon, and the charm-quark excitation. These corrections were determined using a quark-parton model of the nucleon, including QCD fits to the measured structure functions,[14] charm production according to the slow rescaling model,[15] and quark mixing according to the unitarity-constrained Kobayashi-Maskawa matrix.[16,17] The parameters used in the model are listed in TABLE 2, with the corresponding effects on $\sin^2\theta_W$. Higher twist effects are claimed to be small for $E_h > 10\ \text{GeV}$[4,18] and are not included.

The radiative corrections were calculated according to Wheater and Llewellyn Smith.[19] When expressed in the on-shell renormalization scheme,[20] where $\sin^2\theta_W = 1 - m_W^2/m_Z^2$, they result in a change of $\sin^2\theta_W$ by -0.011 ± 0.002. The value of $\sin^2\theta_W$ before radiative corrections is 0.238. After these corrections, the result for $\sin^2\theta_W$ is

$$\sin^2\theta_W = 0.227 \pm 0.005\ (\text{exper}) \pm 0.003\ (\text{theor}) + 0.013\ (m_c - 1.5\ \text{GeV}/c^2).$$

TABLE 2. Effect of the Parameters of the Model on $\sin^2\theta_W{}^a$

Parameter	$\Delta \sin^2\theta_W$
Quark generation mixing	$+0.0031 \pm 0.0003$
$\quad\lvert U_{ud}\rvert^2 = \lvert U_{cs}\rvert^2 = 0.947 \pm 0.006$	
$\quad\lvert U_{us}\rvert^2 = \lvert U_{cd}\rvert^2 = 0.053 \pm 0.006$	
Longitudinal structure function	$+0.0006 \pm 0.0006$
$\quad\sigma_L/\sigma_T = R_{\text{QCD}} \pm R_{\text{QCD}}$	
Nonstrange sea	$+0.0022 \pm 0.0003$
$\quad(\bar{U}+\bar{D})/(U+D) = 0.13 \pm 0.02$	
Strange sea	$+0.0043 \pm 0.0010$
$\quad S/D = 0.45 \pm 0.10$ at $E_h = 30\ \text{GeV}$	
Charm Sea	$+0.0003 \pm 0.0003$
$\quad C/S = 0.15 \pm 0.15$	
Nonisoscalar target (Fe)	-0.0090 ± 0.0009
$\quad D_\nu/U_\nu = 0.39 \pm 0.04$	
Radiative Corrections	-0.011 ± 0.002
$\quad m_{\text{top}} = 45\ \text{GeV}/c^2$	
$\quad m_{\text{Higgs}} = 100\ \text{GeV}/c^2$	
Uncertainties for a fixed m_c	± 0.003
Charm-quark mass[b]	$+0.010 \pm 0.004$
$\quad m_c = (1.5 \pm 0.3)\ \text{GeV}/c^2$	
Total theoretical uncertainty	± 0.005

^{a}The change of $\sin^2\theta_W$ is given when one parameter is varied, leaving all the others at their nominal value. All the corrections being correlated, they were applied at once.

^{b}This value also includes the influence of m_c on the determination of structure functions.

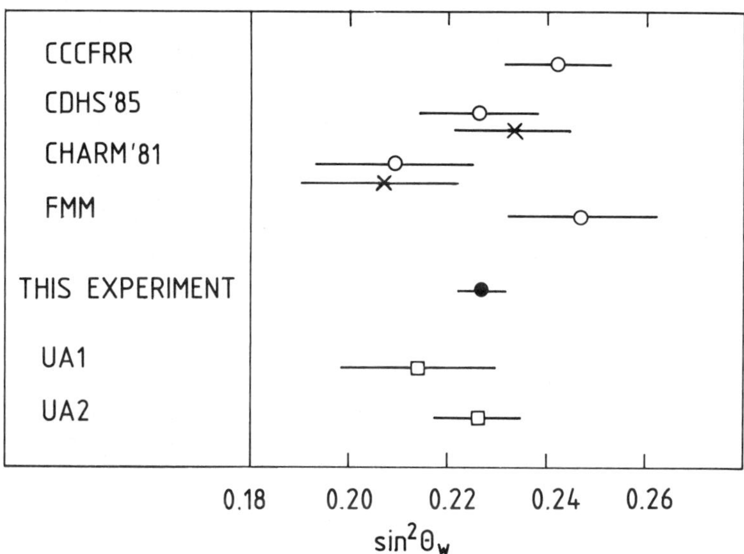

FIGURE 5. Recent measurement of $\sin^2\theta_W$ from semileptonic neutrino scattering[5–8] and from the W mass.[10,11] To allow a direct comparison of all neutrino experiments, the published values (open circles) have been corrected to the same model assumptions as used in this analysis (crosses). In order to compare with the W mass results, a common theoretical error of ± 0.005 is to be added to the neutrino results.

The experimental error results from the statistical and systematic errors on R_ν and r added in quadrature. The theoretical error corresponds to the uncertainties on the parameters of the model used to calculate the corrections; however, this is excluding the uncertainty on the charm-quark mass m_c. The largest uncertainty of the correction is in the choice of m_c, which affects the threshold suppression of charm production in CC events. Hence, the result on $\sin^2\theta_W$ is given as a function of m_c. With a central value of $m_c = (1.5 \pm 0.3)\ \text{GeV}/c^2$, the result is $\sin^2\theta_W = 0.227$, and the total theoretical error increases to ± 0.005.

DISCUSSION AND SUMMARY

The value of $\sin^2\theta_W$ obtained in this experiment is compatible with earlier results obtained by this group.[6,9] It is also compatible with recent measurements from other semileptonic neutrino experiments,[5,7,8,21] as well as with recent determinations of $\sin^2\theta_W$ from the W mass.[10,11] The reported values are listed and are shown graphically in FIGURE 5.

Assuming the validity of the radiative corrections, a precise determination of the ρ parameter, $\rho = m_W^2/(m_Z^2 \cdot \cos^2\theta_W)$, can be obtained. To a very good approximation, ρ^2 is equal to the ratio of the measured value of R_ν to the one predicted for $\sin^2\theta_W = 0.223 \pm 0.008$ as determined from the weighted average of the W mass measurements.

Propagating all errors quadratically, one finds

$$\rho = 0.996 \pm 0.011,$$

which is in good agreement with the minimal standard model value of $\rho = 1$.

In summary, a precise measurement of the NC to CC cross section ratio for neutrino scattering off iron has been performed: $R_\nu = 0.3059 \pm 0.0025$ (stat) ± 0.0020 (syst) for $E_h > 10$ GeV. From this, a value of $\sin^2\theta_W = 0.227 \pm 0.005$ (exper) ± 0.005 (theor) has been derived in the on-shell renormalization scheme. The experimental error of 0.005 in $\sin^2\theta_W$ matches the present theoretical uncertainty that arises mostly from the uncertainties in the charm-quark excitation.

REFERENCES

1. GLASHOW, S. L. 1961. Nucl. Phys. **22:** 579; SALAM, A. 1968. Proc. 8th Nobel Symposium, Aspenäsgården, p. 367. Almqvist & Wiksells. Stockholm; WEINBERG, S. 1967. Phys. Rev. Lett. **19:** 1264; 1972. Phys. Rev. **D5:** 1412.
2. See, for example: LANGACKER, P. 1981. Phys. Rep. **72:** 185.
3. MARCIANO, W. J. & A. SIRLIN. 1984. Phys. Rev. **D29:** 945.
4. LLEWELLYN SMITH, C. H. 1983. Nucl. Phys. **B228:** 205.
5. REUTENS, P. G. *et al.* 1985. Phys. Lett. **152B:** 404.
6. ABRAMOWICZ, H. *et al.* 1985. Z. Phys. **C28:** 51.
7. JONKER, M. *et al.* 1981. Phys. Lett. **B99:** 265.
8. BOGERT, D. *et al.* 1985. Phys. Rev. Lett. **55:** 1969.
9. HOLDER, M. *et al.* 1977. Phys. Lett. **71B:** 222; **72B:** 254.
10. ARNISON, G. *et al.* 1986. Phys. Lett. **166B:** 484.
11. APPEL, J. A. *et al.* 1985. Preprint no. CERN-EP/85-166.
12. PEREZ, P. 1983. Proc. International Europhysics Conf. on High-Energy Physics, Brighton, England, p. 316. Rutherford Appleton Laboratory. United Kingdom.
13. HOLDER, M. *et al.* 1978. Nucl. Instrum. Methods **148:** 235.
14. ABRAMOWICZ, H. *et al.* 1983. Z. Phys. **C17:** 283; BRUMMEL, H. D. 1984. Diplomarbeit. University of Dortmund. Unpublished.
15. BARNETT, R. M. 1976. Phys. Rev. **D14:** 70; GEORGI, H. & H. D. POLITZER. 1976. Phys. Rev. **D14:** 1829.
16. KOBAYASHI, M. & K. MASKAWA. 1973. Progr. Theor. Phys. **49:** 652.
17. KLEINKNECHT, K. & B. RENK. 1983. Phys. Lett. **130B:** 459.
18. FAJFER, S. & R. J. OAKES. 1985. Proc. International Europhysics Conf. on High-Energy Physics, Bari, Italy, p. 190. CASTORINA, P. & P. J. MULDERS. 1985. Phys. Rev. **D31:** 2753.
19. WHEATER, J. F. & C. H. LLEWELLYN SMITH. 1982. Nucl. Phys. **B208:** 27. (Erratum: 1983. Nucl. Phys. **B226:** 547.) Reasonable agreement was also found with the following calculations: BARDIN, D. YU., P. CH. CHRISTOVA & O. M. FEDORENKO. 1982. Nucl. Phys. **B197:** 1; BARDIN, D. Private communication; ABRAMOWICZ, H. CDHS internal note. Unpublished; SIRLIN, A. & W. J. MARCIANO. 1981. Nucl. Phys. **B189:** 442. At the conference, the value $\sin^2\theta_W = 0.219 \pm 0.005$ (exper) ± 0.005 (theor) ± 0.006 (rad. corr.) was given with some reservation concerning the radiative corrections. The corrections of Bardin *et al.* were used, leading to a change in $\sin^2\theta_W$ of -0.017 much larger than the other calculations. In the meanwhile, a bug was found in Bardin's program. Although, after correction, Bardin's program gave results consistent with the other calculations, we have chosen to use the well-tested programs of Wheater and Llewellyn Smith instead.
20. SIRLIN, A. 1980. Phys. Rev. **D22:** 97; STUART, R. G. 1985. Preprint no. CERN-TH 4342/85.
21. PAIN, R. 1985. Proc. International Europhysics Conf. on High-Energy Physics, Bari, Italy, p. 249. European Phys. Soc. Geneva; see also: LANCERI, L. 1987. These proceedings.

Measurements of $\sin^2\theta_W$ from Studies of the Elastic Scattering of Neutrinos by Protons and Electrons

A. K. MANN[a]

Department of Physics
University of Pennsylvania
Philadelphia, Pennsylvania 19104

This talk is intended to be a brief report on studies of the elastic scattering of neutrinos by protons and electrons. Much of this work has been recently published[1-3] or is circulating[4,5] in preprint form. Consequently, a short statement of the motivation and a summary of the experimental method should be sufficient for the purpose here, and attention can be concentrated on the experimental results.

The semileptonic and purely leptonic reactions $\nu_\mu p \rightarrow \nu_\mu p$, $\bar{\nu}_\mu p \rightarrow \bar{\nu}_\mu p$ and $\nu_\mu e \rightarrow \nu_\mu e$, $\bar{\nu}_\mu e \rightarrow \bar{\nu}_\mu e$ are particularly simple reactions (both experimentally and theoretically) with which to study weak neutral currents. These processes have clean, distinctive experimental signatures, and their interpretation is almost completely constrained with the standard electroweak model. Measurements of the ratio of $\bar{\nu}_\mu$ and ν_μ elastic scattering on protons and the corresponding ratio for elastic scattering on electrons minimize systematic experimental errors and lead directly to values of the fundamental parameter of the electroweak interaction, $\sin^2\theta_W$, with minimal ambiguity. Accordingly, our principal motivation in carrying out these studies was the desire to obtain and compare precise values of $\sin^2\theta_W$ from both the semileptonic and leptonic reactions as still another test of the basic validity of the standard electroweak theory.

EXPERIMENTAL METHOD TARGET-DETECTOR

The experiments were done at the Brookhaven National Laboratory Alternating Gradient Synchrotron (AGS). The target-detector in which the ν_μ and $\bar{\nu}_\mu$ interactions occurred and were measured is shown in FIGURE 1. It consisted of three major components: (*i*) a massive high resolution modular target-detector, (*ii*) a shower counter for longitudinal containment of electron and photon showers, and (*iii*) a muon spectrometer, which was used to measure the energy spectra of both majority and minority ν_μ and ν_μ by measuring the angle and momentum of the muon produced in charged current quasi-elastic interactions.

[a]A. K. Mann represents the BNL-Brown-Hiroshima-KEK-Osaka-Pennsylvania–Stony Brook Collaboration consisting of K. Abe, L. A. Ahrens, K. Amako, S. H. Aronson, E. W. Beier, J. L. Callas, P. L. Connolly, D. Cutts, M. Diwan, D. C. Doughty, L. S. Durkin, B. G. Gibbard, S. M. Heagy, D. Hedin, J. S. Hoftun, M. Hurley, S. Kabe, Y. Kurihara, R. E. Lanou, A. K. Mann, M. D. Marx, M. J. Murtagh, S. J. Murtagh, Y. Nagashima, F. M. Newcomer, T. Shinkawa, E. Stern, Y. Suzuki, S. Tatsumi, S. Terada, D. H. White, H. H. Williams, Y. Yamaguchi, and T. York.

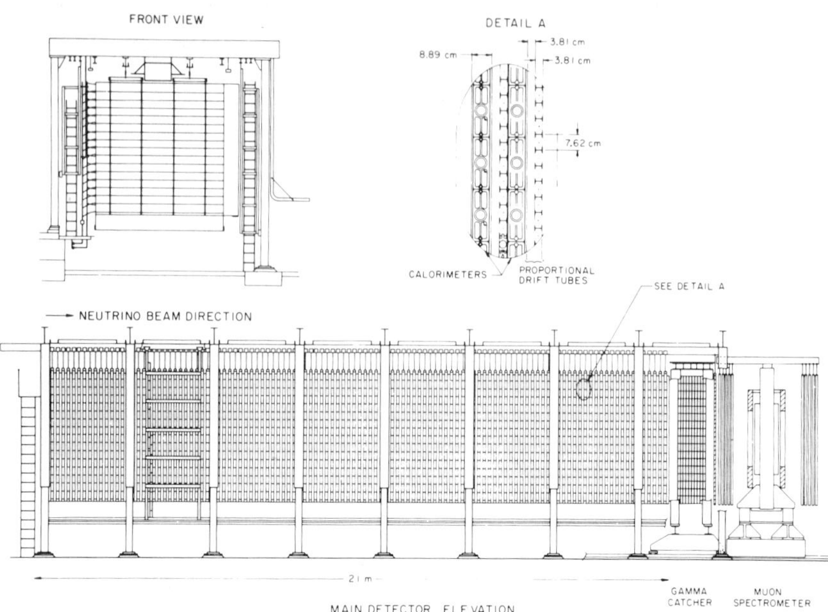

FIGURE 1. A schematic drawing of the BNL-Brown-KEK-Osaka-Pennsylvania–Stony Brook neutrino detector.

The target-detector comprised 112 modules, each of which contained a plane of 16 liquid scintillator cells (each 4m × 25cm × 8cm) and two crossed planes of 54 proportional drift tubes (PDT) (each 4m × 7.6cm × 3.8cm) arranged for x and y track position measurements. Each liquid scintillator cell yielded charge (energy loss) and nanosecond timing information, while each PDT cell yielded charge and 1.5-mm resolution position information. The fine segmentation (1,892 scintillator cells and 12,096 PDT cells) and the pulse height and timing characteristics of the elements provided determination of event topology and discrimination of pions and protons through energy-loss measurements. Eighty percent of the target-detector mass was liquid scintillator. Seventy-nine percent of the target protons were bound in carbon and aluminum, while the remaining 21% were free protons.

The shower counter, located directly downstream of the main detector, consisted of 12 radiation lengths of lead and liquid scintillation counters that contained and measured electromagnetic energy from neutrino interactions occurring in the downstream part of the target-detector. The shower counter was unnecessary in the analysis of neutrino-proton elastic scattering. The muon spectrometer was located downstream from the shower counter. It consisted of a wide aperture dipole magnet instrumented with nine x-y PDT plane pairs to measure the momenta of forward-going muons from ν_μ-induced quasi-elastic interactions. The detector was operated in a triggerless mode that recorded all information within a 10-μsec gate beginning at the AGS proton beam extraction and continuing several muon decay lifetimes after the passage of the neutrino beam.

BEAMS

The horn-focused wideband neutrino beam[6] consisted of protons of 28-GeV kinetic energy incident on a production target that produced pions and kaons that subsequently decayed to yield primarily ν_μ (positive horn focus) or $\bar{\nu}_\mu$ (negative horn focus) beams with mean energy 1.3 GeV and 1.2 GeV, respectively. The ν_μ and $\bar{\nu}_\mu$ fluxes as a function of neutrino energy are shown in FIGURES 2 and 3. In addition to the selected primary helicity neutrinos in a given beam, a small fraction of opposite helicity neutrinos are produced by the decay of defocused mesons. This "wrong helicity" contamination was measured using the muon spectrometer by observing the fraction of quasi-elastic events occurring with the wrong-sign muon for the primary beam. The fraction (0–5 GeV) of antineutrino contamination in the neutrino beam was 0.024 ± 0.005, while the neutrino contamination in the antineutrino beam was 0.087 ± 0.013.

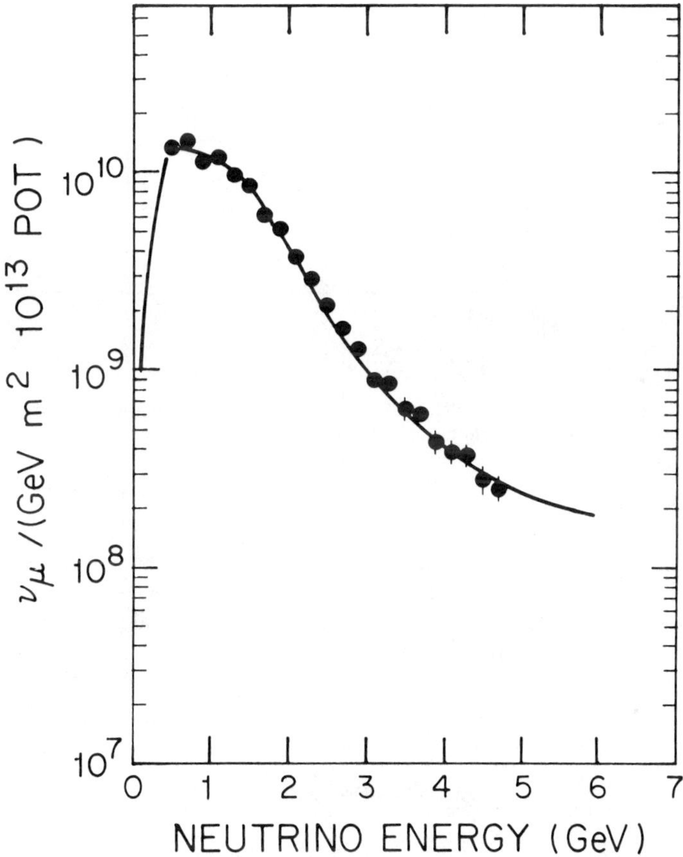

FIGURE 2. The measured ν_μ flux. The error bars represent data from the reaction $\nu_\mu n \rightarrow \mu^- p$. The solid curve is a Monte Carlo beam flux calculation. Details of the flux measurement and Monte Carlo calculations are given in reference 6.

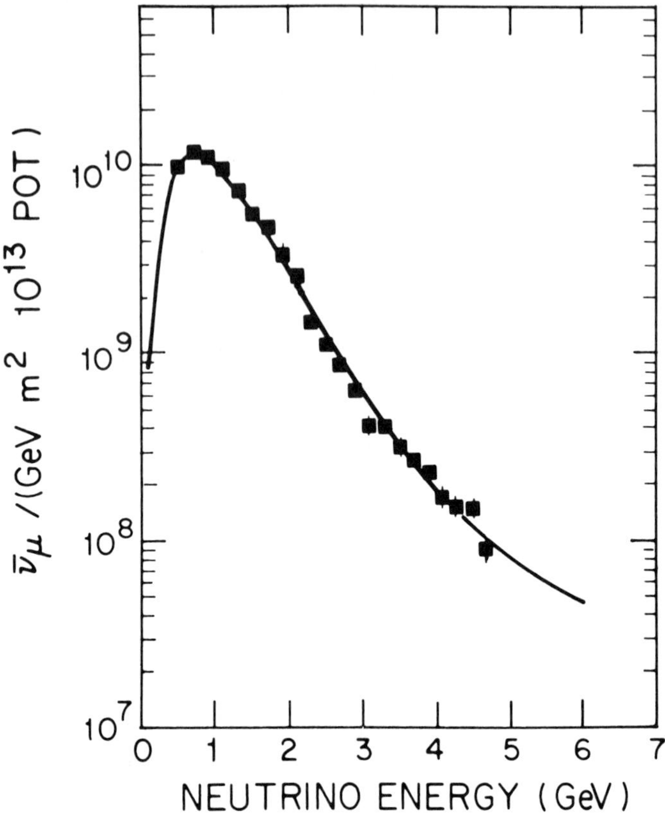

FIGURE 3. The measured ν_μ flux. The error bars represent data from the reaction $\bar{\nu}_\mu p \rightarrow \mu^+ n$. The solid curve is a Monte Carlo beam flux calculation. Details of the flux measurement and Monte Carlo calculations are given in reference 6.

The neutrino beam preserves the internal AGS time (bunch) structure, thus providing neutrinos in 12 distinct bunches of 30 ns width that are spaced 224 ns apart. FIGURE 4 shows a measurement of the time structure using actual neutrino interactions in the detector. Knowing the time of events relative to the AGS extraction and the given structure of the beam, "out of time" (background) events were eliminated from the data.

SIGNAL RECOGNITION

$\nu_\mu p \rightarrow \nu_\mu p$ and $\bar{\nu}_\mu p \rightarrow \bar{\nu}_\mu p$

A typical elastic scattering $\nu_\mu p \rightarrow \nu_\mu p$ candidate is shown in FIGURE 5. In general, the proton track is at a wide angle with respect to the incident neutrino beam

($\langle\theta_p\rangle \approx 45°$), and the kinetic energy of the detected proton set by the experimental detection efficiency lies between about 0.15 and 0.6 GeV. The energy loss per detector element along the tracks of candidate events was used to differentiate between protons and pions (muons). The fine-grained nature of the detector provided up to 9 samples (3 scintillator and 6 PDT) along the shortest tracks in the sample. The particle identification (PID) procedure compared the actual deposited energy in the i-th cell along candidate tracks with that predicted for either a pion hypothesis or a proton hypothesis for the measured range.

Confidence levels were evaluated for each hypothesis and for both detector element types (scintillator and PDT cells) by Monte Carlo integration of appropriate likelihood functions. The confidence levels for proton candidates as determined by scintillator

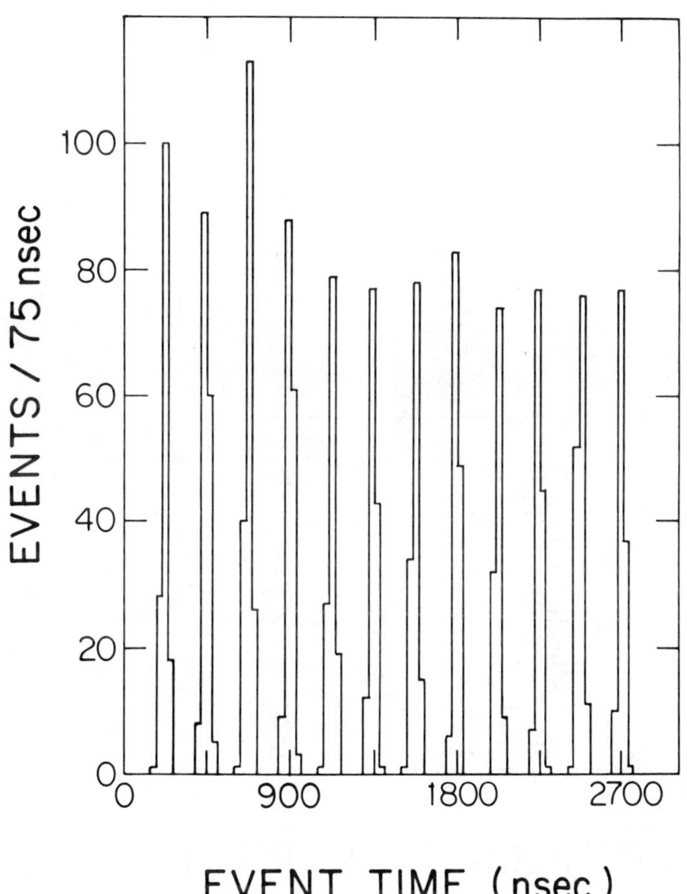

FIGURE 4. The event time distribution of the neutrino-proton elastic scattering sample relative to the start of the AGS beam spill showing the internal AGS time structure. The distribution for the antineutrino sample is similar.

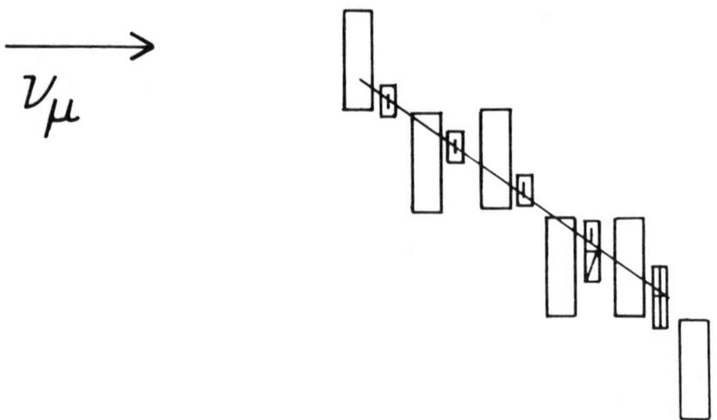

FIGURE 5. The side view of a typical elastic scattering candidate. The large boxes represent scintillator cells and the smaller boxes represent PDT cells hit. The best fit to the PDT drift positions (shown inside the PDT cells) is drawn on the event.

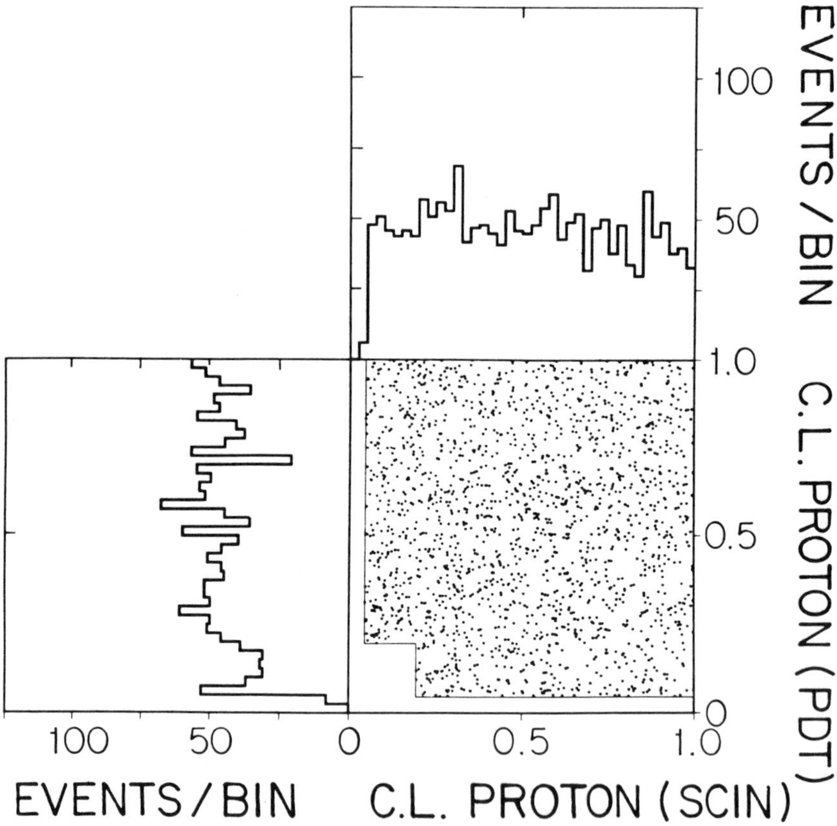

FIGURE 6. Scatter plot for events passing the proton requirements and the topological cuts of the confidence levels (C.L.) for a proton hypothesis using PDT energy deposits versus a proton hypothesis using scintillator energy deposits to exhibit the uniformity of the distribution. The confidence level boundary used to define the protons is indicated in the figure.

and PDT data are shown in FIGURE 6. The distributions and projections are seen to be uniform, which is an indication that the probability distributions have been correctly modeled. After particle identification, the data samples were reduced to 3,015 $\nu_\mu p$ candidates and 2,748 $\bar{\nu}_\mu p$ candidates.

Further criteria were applied to the samples to reject backgrounds from other neutral current channels. At the same time, this helped to minimize the uncertainty in modeling the detection efficiency. These cuts reduced the samples to 1,686 $\nu_\mu p$

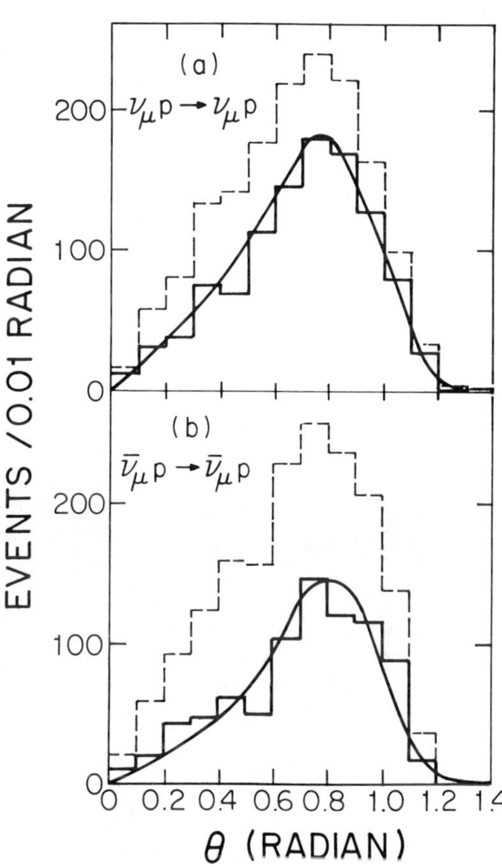

FIGURE 7. Distribution in the angle θ_p of the proton candidates (a) for $\nu_\mu p \rightarrow \nu_\mu p$ and (b) for $\bar{\nu}_\mu p \rightarrow \bar{\nu}_\mu p$. The dashed lines represent the data after PID and elimination of events with extra in-time energy depositions. The solid histograms are the observed elastic scattering signals after the Monte Carlo calculated background subtraction, and the solid curves are the Monte Carlo calculated elastic scattering distributions.

candidates and 1,821 $\bar{\nu}_\mu p$ candidates. The angular distributions of events both before and after background subtraction are shown in FIGURE 7. The distributions in energy deposited in the interaction cell are shown with Monte Carlo calculated background and signal predictions in FIGURES 8 and 9.

The uniformity of the distribution in FIGURE 6 and the good agreement of data and Monte Carlo calculation in FIGURES 7–9, as well as a number of other tests of the experimental method,[1,4] confirm the validity of the event selection procedure for $\nu_\mu p \rightarrow \nu_\mu p$ and $\bar{\nu}_\mu p \rightarrow \bar{\nu}_\mu p$.

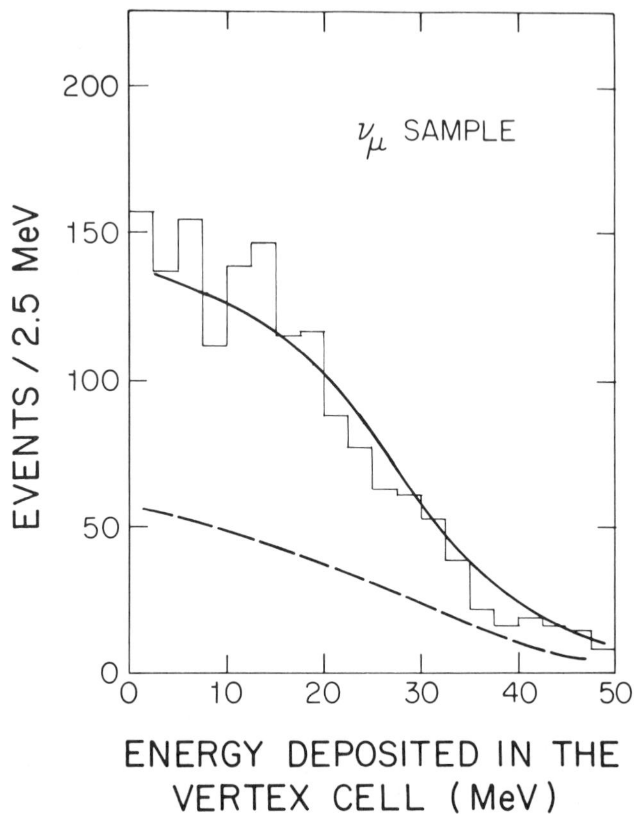

FIGURE 8. The distribution of the energy deposited in the vertex (interaction) cell for the neutrino-proton elastic scattering sample after all cuts. The histogram represents data events, the dashed curve represents the Monte Carlo background prediction, and the solid curve represents the Monte Carlo signal plus background prediction.

$$\nu_\mu e \to \nu_\mu e \text{ and } \bar{\nu}_\mu e \to \bar{\nu}_\mu e$$

The signature of the elastic scattering of a neutrino by an electron is fixed by the kinematics of the reaction that requires that $E_e\theta_e^2 = 2m_e(1 - E_e/E_\nu)$. Hence, for $\langle E_e \rangle \approx 0.5\,\text{GeV}$, appropriate to BNL, $\theta_e \sim 40$ mrad. Backgrounds from neutral current production of π^0 and induced by the contamination ($\sim 1\%$) of ν_e in the incident neutrino beam were reduced by discriminating against (doubly ionizing) photon products in favor of singly ionizing electrons and by the much flatter angular distributions of the backgrounds relative to the signal. The resulting distributions in $E_e\theta_e^2$ are shown in FIGURE 10, where the $\nu_\mu e$ and $\bar{\nu}_\mu e$ signals are clearly evident. The remaining background in the small angle region is most accurately determined and subtracted by means of the distributions in θ_e^2, as shown in FIGURE 11.

NORMALIZATION

The event rates for the elastic scattering reactions were compared to the measured event rates of the quasi-elastic processes $\nu_\mu n \rightarrow \mu^- p$ and $\bar{\nu}_\mu p \rightarrow \mu^+ n$. This indirectly measured the integrated ν_μ and $\bar{\nu}_\mu$ fluxes because the cross sections for the quasi-elastic reactions are well known theoretically and experimentally. To treat ν_μ- and $\bar{\nu}_\mu$-induced reactions as similarly as possible, a topological selection of quasi-elastic events was employed based on isolating single muon candidates at small ($\theta_\mu < 15°$) angles. Additionally, independent confirmation of the number of ν_μ-induced quasi-elastic events was obtained from quasi-elastic events in which both the muon and proton tracks were visible in the detector. These fully reconstructed events yielded a total number of quasi-elastic ν_μ events that was within 3% of the number found from the single muon data.

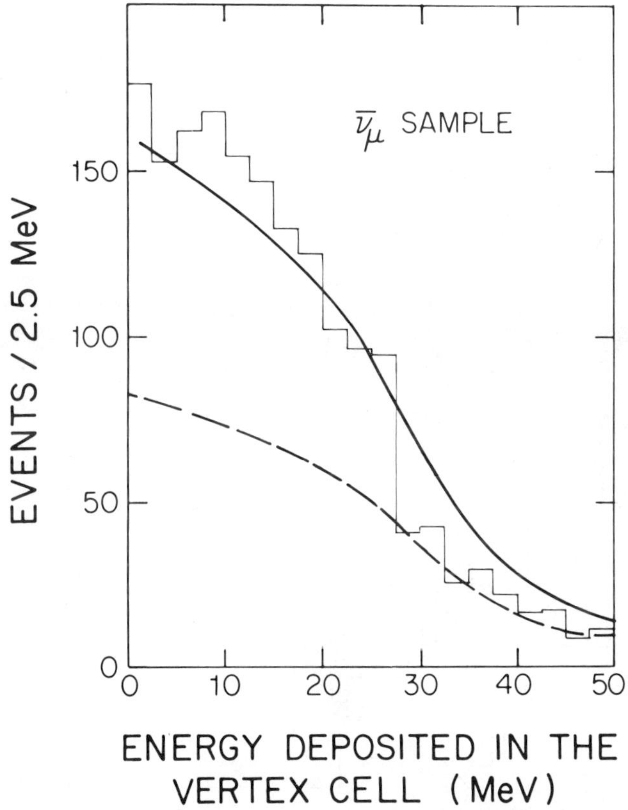

FIGURE 9. The distribution of the energy deposited in the vertex (interaction) cell for the antineutrino-proton elastic scattering sample after all cuts. The histogram represents data events, the dashed curve represents the Monte Carlo background prediction, and the solid curve represents the Monte Carlo signal plus background prediction.

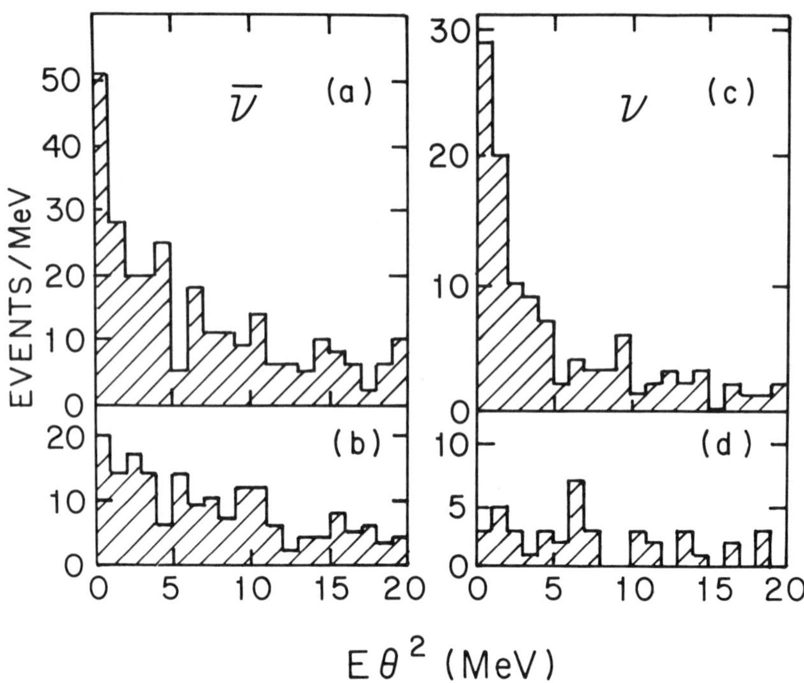

FIGURE 10. Distributions in $E_e\theta_e^2$ for (a) the primary singly ionizing events from the $\bar{\nu}_\mu$ data and (b) the predominantly photon sample from the $\bar{\nu}_\mu$ data; (c) and (d) the corresponding plots for the ν_μ-induced data.

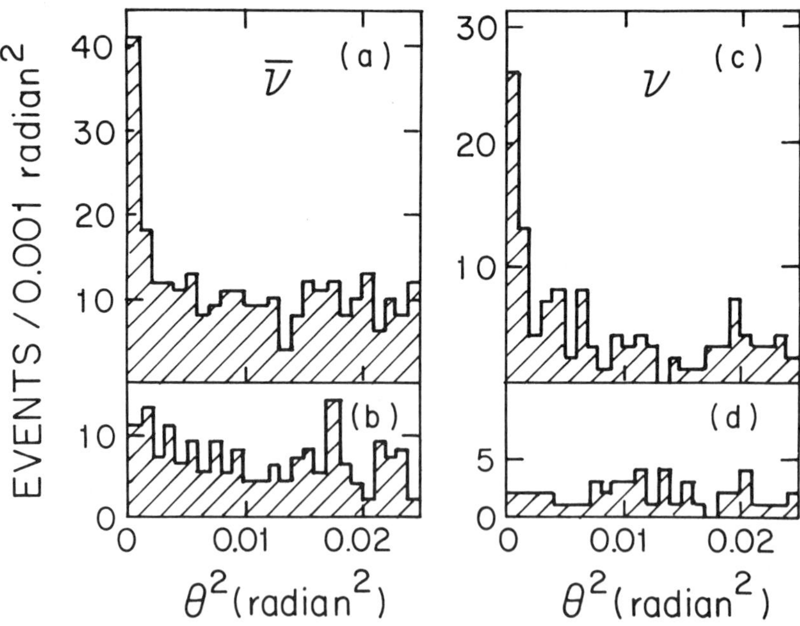

FIGURE 11. Distributions in θ_e^2 for the same event samples as in the corresponding plots of FIGURE 10.

The distribution of the energy deposited in the interaction cell for the neutrino quasi-elastic sample is shown in FIGURE 12. In FIGURE 13, we show the distribution in θ_μ for that sample. The final results after background subtraction and correction for detection efficiency were 5.29×10^4 and 5.12×10^4 quasi-elastic events in the ν_μ and $\bar{\nu}_\mu$ samples, respectively.

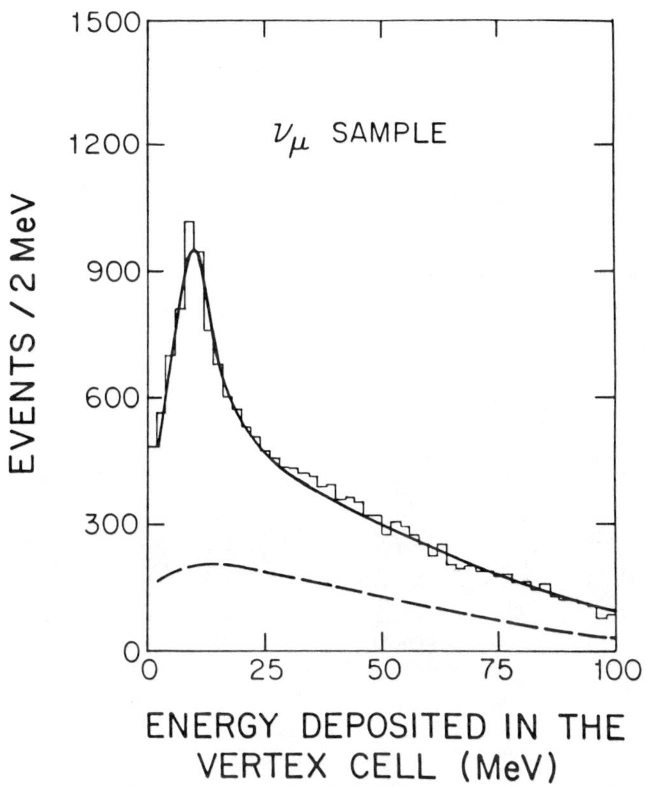

FIGURE 12. The distribution of the energy deposited in the vertex (interaction) cell for the neutrino quasi-elastic sample after all cuts. The histogram represents data events, the dashed curve represents the Monte Carlo background prediction, and the solid curve represents the Monte Carlo signal plus background prediction.

RESULTS

$\nu_\mu\mathrm{p}$ and $\bar{\nu}_\mu\mathrm{p}$

From the corrected data, one obtains the absolute flux-averaged differential cross sections, $d\sigma(\nu_\mu p)/dQ^2$ and $d\sigma(\bar{\nu}_\mu p)/dQ^2$, shown in FIGURE 14, and the total quasi-elastic rates ($Q^2 = 2m_p T_p$, with T_p being the proton kinetic energy). To extract values of the axial vector mass M_A and $\sin^2\theta_W$, corrections to the axial vector form factor were taken to be zero initially and the cross sections in FIGURE 14 were fit over the interval

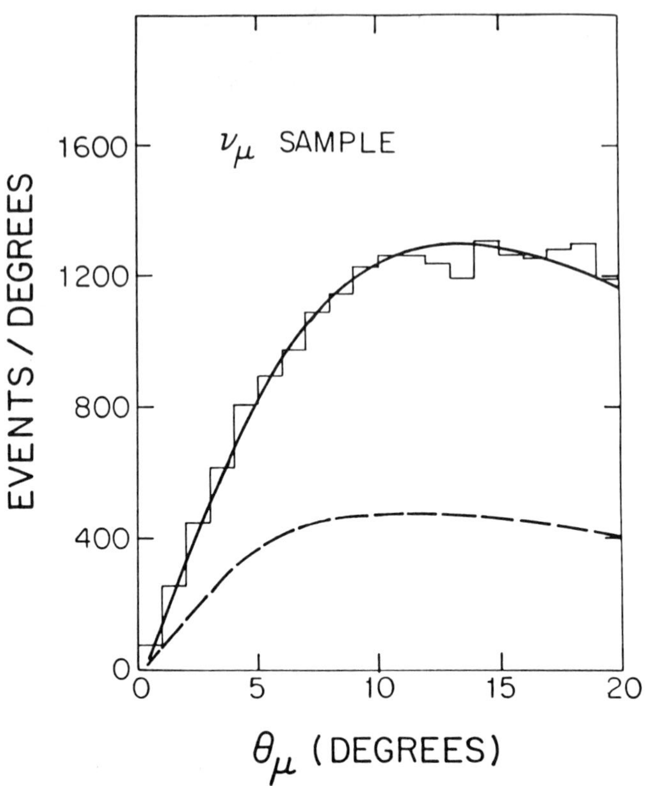

FIGURE 13. The distribution of the muon angle, θ_μ, for the neutrino quasi-elastic sample after all cuts. The histogram represents data events, the dashed curve represents the Monte Carlo background prediction, and the solid curve represents the Monte Carlo signal plus background prediction.

$0.4 < Q^2 < 1.1$ $(\text{GeV}/c)^2$. The simultaneous fit to both differential cross sections yielded

$$\sin^2\theta_W = 0.218 \, {}^{+0.039}_{-0.047}$$

and

$$M_A = 1.06 \pm 0.05 \, \text{GeV}/c^2,$$

where the errors represent a 67% rectangular confidence area. The best-fit curves are shown in FIGURE 14. The value of M_A is in good agreement with the world average value of $M_A = 1.032 \pm 0.036 \, \text{GeV}/c^2$.

A search for additional terms in the axial vector current $G_A(Q^2)$ with $\sin^2\theta_W$ fixed at 0.22 and M_A constrained to the world average value yielded $\eta = 0.12 \pm 0.07$, or equivalently $0.00 < \eta < 0.25$ at 90% C.L., independent of the value assumed for $\sin^2\theta_W$. Here, η represents either a heavy quark contribution to the standard weak axial vector current or a "nonstandard" primitive axial isoscalar current in the parameterization $G_A(Q^2) = (\tfrac{1}{2})g_A(0)(1 + \eta)/(1 + Q^2/M_A^2)^2$.

To exhibit the internal consistency of the data and to extract the most precise value of $\sin^2\theta_W$, M_A was constrained at the world average value and $\sin^2\theta_W$ was calculated in all ways combining ν_μ and $\bar{\nu}_\mu$ data. The resulting most precise value was

$$\sin^2\theta_W = 0.220 \pm 0.016 \text{ (stat)} \, {}^{+0.023}_{-0.031} \text{ (syst)}.$$

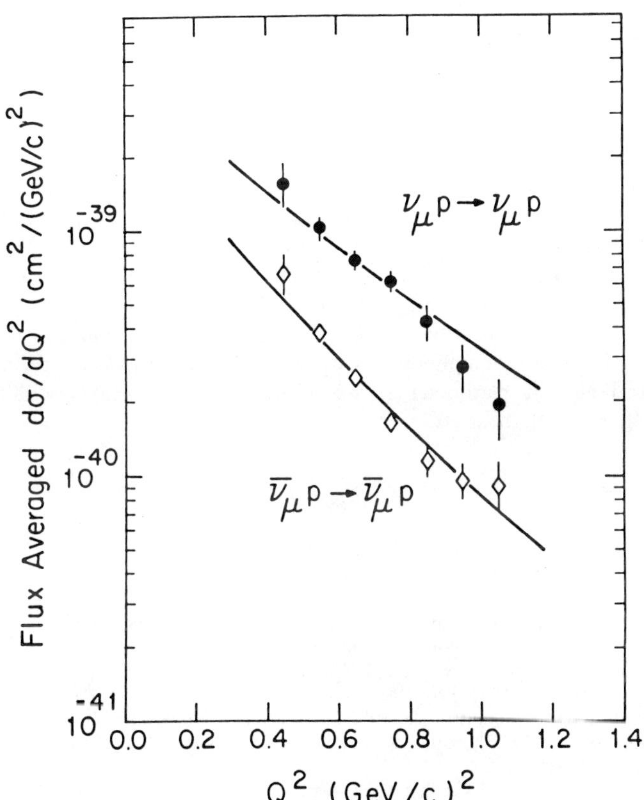

FIGURE 14. The data points are the measured flux-averaged differential cross sections for $\nu_\mu p \rightarrow \nu_\mu p$ and $\bar{\nu}_\mu p \rightarrow \bar{\nu}_\mu p$ from this experiment. The solid curves are best fits to the combined data with the values, $M_A = 1.06$ GeV/c^2 and $\sin^2\theta_W = 0.220$. This fitting procedure imposes adjustments of the solid curves by scale factors of 1.05 for $\nu_\mu p$ and 1.09 for $\bar{\nu}_\mu p$. These values are consistent with the absolute scale uncertainty of approximately 11% in each of the individual cross sections that was included in the fitting procedure. The error bars represent statistical error and also include Q^2 dependent systematic errors.

$\nu_\mu e$ *and* $\bar{\nu}_\mu e$

The fully corrected ratios of the normalized cross sections for $\nu_\mu e$ and $\bar{\nu}_\mu e$ scattering directly yield the ratio

$$\frac{\sigma(\nu_\mu e)}{\sigma(\bar{\nu}_\mu e)} = 1.38 \begin{array}{c} +0.40 \\ -0.31 \end{array} \text{(stat)} \pm 0.17 \text{ (syst)},$$

where the precision is limited by the numbers of $\nu_\mu e$ and $\bar{\nu}_\mu e$ events. The resulting value of $\sin^2\theta_W$ is

$$\sin^2\theta_W = 0.209 \pm 0.029 \text{ (stat)} \pm 0.013 \text{ (syst)},$$

which is in good agreement with the only previously published measurement[7] of $\sigma(\nu_\mu e)/\sigma(\bar{\nu}_\mu e)$ and the value of $\sin^2\theta_W$ obtained from it, specifically, $\sin^2\theta_W = 0.215 \pm 0.032$ (stat) ± 0.013 (syst).

CONCLUSIONS

The values of $\sin^2\theta_W$ determined from the measurements described here of $\nu_\mu p$ and $\bar{\nu}_\mu p$ elastic scattering and $\nu_\mu e$ and $\bar{\nu}_\mu e$ elastic scattering are in good agreement within experimental errors. Furthermore, these values are in good agreement with the values of $\sin^2\theta_W$ determined from the masses of the W and Z bosons,[8] from inelastic electron-deuteron scattering,[9] and from deep-inelastic neutral current neutrino reactions.[10] Therefore, over a wide range of Q^2 [10^{-2} to 10^4 $(\text{GeV}/c)^2$] and with significantly different assumptions and corrections in the various experiments, the weak neutral current parameter $\sin^2\theta_W$ is a universal constant (within present experimental errors of about 10%).

REFERENCES

1. AHRENS, L. A. *et al.* 1983. Phys. Rev. Lett. **51:** 1514.
2. AHRENS, L. A. *et al.* 1985. Phys. Rev. Lett. **54:** 18.
3. ABE, K. *et al.* 1986. Phys. Rev. Lett. **56:** 1107.
4. AHRENS, L. A. *et al.* 1987. Phys. Rev. D. (February 1 issue). To be published.
5. AHRENS, L. A. *et al.* 1987. Nucl. Instrum. Methods. To be published.
6. BERGSMA, F. *et al.* 1982. Phys. Lett. **117B:** 272.
7. AHRENS, L. A. *et al.* 1986. Phys. Rev. **D34:** 75.
8. RUBBIA, C. *In* Proceedings of the 1985 International Symposium on Lepton and Photon Interactions at High Energies, Kyoto, Japan. M. Konuma & K. Takahashi, Eds.: 242; DiLELLA, L. Ibid, p. 280.
9. PRESCOTT, C. Y. *et al.* 1978. Phys. Lett. **77B:** 347; 1979. Phys. Lett. **84B:** 524.
10. See, for example: FOGLI, G. L. 1985. Nucl. Phys. **B260:** 593.

Underground Muons from Cygnus X-3[a]

K. RUDDICK

School of Physics and Astronomy
University of Minnesota
Minneapolis, Minnesota 55455

During 1985, the University of Minnesota–Argonne group reported the observation of underground muons coming from the direction of Cygnus X-3, a binary star system thought to consist of a neutron star in very close proximity to a large companion star, and which is probably the most energetic source of high energy cosmic rays in the galaxy.[1,2] These data were obtained during 1982–83 in the Soudan I nucleon decay detector located at a depth of 1,800 mwe (meters water-equivalent) in the Tower-Soudan iron mine in northern Minnesota. Confirmation of the signal has been reported from the NUSEX detector at a depth of 5200 mwe under Mont Blanc.[3] The effect has not yet been established in any other detectors.[4]

The data were obtained by using the known periodicity of the source, which was taken from the very accurate X-ray observations of Cygnus X-3.[5] In the Soudan experiment, the total flux of muons was 60 ± 17 during 0.96 years live-time, corresponding to an average flux of 7×10^{-11} cm^{-2} s^{-1}. The NUSEX experiment observed 19 ± 4 events, which is a flux approximately ten times less than at Soudan. The emission of the muons occurs during only a short part of the 4.8-hour period of the source, centered at a phase of approximately 0.7 to 0.9. This same feature occurs in the observation of TeV cosmic rays from the object (assumed to be gamma rays) by Cherenkov detectors on the surface of the earth. It is assumed to arise from a beam of ions that is accelerated in the region of the neutron star and that interacts in the limb of the companion star.[6]

In order to reach the depths of the Soudan and NUSEX detectors, the energies of muons produced in the atmosphere must be at least 0.65 and 5.0 TeV, respectively. However, the muon fluxes at both Soudan and NUSEX are approximately the same as the gamma fluxes observed by the surface detectors at these energies. It then follows that the source of the underground muons cannot be photons: this would require each incident photon to produce a secondary muon of energy similar to the primary photon. Calculations have shown that the expected flux of muons from incident gamma rays is reduced by a factor of approximately 200.[7]

If these underground muon observations are correct, then no conventional particles can explain them. Neutrons have too short a lifetime. Muons produced by neutrinos would be observed with an approximately uniform zenith angular distribution, while the data would show a zenith dependence indistinguishable from the background muons (produced by charged nucleons incident on the atmosphere) at the present level

[a]The Soudan collaboration consists of H. Courant, K. Heller, S. Heppelmann, T. Joyce, M. Marshak, E. Peterson, K. Ruddick, and M. Shupe of the University of Minnesota, and D. Ayres, J. Dawson, T. Fields, E. May, and L. Price of the Argonne National Laboratory. The experiment was supported by the U.S. Department of Energy.

of statistical accuracy. Neutral atoms would be stripped of their electrons by material in the galaxy. Charged primaries coming from Cygnus X-3 thus retain no memory of their initial directions due to bending in the galactic magnetic field.

Before the discovery of these muons, the Soudan I detector had been turned off. This was done because it had become noncompetitive in the search for nucleon decay at a level of two orders of magnitude below that for which it was designed. It was reactivated at the beginning of 1985. Up to October 1985, no Cygnus signal was observed—either refuting the original data, or indicating a reduction in the flux by at least a factor of three over the period of one to two years since Cygnus was previously observed. The source is indeed known to be highly variable: it has been previously noted that the photon flux observed by the surface detectors may be decaying by a factor of two per year.[8]

During October 1985, radio astronomers observed a very large outburst from Cygnus X-3, with the flux increasing by approximately three orders of magnitude. FIGURE 1 shows the muon signal in the Soudan I detector during this period. The muons appear to precede the radio signal by approximately one week. They arrive within a very sharp phase interval, 0.725 to 0.745, which is much narrower than the previously observed 0.6 to 0.9.

The data were obtained by requiring that the muon directions lie within 5 degrees of the direction of Cygnus X-3. The previous Soudan data were taken from a cone of half-angle 3 degrees (chosen to optimize the signal-to-background ratio), while the NUSEX data were from a 10-by-10 degree region (in celestial coordinates) about Cygnus X-3. The greater signal-to-background ratio during the starburst enables this finite angular dispersion of the Soudan muons to be investigated in more detail. FIGURE 2 shows the angular spread of the muons at both Soudan (during the burst) and at NUSEX. The angular resolutions of the detectors are less than 1 degree, and multiple Coulomb scattering causes less than about 0.5-degree dispersion in the directions. We conclude that this finite spread in the directions of the muons is a real effect.

Several suggestions have been made about the nature of the possible particles (and of their interactions) that might explain the underground muons. An upper limit on the mass of the primary particle originating at Cygnus X-3 and traveling to Earth can be

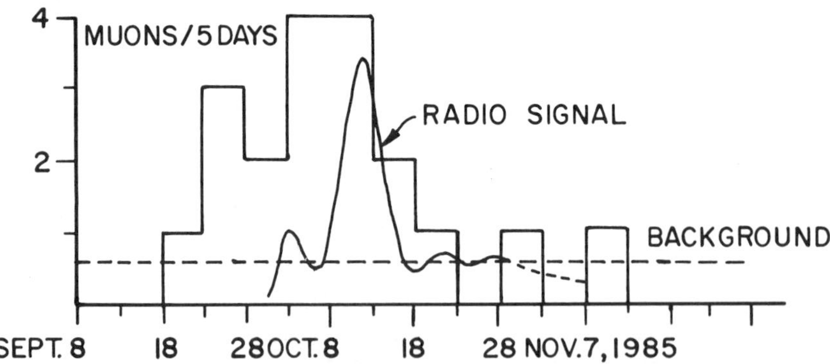

FIGURE 1. Muon flux at Soudan I during radio burst from Cygnus X-3 (muons within 5° of Cygnus direction, $0.725 < \phi < 0.745$).

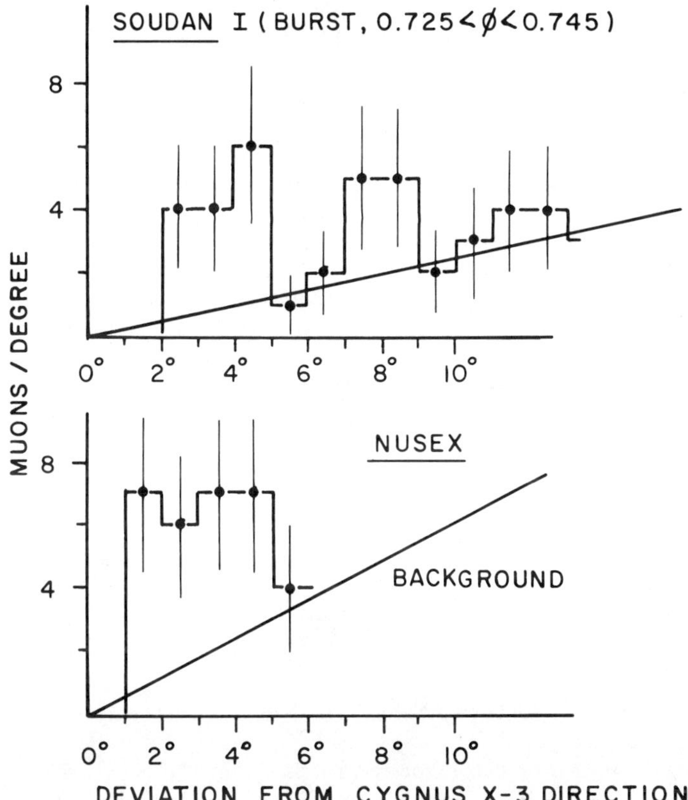

FIGURE 2. Angular distributions of muons associated with Cygnus X-3. (NUSEX data have been converted from celestial to local coordinates.)

obtained from the width of the peak (Δt) in the phase distribution: in order to maintain this peak over the transit time ($T = 10^{-12}$ sec) from Cygnus, the particle velocities must lie in a range $\Delta\beta/\beta$ less than $\Delta t/T$. For particles of mass m and energies above some minimum E, then

$$\frac{\Delta\beta}{\beta} = \frac{\Delta t}{T} \approx \frac{M^2}{2E^2}$$

which corresponds to masses less than about 30 MeV for Δt about 0.02×4.8 hours and $E = 1$ TeV.

None of the previous suggestions for the phenomenon have addressed the cause of the angular spread in the data. The width of the angular distributions are suggestive of the production and decay of a massive particle by the incoming primary. In order to explain this apparent angular size of the source, the relative rates observed at Soudan and NUSEX, and the observed zenith dependence of the signal, we shall show that fairly stringent limits can be placed on the mass of this secondary particle and on its production cross section. Its mass must be less than 40 GeV and the production cross section must lie in the range of 10–20 microbarns. Production cannot occur off electrons.

A particle of mass M and energy E decays to light particles at a median angle M/E. Because the angles observed at Soudan and NUSEX are similar, then the energies of both the primaries and the secondary muons producing the signal must be similar. The lower flux observed at NUSEX compared to Soudan can then only be explained by assuming that the mass M is produced in the rock above the detectors; the lower NUSEX rate is due to the attenuation of the primary beam. The average difference in slant depth between Soudan and NUSEX (3,000 to 4,000 mwe) must correspond to 2 to 3 interaction lengths, which is equivalent to a primary interaction cross section in the range of 10–20 microbarns. Such a cross section would also produce the observed zenith angle dependence of the data.

Limits on the possible mass of the produced particle can be obtained by simple kinematic arguments.[9] First, consider a primary particle of energy E interacting resonantly with a target particle of mass m to produce a secondary of mass M, which we assume is large. Thus, the total CM energy squared (s) is given by

$$s = 2\,Em = M^2.$$

The mass M then decays to a muon plus anything with the median angle of the muon given by

$$\theta = \frac{M}{E} = \frac{2m}{M}.$$

Typical muon energies will be on the order of $0.5E$ for two-body decay; they will be less for multiparticle decays.

Taking $O = 0.05$–0.1 radian (as observed at Soudan and NUSEX), we find M to lie in the range of 10–20 MeV for an electron target, with typical muon energies of hundreds of MeV. All such muons would stop in the detectors, which is not the case. We conclude that resonant processes off electrons cannot explain the effect, even when taking into account the finite velocities of electrons in the heaviest of atoms. For nucleon targets, the corresponding masses lie in the range of 20–40 GeV, with muon energies of hundreds of GeV, which can be accommodated by the observations.

This simple kinematic analysis is only slightly modified if we consider the possibility of production of the mass M in an n-body final state. For a primary differential energy spectrum falling as E (or steeper) and for a production cross section parameterized as

$$\sigma = \sigma_0 \left(\frac{M^2}{s}\right)^p \left(1 - \frac{M^2}{s}\right)^q, \qquad p = 0, 1, \quad q = 1, 3,$$

the most probable production occurs at $s = kM^2$, where k is typically less than four. The previous analysis is simply modified to yield

$$E_M = \frac{M^2}{2m}\sqrt{k},$$

$$\theta = \frac{M}{E_M} = \frac{2m}{\sqrt{k}}.$$

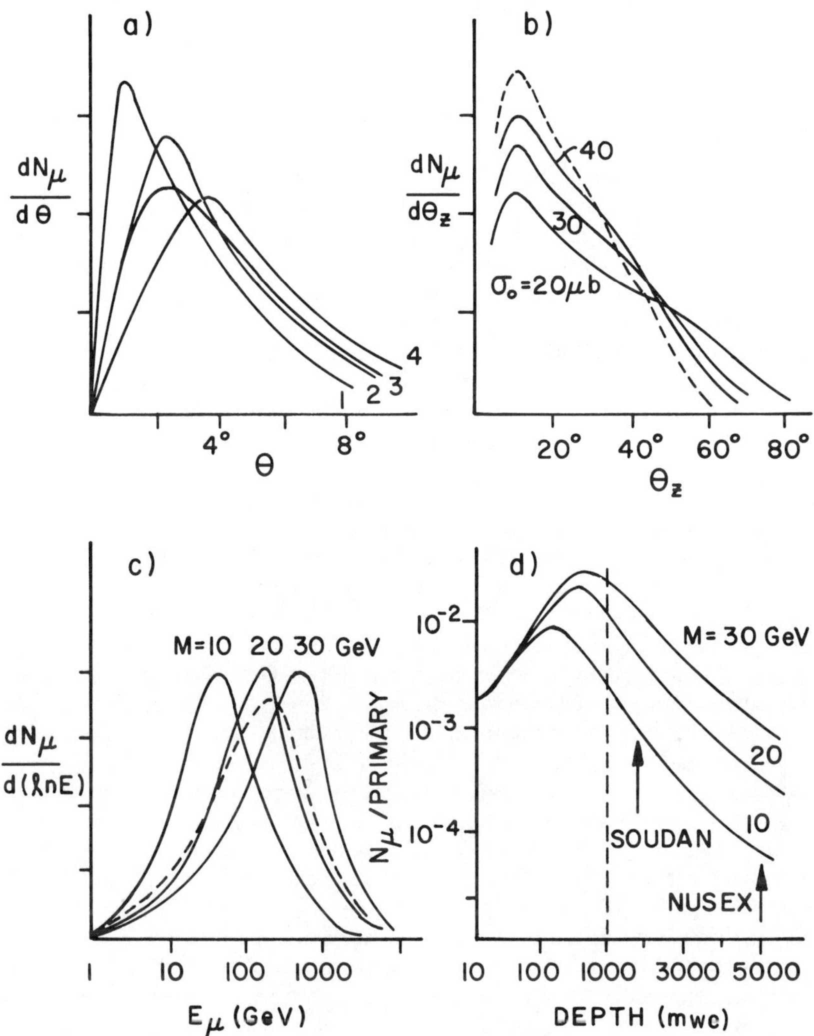

FIGURE 3. Typical Monte Carlo results: (a) Angular distributions at Soudan for $M = 15$ GeV, $\sigma = 30$ μb $(1 - M^2/S)$. The curves 1–4 are for decay at rest in CM, backward production with isotropic decay, primary spectrum E^{-3}, and spectrum E^{-3} with backward production, respectively; (b) Zenith distributions for $M = 15$ GeV, $\sigma = \sigma_0(1 - M^2/S)$. The dotted line is background; (c) Energy distributions of muons at Soudan assuming two-body decay of the mass M. The dotted line is background; (d) Depth dependence assuming $\sigma = \sigma_0(1 - M^2/S)$; E^{-2} is the primary spectrum.

Thus, for a nucleon target, the secondary particle mass must lie in the range of 10–40 GeV, and muon energies are several hundred GeV.

This simple analysis is adequate for depths less than approximately one interaction length of the primary; at greater depths, the energy dependence of the interaction cross section will distort the energy spectrum. A more careful analysis requires a Monte Carlo calculation. FIGURE 3 shows the results of such a calculation. Some typical muon angular distributions are shown in FIGURE 3a. A feature of the predictions is that the median muon angle at NUSEX can be greater or less than at Soudan, depending upon whether the primary interaction cross section rises monotonically from threshold: a continually rising cross section results in lower average interaction energies at greater depths, and thus to a greater median decay angle. FIGURES 3b and 3c show typical predicted zenith angle distributions and muon energy spectra (assuming two-body decay) at Soudan, along with the background shapes. These are clearly indistinguishable with the present data.

We have also considered diffractive-type interaction cross sections, which are attractive because of the possibility of an enhanced coherent cross section. These cross sections rise more slowly than those previously considered and have effective thresholds much higher. Solutions are obtained for secondary masses in the range of 1–4 GeV with possible muon energies of typically 5–20 GeV. Up to 20% of these muons should stop in the detectors, and with only slightly better data than presently exists, they could be excluded as possible explanations of the phenomenon.

Such models as that presented here, with a primary interaction in the rock as opposed to the atmosphere, obviously predict a very different depth dependence of the signal. The optimum depth for observation of the signal is at approximately the average range of the decay muons. This is illustrated in FIGURE 3d. At the surface of the earth, the maximum signal occurs at very large zenith angles due to the large interaction length of the primaries.

In summary, we have demonstrated that the production of an intermediate particle of mass less than approximately 40 GeV (and which subsequently decays to at least one muon) is capable of explaining all of the features of the underground muons associated with Cygnus X-3. Unfortunately, the typical muon energies and zenith angular distributions predicted by the model are very similar to those of the background; considerably more data will be required to prove or disprove the model. Until the very large detectors—Frejus and especially Soudan II—are able to detect a signal, the most likely early verification of the model would come from observation of large zenith-angle muons by detectors situated at, or near, the surface of the earth. The fluxes there will likely be smaller than those in Soudan I, and thus large detectors will be necessary.

REFERENCES

1. MARSHAK, M. *et al.* 1985. Phys. Rev. Lett. **54:** 2079.
2. MARSHAK, M. *et al.* 1985. Phys. Rev. Lett. **55:** 1965.
3. BATTISTONI, G. *et al.* 1985. Phys. Lett. **155B:** 465.
4. OYAMA, K. *et al.* 1986. Phys. Rev. Lett. **56:** 991.
5. VAN DER KLIS, M. & J. M. BONNET-BIDAUD. 1981. Astron. Astrophys. **95:** L5.
6. HILLAS, A. M. 1984. Nature **50:** 312.
7. STANEV, T., T. GAISSER & F. HALZEN. 1985. Phys. Rev. **32:** 1244.
8. HALZEN, F. 1985. Int. Conf. on HEP, Bari, Italy.
9. This is also remarked upon in reference 8.

Cosmic Accelerators

FRANCIS HALZEN

Physics Department
University of Wisconsin
Madison, Wisconsin 53706

INTRODUCTION: γ-RAYS FROM COSMIC SOURCES AND THEIR DETECTION

An incomplete compilation of fluxes measured by experiments[1] observing γ-rays emitted by the source Cygnus X-3 is shown in FIGURE 1. This source is observed in radio, infrared, MeV γ-ray, and X-ray experiments and emits photons of energies all the way up to 10^5 TeV, with the Haverah park observations (open circles) suggesting a cutoff at that energy. The integral flux of γ-rays above 1 TeV is well fitted by a normal emission spectrum:

$$F(> E) \text{ [in particles cm}^{-2}\text{ sec}^{-1}] = \frac{4 \times 10^{-11}}{E \text{ [in TeV]}}. \tag{1}$$

The experiments in FIGURE 1 measured fluxes averaged over long periods of time (days to years), but bursts have been observed where the flux periodically increases. About once a year, when radio bursts of a duration of a few weeks are observed, the flux may increase by over two orders of magnitude. Some TeV-experiments have identified epochs of increased activity of a few minutes of duration. The emission of the Cygnus beam has a very rich time structure that has not been deciphered. Periodicities of 4.8 hours, 19 days, and 1 year have been suggested. Only the 4.8-hour bunching of the beam is established, although we know that the actual high energy emission only occurs very sporadically in these 4.8-hour periods.

The primary flux of photons given by equation 1 is not directly observed. Photons originate from an electromagnetic shower in the atmosphere that is 25 radiation lengths λ_R thick, with $\lambda_R \simeq 37$ g/cm^2. A primary photon converts to an e^+e^- pair after 1 radiation length near the top of the atmosphere. In subsequent radiation lengths, the electromagnetic particles lose further energy by bremsstrahlung $e^\pm \rightarrow \gamma e^\pm$ or pair production $\gamma \rightarrow e^+e^-$ (see FIGURE 2). The particles lose approximately one-half of their energy per radiation length. It is straightforward to calculate the flux of photons as a function of depth in the atmosphere using linear shower theory.[2] For the primary E^{-1} spectrum (equation 1) at the top of the atmosphere, we obtain

$$N_\gamma(> E, z) \simeq \tfrac{1}{2} A E^{-1}. \tag{2}$$

Here, $A = 4 \times 10^{-11}$ in the units of equation 1, and z is the depth in the atmosphere described in terms of linear column density. The number of photons of a given energy E is independent of depth. This somewhat surprising result can be understood as follows: while photons lose energy with depth, others of energy E are generated by showers with $E' > E$. Energy loss and feed-down are in equilibrium for a "normal" E^{-1} spectrum;

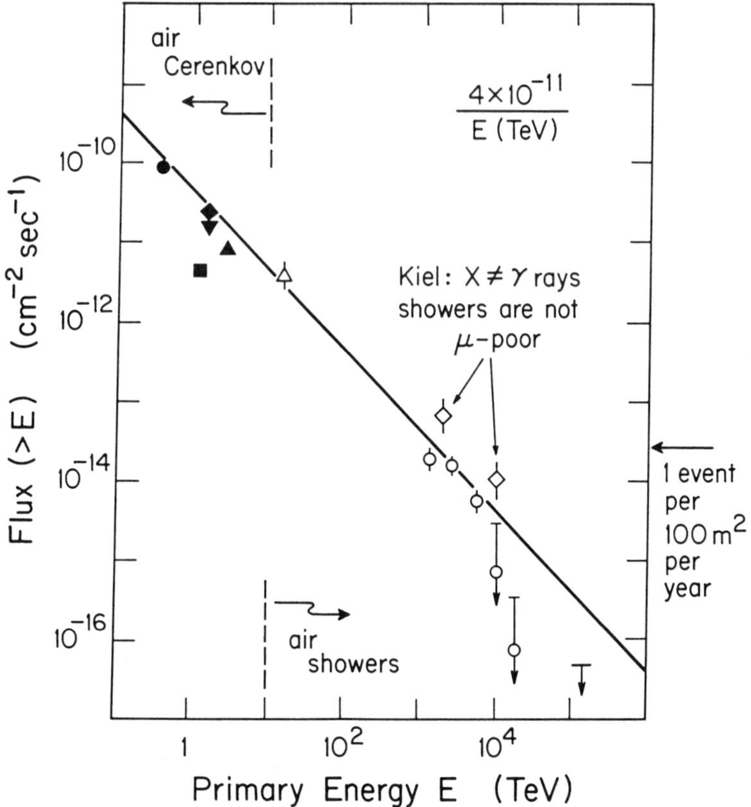

FIGURE 1. Integral flux of very high energy particles (X) from the X-ray binary Cygnus X-3. We will call these particles γ-rays, although the Kiel experiment challenges this identification. Note the flatter E^{-1} dependence compared to the E^{-1} falloff of the cosmic ray flux.

hence, the z-independent result of equation 2. Every radiation length of atmosphere contains the same number of photons of energy E. The number of radiation lengths, of course, is limited to some number n_{max} that is determined by the maximum energy of the source (10^5 TeV for Cygnus; see FIGURE 1) and by the fact that a photon will lose one-half of its energy per radiation length; therefore,

$$n_{\text{max}}\ (\text{radiation lengths}) \simeq \frac{\ln\,(E_{\text{max}}/E)}{\ln\,(2)}.\tag{3}$$

For example, $n_{\text{max}} \simeq 15$ for 1-TeV γ-rays. It is also important to remember that roughly equal numbers of electrons are present in each radiation length (see FIGURE 2).

The atmospheric experiments view this shower as a pancake of electromagnetic radiation $10^2 \sim 10^3$ m^2 in area, a few nanoseconds thick, moving down the atmosphere at the speed of light. The enhancement of emission of γ-rays in source direction over the isotropic background of hadron-induced cosmic ray showers is established by timing or on/off-source subtraction.

The most prominent feature to set the on-source γ-rays apart from background hadron showers is their low muon content. This is not exploited in present experiments.[3] The number of muons in a γ-shower is typically a few percent of that in a hadron shower in which muons are abundantly generated by the decay of the produced $\pi^{\pm}$. In γ-initiated showers, processes resulting in muons are characterized by small cross sections: π photoproduction followed by $\pi \rightarrow \mu\nu$ decay, production and subsequent decay of charm quarks, and $\gamma \rightarrow \mu^{+}\mu^{-}$ pair production, which is suppressed by a large factor $(m_e/m_\mu)^2$ relative to $\gamma \rightarrow e^{+}e^{-}$. The muon production through these channels has been the subject of a Monte Carlo study.[4] An illustrative result is shown in FIGURE 3, where the number of muons with energy in excess of 1 GeV is plotted as a function of shower size (number of electrons N_e) for γ- and nucleon-initiated showers. However, the result is hardly surprising: muons are the progeny of hadrons, and photons are hadronic at $O(\alpha)$ and electrons are hadronic at $O(\alpha^2)$; therefore, the number of muons in an electromagnetic shower is a few percent of the number in a proton shower. The fluctuations in the number of muons for a fixed primary photon

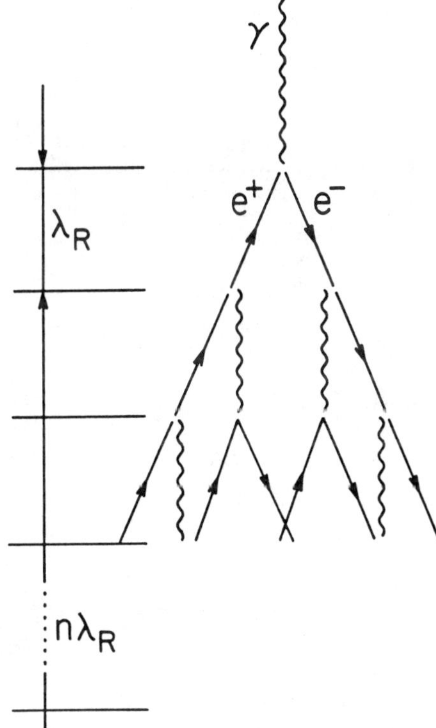

FIGURE 2. The γ-ray initiated cascade in the atmosphere. Each layer represents a radiation length. The number of layers is limited by the energy cutoff of the source around 10^5 TeV (see FIGURE 1).

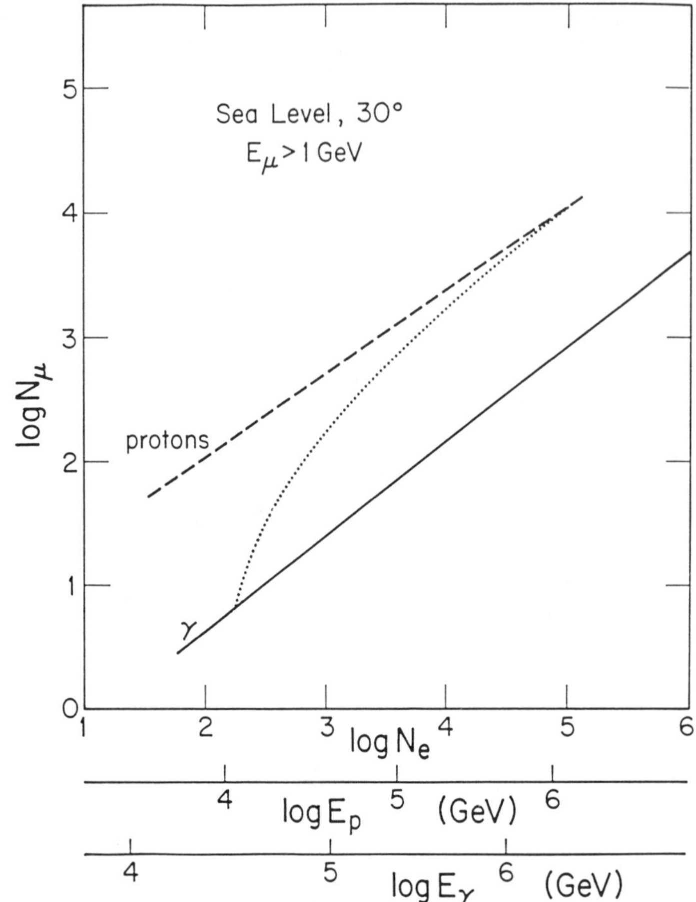

FIGURE 3. Muon number (N_μ) versus shower size (N_e) for γ- (solid line) and proton- (dashed line) showers.[4] The dotted line shows the transition of γ-showers to hadron showers due to a strong coupling process. The energy scales show the conversion of shower size to proton and γ-ray energy.

energy are very large because photoproduction is a rare event. This is to be contrasted with the muon number in a hadron shower, which is relatively stable.

THE COSMIC ACCELERATOR AND ITS IMPACT
ON THE COSMIC RAY PUZZLE

A useful visualization of the Cygnus X-3 accelerator is shown in FIGURE 4. Cygnus is a binary system of a compact star and a regular main sequence star companion. The binary period is 4.8 hours. The compact star is most likely a pulsar (a 12-msec period has been tentatively identified[5]) that accelerates particles up to 10^5 TeV in the 10^{12} gauss magnetic fields near its polar caps. Twice per binary cycle will this beam interact with the atmosphere of the companion as shown in FIGURE 4. Twice per cycle will the system turn into a beam dump experiment with

$$p \longrightarrow \pi^0 \longrightarrow \gamma,$$

$$p \longrightarrow \pi^\pm \longrightarrow \nu.$$

Neutral γ's and ν's—decay products of π^0 and $\pi^\pm$ produced by the proton beam dumped into the atmosphere of the companion—are beamed towards earth. Cygnus is at the edge of our galactic plane, more than 10^4 parsecs away from earth. Charged secondaries produced in the dump are inevitably deflected by the intergalactic magnetic field.

In all but a few experiments, the signal can only be established by exploiting the fact that it is a 4.8-hour pulsed signal on top of a continuous background. It is customary to define a phase angle from 0 to 1 that traces the binary revolution as shown in FIGURE 4. The 0 corresponds to the minimum in X-ray emission that results from the eclipse of the pulsar by the companion star. The "beam-on" positions in the model of FIGURE 4 roughly correspond to phases of 0.2 and 0.8.[6]

To a particle physicist, beams of 10^5 TeV might sound out of the ordinary, but the

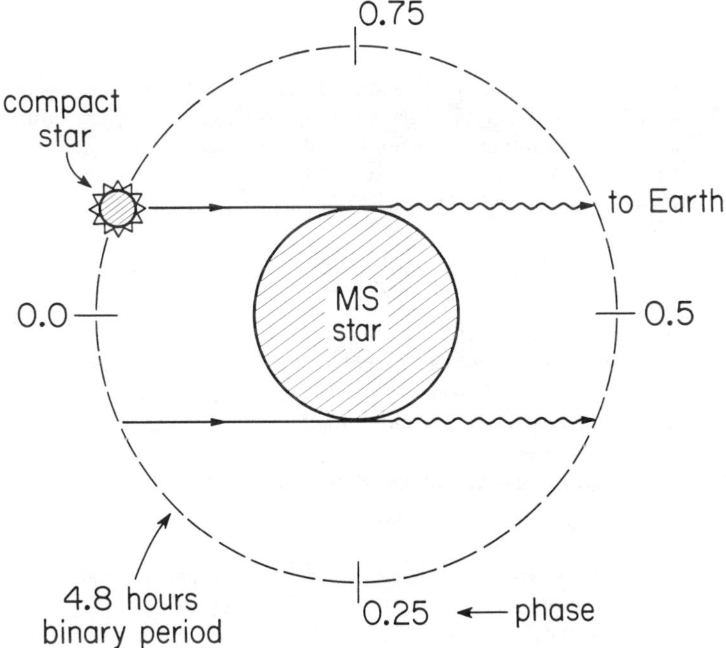

FIGURE 4. The Cygnus accelerator beams particles to earth produced in the beam dump at phases of 0.2 and 0.8 of the pulsar (HILLAS, A. M. 1984. *Nature* **312:** 50, and references therein).

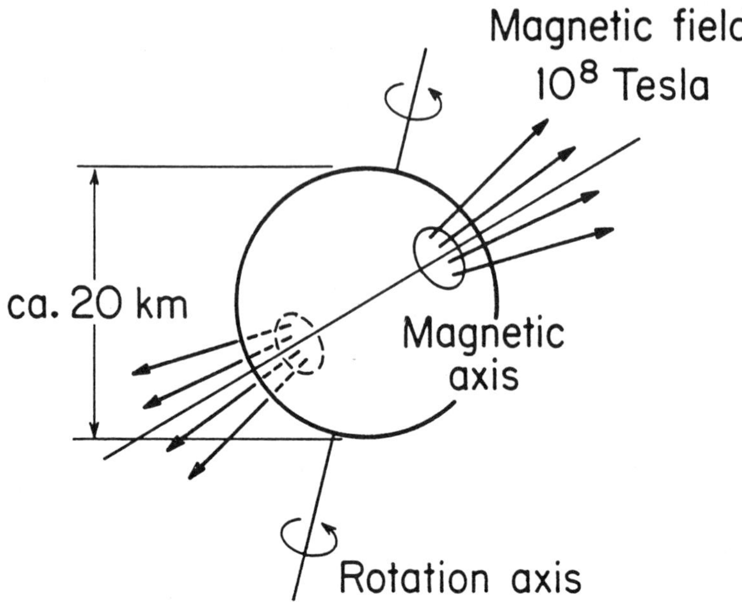

FIGURE 5. The typical pulsar.

basic reason why such energies are achieved can be understood on the basis of dimensional arguments. The typical pulsar shown in FIGURE 5, 10 km in diameter, emits particle beams like a lighthouse along the dipole magnetic axis. Field values can reach 10^{12} gauss. It spins with pulsar periods of $10^{-3} \sim 1$ sec around an axis that does not coincide with the magnetic axis. The EMF of such a system is

$$\mathcal{E} = Blv = Bl\frac{l}{t} = 10^{16} \sim 10^{19} \text{ eV}. \tag{4}$$

More intriguing is the energy output of the system. It can be calculated from the flux observed in our earth-based apparatus by making the following corrections: (i) we only catch a fraction of the $4\pi R^2$ emission; (ii) we have to take into account the duty cycle of the accelerator (see FIGURE 4); (iii) only a fraction of the energy goes into $p \rightarrow \pi^0 \rightarrow \gamma$; and (iv) γ's are absorbed on the 3 K background along the way. Many of these corrections are model-dependent, but one conservatively estimates that

$$L > 10^{39} \text{ ergs sec}^{-1}, \tag{5}$$

that is, more than 10^6 times the total energy output of the sun. The data violates the Eddington limit for acceleration by electromagnetic processes.

The discovery of this point source in the very high energy spectrum might have solved the fifty-year-old problem of the origin of cosmic rays. The cosmic ray spectrum can be understood up to 10^2 TeV in terms of shock wave acceleration in supernova remnants. Although the spectrum shows a kink at 10^2 TeV (see FIGURE 6), cosmic rays with higher energies are observed and cannot be accounted for by this mechanism. The flux in FIGURE 1, however, is of the same order of magnitude as the total cosmic ray flux above 10^2 TeV. Therefore, Cygnus (and a few other sources emitting TeV γ-rays such as Hercules X1, the X-ray binary 4U 0115 + 63, and the Crab pulsar, PSR 1953 + 29) could very well be responsible for all high energy cosmic rays not accounted for by supernovas.[7]

PARTICLE PHYSICS WITH COSMIC ACCELERATORS?

Whereas Cygnus X-3 is possibly closing a chapter in cosmic ray physics, it might be opening a new one in particle physics. As FIGURE 2 suggests, Cygnus is truly a HERA in the sky.[8] The direction and characteristic time structure of the radiation from such sources can be used to tag a beam of known composition (γ-rays) with well-understood interactions (described by QED) with the target atmosphere. An experiment using a tagged cosmic photon beam, therefore, overcomes the classic hurdles in interpreting cosmic ray data. If a new physics threshold is observed in the experiment, it cannot be due to some "mundane" change in the chemical composition of the cosmic ray primaries or of their hadronic interaction with atmospheric nuclei.

The accelerator is free. The question is what the detector requirements are to achieve statistics and signal/noise to possibly probe the much anticipated new structure in particle physics at a scale of $(\sqrt{2}\, G_F)^{-1/2} \simeq 0.25$ TeV. As we have seen, the most discriminating signature of the photon-induced shower against the hadron-

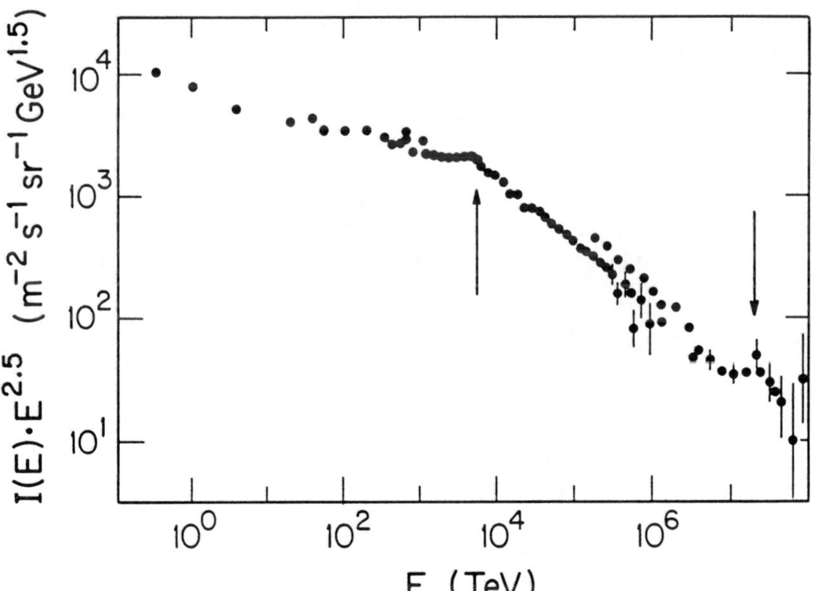

FIGURE 6. Cosmic ray spectrum compiled by Hillas.

induced cosmic ray background showers is the scarcity of muons. The two essential components of a tagged photon experiment are therefore obvious:

(*i*) an array of scintillation counters with fast timing or a lattice of simple single mirror Čerenkov telescopes giving an observation of the electromagnetic component of the shower with good angular resolution on the arrival direction;

(*ii*) an array of shielded counters identifying the content of (GeV) muons in the shower.

Both goals could be achieved by having vertically stacked counters in coincidence with different shielding. The array can be triggered on specific timing sequences of the counters that correspond to showers from one or even multiple source directions. From equation 1, we can see that at 1-TeV energy, even a "modest" $(100 \text{ m})^2$ array can accumulate 10^5 γ-rays. The number of electrons/muons at sea level, however, is insufficient to perform a realistic experiment. With a 100-TeV threshold, this problem is solved, but it now takes a $(1 \text{ km})^2$ array to accumulate the 10^5/year statistics required to take a detailed look at photoproduction under difficult experimental circumstances. Some relevant information is summarized in TABLE 1. Such a facility[9] would still constitute a not-too-far extension of present facilities such as the Akeno array in Japan. One should, however, be aware of the possibly severe punch-through problems when identifying GeV muons. Not just hard electromagnetic particles contribute; softer ones have a low probability to punch through, but they have very large multiplicity. For a detailed discussion, see Gaisser *et al.* in reference 4. What about signal-to-background?

Present detectors (e.g., Čerenkov) telescopes achieve a nominal

$$\frac{S}{N} = 10^{-2}. \tag{6}$$

This is clearly insufficient if we hope to detect traces of new particle thresholds in the tagged photon beam. This number can be greatly improved in the future when the detailed emission time structure of sources like Cygnus X-3 is well known. The binary 4.8-hour and candidate 19-day, 12-month (?), and 5-year burst (?) repetition rates of the source can be used to enhance the S/N possible by 10^2 using phase information. The μ-poor property of γ-showers further enhances the signal by another factor of 10^2, but not more, because at that level, γ-showers photoproducing a π in the primary interaction are virtually indistinguishable from a hadron-induced background cosmic ray cascade. With possible improvements in angular resolution, $S/N \simeq 10^2 \sim 10^3$ can possibly be achieved. However, preliminary data[10] from the Haleakala Čerenkov

TABLE 1. Properties of Hadron- and Photon-Induced Air Showers at Akeno[a]

Primary Energy (TeV)	1	10^2
γ-flux from Cyg X-3 using equation 1 (km^{-2} year^{-1})	10^7	10^5
Number of electrons in hadron shower	—	3,800
Number of muons in hadron shower ($E > 1$ GeV)	30	1,525
Number of muons in γ-shower (see also FIGURE 3)	—	40

[a]Depth = 920 g/cm^2.

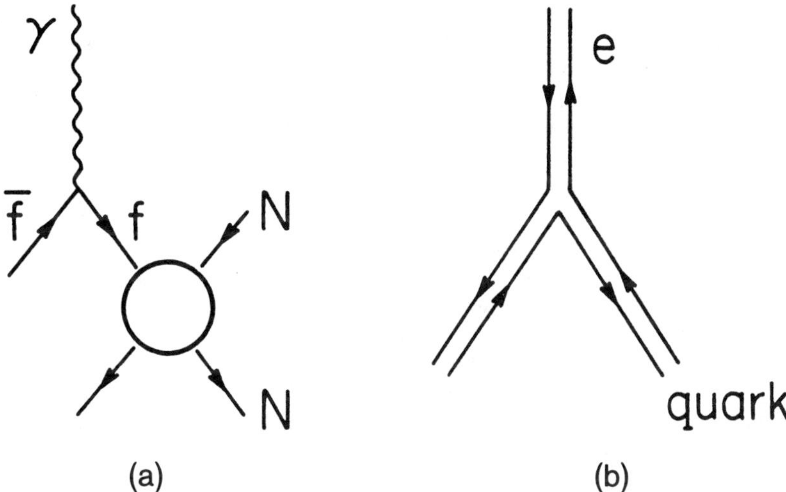

FIGURE 7. (a) Photon interacting with matter via a new fermion f. (b) Colored preons converting leptons to quarks.

telescope suggest emission from Cygnus in bursts with a typical duration of one minute. During these bursts, $S/N \simeq 1$ rather than 10^{-2}. Tagging such bursts could therefore yield data at the $10^4 \sim 10^5$ signal-to-noise level.

It is exciting to contemplate the feasibility of such experiments as a not too farfetched extrapolation of present facilities (e.g., Akeno). Even if no other than routine particle physics results are ever obtained from such experiments, imagine their power as a telescope. It is also very easy to convince oneself that deep underground experiments observing TeV-muons rather than GeV-muons would not be competitive.[11]

New physics can reveal itself in the experiment in two distinct ways: anomalous interactions of the photons or electrons in the atmospheric cascade (see FIGURE 2), or admixture in the γ-beam of new neutral particles produced in the cosmic beam dump where the primary energies reach 10^5 TeV (see FIGURE 4). We illustrate both possibilities. In FIGURE 7, examples are shown on new interactions of photons and electrons with matter. In FIGURE 7a, the photon couples to nucleons via new fermions f. Such interactions are certainly associated with heavy quark photoproduction with $f = c, b, t, \ldots$, and $\alpha_f - \alpha_s$. They have already been included in the calculations of the muon content of γ-showers shown in FIGURE 3 and would be observable in the more ambitious versions of the experiment previously described. Speculation that photon interactions become strong at high energies[12] (though unpopular in the context of gauge theories) could be easily investigated. If at some energy, $\alpha_f \simeq 1$, then photon showers would eventually resemble hadron showers. The dotted line in FIGURE 3 shows the result of a Monte Carlo study of the muon number dependence on energy for a new strong coupling process with 0.25-TeV threshold.

It is important to realize that electromagnetic cascades contain roughly equal numbers of electrons and photons. Composite models of quarks and leptons provide us with an example where new physics in the electron-nucleon interaction will also affect the development of the γ-initiated cascade. At very high energy, electrons have a finite size in which pointlike preons move around. They can interact with quarks (hadrons) via the exchange of preons. If these are colored, electrons can turn into quarks and they begin to interact like hadrons (see the preon diagram of FIGURE 7b). A drastic increase of the muon yield will result.

In both of the examples shown in FIGURE 7, photon showers will turn mixed electromagnetic–hadronic at high energy, with enhanced muon production signaling the onset of the hadronic component. The Kiel experiment,[13] showing an increase of the muons in the Cygnus showers by at least a factor of ten over the QED calculation, could be interpreted along these lines.

The photino ($\tilde{\gamma}$) or gluino ($\tilde{g}$) supersymmetric partners of the photon and gluon, respectively, provide us with postulated particles that should be produced by cosmic accelerators and that could be detected as admixtures in the γ-beam. Calculations are straightforward and are relatively insensitive to the detailed properties of the binary system.[14] Given the observed γ-emission, their relative abundance is dictated by particle physics. One can show[14] that

$$\frac{dN\tilde{g}/dE}{dN\gamma/dE} \simeq \frac{\sigma(p \to \tilde{g})}{\sigma(p \to \pi)} \lesssim 10^{-2}, \tag{7}$$

which corresponds to a one-percent gluino admixture for O (GeV) masses. For heavier gluinos, the flux is suppressed following equation 7. If the photino is the lightest stable supersymmetric particle, the gluinos will decay by $\tilde{g} \to \tilde{\gamma}q\bar{q}$, and the resulting $\tilde{\gamma}$ flux will be further suppressed. Signatures of supersymmetric particles will be discussed further on. To a good approximation, however, a stable gluino or photino would, respectively, interact like a hadron and neutrino with the atmosphere. Heavy squarks, though, could appear in the final state, thus making detection of small supersymmetric admixtures possible. We shall have more to say about that later.

MUONS FROM CYGNUS X-3: WHO ORDERED THAT?

By last count, 15 different experiments[1,10] have identified air showers in direction and with the characteristic time structure of the source Cygnus X-3. Some have made repeated observations and have found signals from astronomically similar binaries, for example, Hercules X-1. Although a wealth of puzzling questions remain to be resolved, a signal at the flux level given by equation 1 has solid experimental support. In addition, even though γ-rays are ineffective at producing muons, the very high energy ones in the spectrum shown in FIGURE 1 will generate TeV muons, and these will produce a calculable signal in underground (proton decay) detectors. A proton decay detector, therefore, is also an underground muon telescope. Deep underground muons remember the original γ-ray direction with an angular precision of

$$\theta(\text{in mrad}) \simeq \frac{\pi}{\sqrt{E_\mu(\text{in TeV})}}. \tag{8}$$

We computed[15] this signal for the Soudan and NUSEX detectors. Their experimental thresholds are such that they are mostly sensitive to primary particles with roughly 5 TeV and 35 TeV energies, respectively. The calculated muon fluxes are shown in FIGURE 8 (labeled $X = \gamma$), where they are compared with the parent γ-flux and the published observations[16] of in-direction and in-phase muons from Cygnus X-3. The disagreement between calculated and observed flux is roughly three orders of magnitude for each experiment. It is possible that underground detectors rediscovered an old puzzle. The Kiel air shower array experiment,[13] after detecting a 4σ enhance-

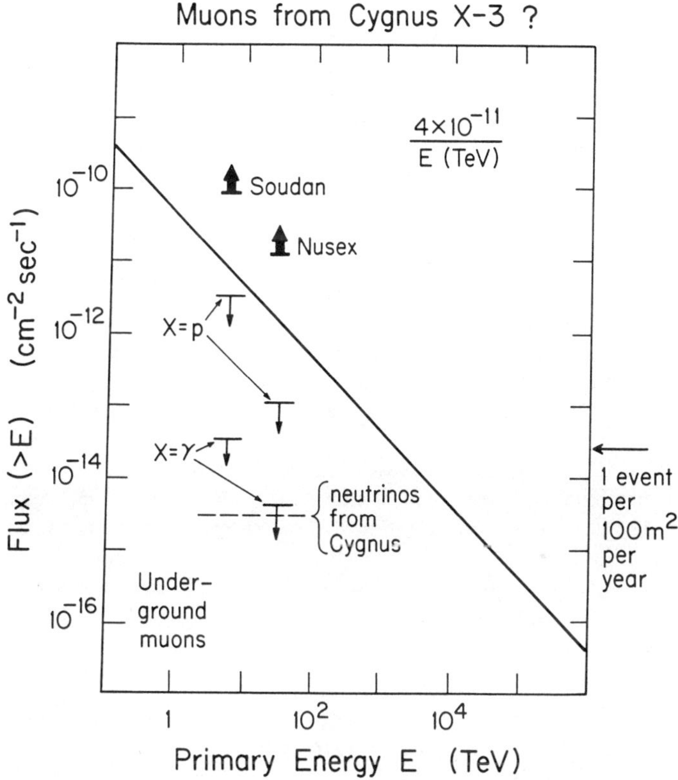

FIGURE 8. Observed and calculated underground muon fluxes assuming that the source of the muons is γ-rays or hadron showers in the atmosphere or neutrinos directly from Cygnus.

ment of the cosmic ray flux in the direction of Cygnus X-3, performed two tests to confirm the signal. They checked that on-source showers indeed remember the 4.8-hour binary period, but they also found that the showers are not μ-poor as expected from γ-ray emission. The expected 2% muon content relative to the hadron-induced background was observed to be 70%. Wdowczyk and collaborators found a similar result[17] for showers from the Crab pulsar. Is our straightforward assumption that FIGURE 1 represents the high energy tail of the electromagnetic emission spectrum to be questioned? Do cosmic accelerators emit particles (referred to as X from now on) other than photons?

What if X were a neutral hadron, for example, a neutron? We can repeat the previous calculation assuming the atmospheric signal (equation 1) is due to hadrons (remember that atmospheric experiments identify showers and not the nature of the primary particle). As expected, about 10^2 times more muons are predicted, but the assumption that X is a hadron falls short of accommodating the data by more than one order of magnitude for Soudan and two for NUSEX (see FIGURE 8). A detailed discussion of these calculations can be found in reference 15. Let me just say here that such shower calculations have been calibrated with respect to a variety of routine cosmic ray measurements. Their margin of error is possibly "a factor," but it is difficult to conceive them to be in error by an order of magnitude.

An alternative candidate for a "conventional" explanation of the underground muon signal is that X is a neutrino. As discussed in relation to FIGURE 4, Cygnus X-3 is expected to emit neutrinos. When arriving on earth, neutrinos basically ignore the atmosphere and interact with the rock above an underground muon detector. This is a direct result of their weak interaction cross sections with matter. However, after interacting, they produce muons by conventional $\nu \rightarrow \mu$ charge current interactions that yield a signal in the underground detector. The neutrino and (associated muon) flux can be readily calculated in the beam dump model of FIGURE 4 as their $\pi^\pm$ parent flux should just be twice the π^0 parent flux of the "photons." The result of such a calculation[18] is shown in FIGURE 8. Neutrinos fall short of the data by about 10^4. One could conceivably modify this calculation by increasing the beam luminosity of the compact star in FIGURE 4. This could increase the $\pi^\pm \rightarrow \nu$ duty cycle of the source without increasing the $\pi^0 \rightarrow \gamma$ yield if there is absorption of the γ's in the companion's atmosphere. One has to be careful, however, because a too-intense beam will critically heat the companion when the compact star eclipses the beam in 0-phase position in FIGURE 4. The heating is the result of neutrinos depositing energy in the core of the companion. In the end (see reference 19), it is still impossible to explain the underground signal in terms of neutrinos. Given the Kiel and underground muons results, the situation is now desperate. How desperate it is will be illustrated next by proving that the particle X cannot exist.

EXPLAINING CYGNUS X-3 MUONS IN TERMS OF A NEW PARTICLE: NO-GO THEOREM

Let us start by reiterating that we have already excluded that X is a particle with regular hadronic interactions (proton, neutron, nucleus, . . .), a photon, or a neutrino. Let us backtrack and list the properties of X implied by the observations:

(i) X is neutral: charged particles forget direction and time (phase) in the 3-μ gauss intergalactic field. The 10^4-parsecs distance represents more than 10^4 gyroradii for particles with rigidity less than 10^3 TeV. This would exclude nuclei and protons anyway.

(ii) The 4.8-hour bunching of the Cygnus beam would be lost after $L^* = 10^4$ parsecs unless the γ factor is large enough. The time delay between the arrival of two particles with velocities v_1 and v_2 that left Cygnus X-3 at the same time is

$$\delta = \frac{L^*}{v_1} - \frac{L^*}{v_2}$$

or

$$\delta t \simeq \frac{L^*}{2c}\left[\frac{1}{\gamma_1^2} - \frac{1}{\gamma_2^2}\right] \lesssim \frac{1}{2}\text{ hour.} \tag{9}$$

In equation 9, we imposed the condition that the muons arrive within a rather narrow time window—approximately, a half hour as observed by the experiment.[16] From equation 9, we conclude that the muon parents must be nearly monoenergetic (which is inconceivable in the type of models sketched in FIGURE 4) or that the Lorentz factors must satisfy the bound

$$\gamma = \frac{E}{M(X)} > 10^4 . \tag{10}$$

Therefore,

$$M(X) < 10^{-4}\, E. \tag{11}$$

As for Soudan, $E \simeq 10^4$ GeV, so we conclude conservatively that

$$M(X) \lesssim \text{a few GeV.} \tag{12}$$

(iii) The lifetime of X must be sufficient to cover the 10^4-parsecs distance; therefore,

$$\tau(X) \gtrsim 10^8 \text{ sec.} \tag{13}$$

This is about 10^5 neutron lifetimes.

(iv) The observations also restrict the interaction cross section of X with matter. An upper limit is obtained from our previous observation that the muons do not originate in conventional air showers. A way out of this is to arrange the interaction length of X to be comparable to or greater than the thickness of the atmosphere so that production of the signal occurs too low for regular air shower production. This requires $\sigma(X\text{-nucleon}) \lesssim 1$ mb. However, if the cross section is made too small, the zenith angle distribution becomes very different from that of muons originating in atmospheric showers and the X particles penetrate so deep that they can interact inside the detector to result in "contained" events. Both of these results disagree with observations. Calculations[15] relevant to the Soudan detector are shown in FIGURE 9. A comparison with the data suggest that the cross section cannot be made smaller than 10 μb. We therefore conclude

$$10\,\mu\text{b} < \sigma(XN) < 1 \text{ mb.} \tag{14}$$

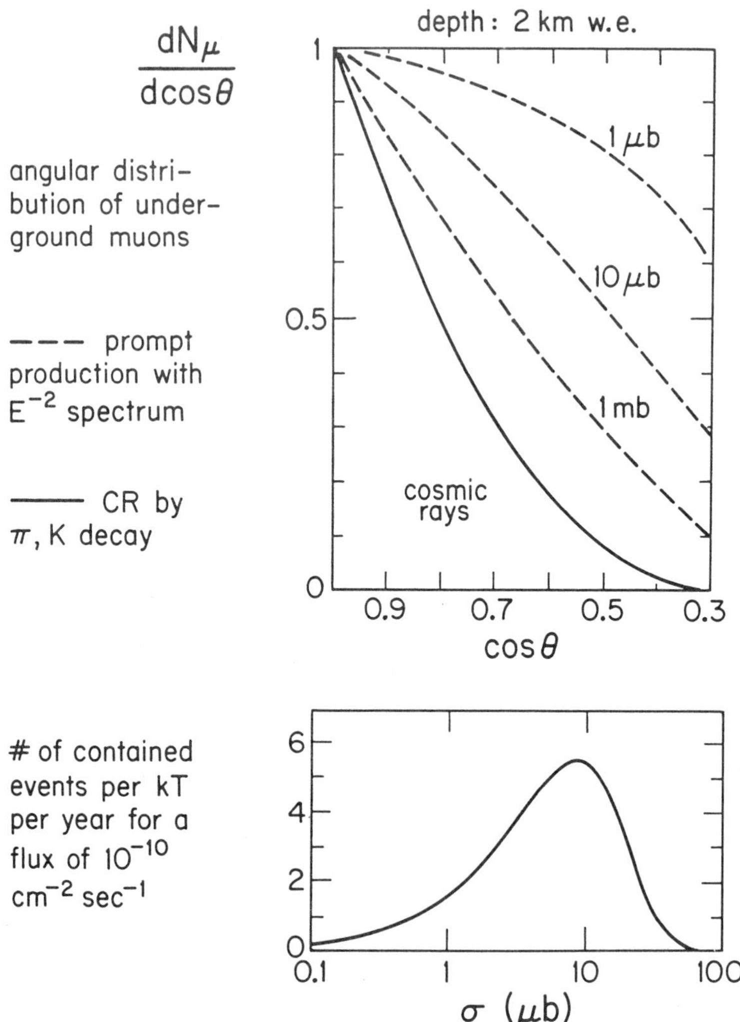

FIGURE 9. Calculations relevant to the Soudan detector.

Also, this discussion eliminates the possibility that neutrinos (or photinos) are the underground muon parents.

This concludes the theorem: a particle with the properties (i)–(iv) should have been discovered by accelerators. If you can imagine that such a particle has been overlooked, consider this as a challenge.

The previous no-go theorem does not even exploit another puzzling feature of the muon data: the angular spread of the muon's arrival directions is roughly 3° around Cygnus in both experiments.[20] This cannot be understood on the basis of any astronomical arguments or in terms of detector resolution, which is roughly 1°. The decay of a heavy particle or large transverse momentum must be the source of the angular spread. Suppose a new particle (a gluino or photino) interacts with quarks in the atmosphere or rock via the resonance production $\tilde{\gamma} q \to \tilde{q}$ (see FIGURE 10). This is the supersymmetric analogue of the Glashow resonance $\bar{\nu}_e e \to W$ suggested to detect a $\bar{\nu}_e$-beam interacting with atomic electrons in the atmosphere. The squark resonance peaks at

$$ E \simeq \frac{M_{\tilde{q}}^2}{\xi_q M_p}. \tag{15} $$

Here, M_p and $M_{\tilde{q}}$ are the proton and squark masses, respectively, and ξ_q is the average momentum of quarks in a proton. The peak is further broadened by the $\tilde{\gamma}$ spectrum. Muons resulting from the decay of the heavy squark would have an angular spread of

$$ \theta \lesssim \frac{M_{\tilde{q}}}{E}. \tag{16} $$

For $M_{\tilde{q}} \simeq M_W$ and $E \simeq 10$ TeV (as in the experiments), equation 15 is satisfied and equation 16 predicts an angular spread of several degrees as observed. The bad news is that the cross section associated with the production of such a heavy mass,

$$ \sigma \simeq \frac{g^2}{M_{\tilde{q}}^2}, \tag{17} $$

is very small and violates equation 13. In equation 17, $g^2 \simeq e^2$ is a typical standard model coupling. Reducing $M_{\tilde{q}}$ does not help as the angular spread θ is reduced accordingly (see equation 16). This is another version of the no-go theorem.

For those not totally discouraged, FIGURE 11 shows two possible contrasting road maps to solutions of the muon puzzle. Yet, it is clear that no one has succeeded at present in explaining every feature of every experiment. In FIGURE 11a, the particle interacts in the atmosphere with $O(\text{mb})$ cross section. The deeper detector samples higher incident energies and therefore measures a reduced flux. This is a feature of the observations.[16] However, by equation 16, the angular spread should be reduced at the deeper observation level. This is not borne out by the data. In FIGURE 11b, the incident particle interacts in the rock with $O(10\,\mu\text{b})$ cross section. The energies sampled by both detectors are similar, and the reduced flux seen in the deeper detector is the result of absorption in the rock. Now, the angular spread could actually be larger in the deeper detector if high energy particles are preferentially absorbed (threshold behavior).[21]

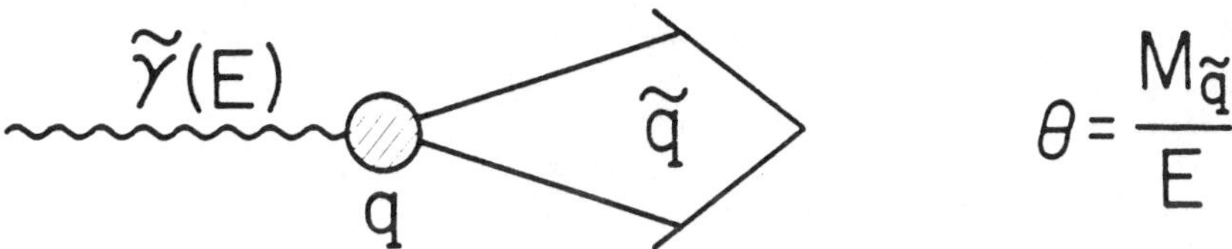

FIGURE 10. "Glashow" resonance photino + quark $\to$ squarks.

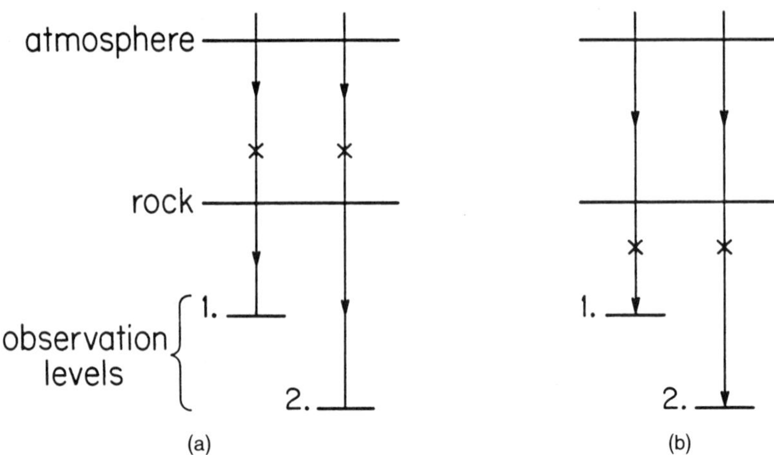

FIGURE 11. Alternative scenarios where muons are produced in the atmosphere or rock above two underground detectors at different depths. See text.

Only the low energy ones dribble down to the deep detector to result in large θ by equation 16. This will only leave Kiel unexplained.

COMPOSITE EXPLANATIONS: IS CYGNUS A QUARK STAR?

The most puzzling feature of the muon data is, of course, the constraint on the particle mass given by equation 12. It is important to realize, however, that the constraint is on the γ-factor; therefore, equation 10 is also satisfied for

$$\gamma = \frac{AE}{AM(X)} > 10^4 . \tag{18}$$

For example, a nucleon, as well as a nucleus, satisfies the constraints resulting from the fact that the Cygnus's radiation carrier keeps time. The observed muon parents are now fragments of a larger system with correspondingly higher energy and the constraint on the mass via the γ factor is relaxed. Pursuing this approach, one possibility is that the muon parents are neutral bits of quark matter.[22] There are theoretical reasons[23] for believing that such objects might exist and might be stable for certain ranges of mass. Furthermore, they might be produced in high energy processes around the compact star. It is even conceivable that the heavy star is a quark star including the crust. Neutron stars, turned quark stars as a result of a phase transition at large density, are made up of u, d quarks. Such a star can stabilize itself (lower its Fermi level) by changing identical u or d into s quarks by weak interaction $u \rightarrow sev$. In the limit that the strange quark mass is neglected, we expect equal numbers of u, d, and s quarks so that stable bits would tend to be neutral. Only charged nuggets could be accelerated, so the neutral beam is produced by charge-changing interactions in the ambient matter of the source after acceleration. Quark globs of the right mass and energy can penetrate the atmosphere. A quark nugget interacting with an atmospheric nucleus heats up and subsequently cools by meson or baryon evaporation. Assuming that a nugget is a Fermi droplet of quarks with $k_F \simeq 1$ GeV, it has been estimated[24] that the energy loss would be a few percent per collision, thus implying a range of

$$R \text{ (in g/cm}^2) \simeq 10^3 \ln \gamma. \tag{19}$$

Nuggets, therefore, can conceivably reach sea level and interact in the rock. As their stability is determined by the strangeness/baryon ratio, they might, however, explode sooner by exchanging baryons with atmospheric nuclei.[25] Some presumably explode to give rise to Centauro events. Finally, the high content of strange quarks would lead to enhanced Λ or K production and thus to a relatively high yield of muons, possibly multimuons. A specific version of a stable ensemble of quarks was suggested[26] some

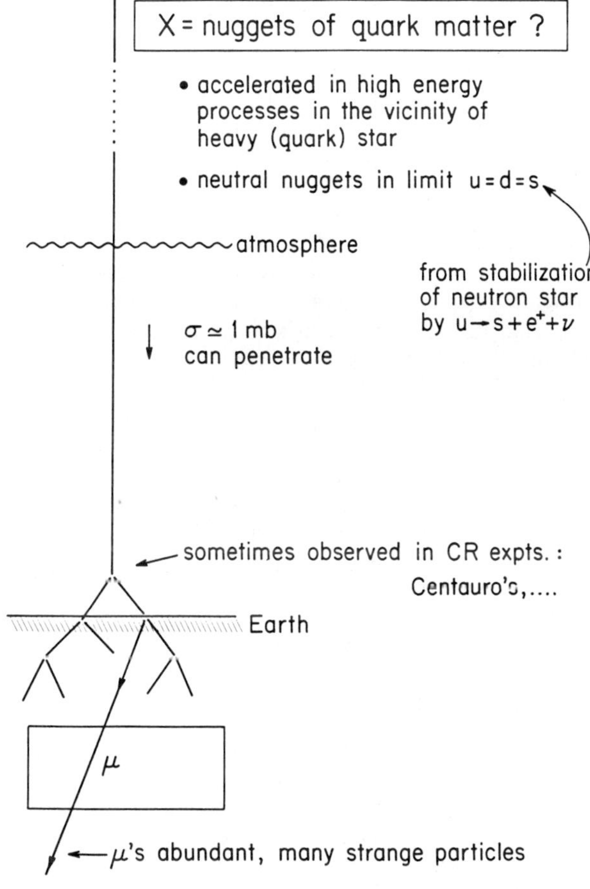

FIGURE 12. A summary of the scenario where the radiation from Cygnus X-3 contains bits of stabilized quark matter.

time ago that could be relevant in the context of underground signals from Cygnus X-3 (but not Centauro events). This is the $(\Lambda\Lambda)$ bound dihyperon state of $2u$, $2d$, and $2s$ quarks. A summary of the scenario where the radiation from Cygnus X-3 contains bits of stabilized quark matter is shown in FIGURE 12. The quark matter beam, of course, would be accompanied by in-phase photon and neutrino radiation produced in the same high energy processes that are the source of the quark matter. Quark nugget explosions are associated with high transverse momenta.[25] It is doubtful, however, that this can ever explain the observed angular spread. The quark star solution has many appealing features that were all anticipated. The solution at present deserves the W. Pauli treatment: "I can paint like Titian, only the details are missing."

CONCLUSIONS

Nonaccelerator experiments now have their great accelerator(s) in the sky. The field undoubtedly has a long and promising future whether the Cygnus muon puzzle survives further scrutiny or not.

Cygnus muons are at present the only challenge to the standard model of quarks and leptons. One should, however, keep in mind the possibility of a "garbage" solution, of which quark matter is an illustration, as an alternative to new particles or interactions. The highest priority is to confirm the data. People who have taken more than a cursory look at the data agree that systematics, not statistics, is the problem. In that sense, the Soudan observation[20,21] of Cygnus during the recent radio burst is of central importance. Little analysis is involved in the two-week observation. We should learn an important lesson from air Čerenkov experiments, which is a field littered with contradictory results: experiments taken in different locations and depths or at different times are difficult to compare because we do not know the details of either the accelerator's schedule or the footprints of its radiation interacting with the earth.

This talk does injustice to the importance of Cygnus X-3 for astronomy. The most common phase observed in the TeV γ-ray experiments is 0.6. How does this square with the model of FIGURE 4? The binary period of Cygnus X-3 increases. In normal binaries, it decreases as the binary system decays under the emission of gravitational waves. This is a test of general relativity. Is the pressure of the accreting matter enormous? Is the compact star a pulsar; a black hole? Is there a correlation between radio bursts and high energy emission as suggested?[27] What is the explanation?

ACKNOWLEDGMENTS

I thank my collaborators, M. Barnhill, T. Gaisser, K. Hikasa, and T. Stanev.

NOTES AND REFERENCES

1. For a complete review, see: WATSON, A. A. 1985. Rapporteur paper presented at the 19th International Cosmic Ray Conference, La Jolla, California. To be published. See also: LEARNED, J. 1985. University of Hawaii internal report. Unpublished; ELBERT, J. W., R. C. LAMB & T. C. WEEKS. 1986. *In* Proceedings of New Particles '85. V. Barger, D. Cline & F. Halzen, Eds. World Scientific. Singapore; VLADIMIRSKII, B. M., A. M.

GAL'PER, B. I. LUCHKOV & A. A. STEPANYAN. 1985. Usp. Fiz. Nauk **145:** 255 (Sov. Phys. Usp. **28:** 153).

2. See, for example: ROSSI, B. 1952. High Energy Particles. Prentice–Hall. New York.

3. Only the Akeno experiment (reference 1) puts a cut on the muon content of the shower. Their cut-value is ten times smaller than what it should be by calculation. It is not clear therefore what their result means. They could pick up photons or fluctuations from showers of another nature.

4. For a detailed discussion, see: HALZEN, F., K. HIKASA & T. STANEV. 1986. Phys. Rev. **D34:** 2061; STANEV, T., T. K. GAISSER & F. HALZEN. 1985. Phys. Rev. **D32:** 1244; STANEV, T. & CH. P. VANKOV. 1985. Phys. Lett. **158B:** 75; STANEV, T., CH. P. VANKOV & F. HALZEN. 1985. 19th International Cosmic Ray Conference Papers, La Jolla, California (NASA, Washington, D.C.). Vol. 7, p. 219; STANEV, T. & CH. P. VANKOV. 1979. Comments Phys. Commun. **16:** 363.

5. TURVER, T. 1985. 19th International Cosmic Ray Conference, La Jolla, California (NASA, Washington, D.C.).

6. Few experiments identify both phases. The most common phase observed is actually 0.6, which is a value difficult to accommodate in this temptative description.

7. The very highest energy cosmic rays are extragalactic. The extragalactic source LMC X-4 has been identified as a TeV γ-emitter (see reference 1).

8. HERA is DESY's electron-proton collider.

9. I am aware of studies to build such detectors in Japan and the U.S. (Chicago, Maryland, Michigan, Wisconsin, and others). For a discussion, see: HALZEN, F., K. HIKASA & T. STANEV. Reference 4.

10. ATHENS, HAWAII, PURDUE, WISCONSIN COLLABORATION. Private communication; FRY, J. 1986. Proceedings of the Aspen Winter Conference, Aspen, Colorado.

11. STANEV, T. 1986. Bartol preprint no. BA-86-7.

12. See, for example: OCHS, W. & L. STODOLSKY. 1985. Max-Planck Institute (München) preprint no. MPI PAE/PTh 54/85.

13. SAMORSKI, M. & W. STAMM. 1983. Astrophys. J. **268:** L17.

14. Calculations from: HALZEN, F. *et al.* Reference 4; STENGER, V. J. 1985. Nature **317:** 411; 1984. Astrophys. J. **284:** 810; ROBINETT, R. W. 1985. Phys. Rev. Lett. **55:** 469; AURIEMMA, G., L. MAIANI & S. PETRARCA. 1985. Phys. Lett. **164B:** 179; BEREZINSKY, V. S. & B. L. IOFFE. 1985. Moscow preprint no. ITEP-127; MIDORIKAWA, S. & S. YOSHIMOTO. 1985. University of Tokyo preprint no. INS-Rep.-560; HAGIWARA, K. Private communication.

15. BARNHILL, M. V., III, T. K. GAISSER, T. STANEV & F. HALZEN. 1985. University of Wisconsin preprint no. MAD/PH/243. Unpublished; Nature **317:** 409; 19th International Cosmic Ray Conference Papers, La Jolla, California (NASA, Washington, D.C.). Vol. 1, p. 99; BAYM, G., E. W. KOLB, L. McLERRAN, T. P. WALKER & R. L. JAFFE. 1985. Phys. Lett. **160B:** 181; DAR, A., J. J. LORD & R. J. WILKES. 1986. Phys. Rev. **D33:** 303. For reviews, see for example: DE RÚJULA, A. (CERN-TH.4267/85), F. HALZEN & L. MAIANI (CERN-TH.4326/85). 1985. *In* Proceedings of the International Europhysics Conference on High Energy Physics, Bari, Italy. L. Nitti & G. Preparata, Eds.; STANEV, T. 1986. *In* Proceedings of New Particles '85. V. Barger, D. Cline & F. Halzen, Eds. World Scientific. Singapore.

16. BARTELT, J. *et al.* (SOUDAN I COLLABORATION). 1985. Phys. Rev. **D32:** 1630; MARSHAK, M. L. *et al.* 1985. Phys. Rev. Lett. **54:** 2079; **55:** 1965; BARRISTONI, C. *et al.* (NUSEX COLLABORATION). 1985. Phys. Lett. **155B:** 465; BERGER, C. (FREJUS COLLABORATION). 1985. *In* Proceedings of the International Europhysics Conference on High Energy Physics, Bari, Italy. To be published. See, however: APRILE, E. *et al.* (HPW COLLABORATION). 1985. *In* Proceedings of the International Europhysics Conference on High Energy Physics, Bari, Italy. To be published; OYAMA, Y. *et al.* (KAMIOKANDE COLLABORATION). 1985. University of Tokyo preprint no. UT-ICEPP-85-03. See also: PRICE, L. E. 1986. Argonne preprint no. ANL-HEP-CP-85-117. *In* Proceedings of Annual Meeting of the Division of Particles and Fields of the American Physical Society. R. C. Hwa, Ed. World Scientific. Singapore; TOTSUKA, Y. 1985. University of Tokyo preprint no. UT-ICEPP-85-05. *In* Proceedings of 1985 International Symposium on Lepton and Photon Interactions at High Energies. M. Konomu & K. Takahashi, Eds.

17. DZIKOWSKI, T. *et al.* 1985. *In* Proceedings of the 19th International Cosmic Ray Conference, San Diego, California.
18. GAISSER, T. K. & T. STANEV. 1985. Phys. Rev. Lett. **54:** 2265; KOLB, E. W., M. S. TURNER & T. P. WALKER. 1985. Phys. Rev. **D32:** 1145; 1986. Phys. Rev. **D33:** 859(E); WALKER, T. P., E. W. KOLB & M. S. TURNER. 1986. *In* Proceedings of New Particles '85. V. Barger, D. Cline & F. Halzen, Eds. World Scientific. Singapore; STECKER, F. W., A. K. HARDING & J. J. BARNARD. 1985. Nature **316:** 418; DAR, A. 1985. Phys. Lett. **159B:** 205.
19. STECKER, F. W. *et al.* See reference 18.
20. MARSHAK, M. Private communication; 1986. Science **231:** 336.
21. RUDDICK, K. 1986. Proceedings of the Aspen Winter Conference, Aspen, Colorado.
22. BARNHILL, M. V. *et al.* See reference 15; BAYM, G. *et al.* See reference 15; SHAW, G. L., G. BENFORD & D. SILVERMAN. 1985. University of California-Irvine technical report no. 85/14.
23. FARHI, E. & R. L. JAFFE. 1984. Phys. Rev. **D30:** 2379; WITTEN, E. 1984. Phys. Rev. **D30:** 272; LIU, H. C. & G. L. SHAW. 1984. Phys. Rev. **D30:** 1137; HALZEN, F. 1984. *In* Proceedings of Workshop on Cosmic Ray and High Energy Gamma Ray Experiments for the Space Station Era, Louisiana State University, Baton Rouge (October 1984); HALZEN F. & H. C. LIU. 1985. Phys. Rev. **D32:** 1716.
24. BJORKEN, J. D. & L. L. MCLERRAN. 1979. Phys. Rev. **D20:** 2353.
25. HALZEN, F. & H. C. LIU. See reference 24.
26. FAHRI, E. & R. L. JAFFE. See reference 24.
27. For a discussion, see: HALZEN, F. Reference 15.

Particle and Astrophysical Results from Underground Experiments[a]

JOHN M. LoSECCO

Department of Physics
University of Notre Dame
Notre Dame, Indiana 46556

INTRODUCTION

Underground physics experiments have emerged in the last five years as unique tools to study a wide variety of problems in physics and astrophysics. The scale and location of these experiments was initially motivated by a search for proton decay.

From prior experiments, the proton was known to live at least 10^{29} years. A number of theories suggest a value for the lifetime in the range of 10^{30} years. To study such lifetimes in a reasonable period requires detectors in the range of 1000 metric tons. A metric ton contains 6.02×10^{32} nucleons, and thus a lifetime of 10^{31} years would yield about 60 proton and bound neutron decays per year. Optimistically, one could hope to do an experiment with less matter if the lifetime were shorter, but errors in the lifetime calculation indicated that values as high as 10^{32} years would still be possible.

In conducting such a sensitive experiment, one must take great care in shielding the detector from spurious signals. Such background could be mistaken for proton decay and, at the very least, this makes the task of finding candidates time-consuming because of all the background that must be rejected. In general, an experiment's sensitivity is determined by its mass and by its ability to reject background. Smaller detectors with good background rejection may be competitive with larger, coarser detectors. This statement is decay mode dependent because some modes such as $P \rightarrow e^+\pi^0$ produce spectacular, unique signatures, while modes with missing neutrals such as $P \rightarrow \nu K^+$ have much greater ambiguity.

To eliminate the hadronic and electromagnetic component of cosmic rays, proton decay detectors are located deep underground. The surviving components of cosmic rays are very energetic muons that penetrate from the surface, along with neutrinos formed by cosmic ray interactions in the atmosphere that propagate unattenuated through the earth. The neutrinos form the ultimate background to proton decay.

One expects about 120 neutrino interactions in a 1000-ton detector in one year of operation. To a large extent, these events are distinguishable from proton decay. Neutrinos have a momentum equal to their energy. Any interaction will conserve the momentum, so any neutrino event with enough energy to be confused with proton decay ($E = 938$ MeV) will have a momentum much higher than could be attributed to a nucleon at rest or with Fermi motion ($P \approx 250$ MeV/c) bound in a nucleus. In practice, ambiguities can arise. Protons can decay into modes with missing neutrals,

[a]This work was supported in part by the National Science Foundation and the U.S. Department of Energy.

and neutrinos can interact and produce unobserved particles that carry off the momentum. Both of these effects imply that the momentum can only be approximately measured even in the best detector.

One of the major tasks of proton decay experiments is to make neutrino background estimates that are consistent with known physics and the behavior of their detector. In addition, as we shall see in the fourth section below, the environment required for proton decay experiments is ideal for a large number of other physics objectives.

CLASSES OF EXPERIMENTS

In general, there have been two approaches to the above problem. One can build a very large detector with modest spatial and momentum resolution, or one can construct a fine-grain detector where the ultimate size is limited by costs. Both classes of detectors exist.

The most massive detectors, in excess of 1000 tons, are water Cherenkov detectors. Charged particles with $E > 1.52mc^2$ will emit Cherenkov light in water. The light is emitted in a cone along the track direction. This light can be collected on arrays of photomultipliers and the event can be reconstructed from the structure of the light output. The technique cannot see neutral particles or particles below the Cherenkov threshold. However, photons will readily convert and produce a shower that represents the photon energy very well.

Water Cherenkov detectors have sizes of 10 to 20 meters on a side. The attenuation of light is not a problem on this scale. The spatial resolution is about 30 centimeters and it depends on the topology of the event. Angular resolution for long tracks traversing the detector is from 2° to 3.5° on the average.

Fine-grain detectors employ flash chamber or ionization technology to measure tracks. The bulk of material comes from iron plates placed between these counters. The spatial resolution is from millimeters to 4 cm, depending on the counter size. To some extent, the resolution is limited by the thickness of the steel plates and the effect of multiple scattering on particles of modest momentum. The angular resolutions of these devices is <1° for long tracks. In general, the masses range from 100 to 912 tons.

CURRENT EXPERIMENTS

TABLE 1 summarizes some information about current underground experiments. The largest detector operating is the 3300-metric ton water IMB detector. The largest of the iron calorimeter type of detectors is the 912-ton Frejus detector that has been in full operation for less than a year. Both the smaller KGF and Nusex detectors are of the iron calorimeter types and have been operating for years. These detectors gave a quick early look at the lifetime of greatest interest ($\sim 10^{30}$ years), but they do not have the mass to be competitive with latter detectors for lifetimes on the order of 10^{31}–10^{32} years. The 880-ton water Kamioka detector, while smaller than the two largest detectors, has better light collection than IMB and a much longer exposure at full mass than Frejus.

The depths of the detectors are listed to give a crude comparison of the energy of muons penetrating from the surface. For IMB, this is $E_\mu > 400$ GeV. For Nusex, this is $E_\mu > 2$ TeV.

The geomagnetic latitude is listed because there is a small dependence of the low energy neutrino flux on the local magnetic field. The flux is lower near the equator than near the poles because low energy primaries are deflected away from the earth near the equator.

The exposures listed in the table do not reflect the total exposure of the detector. The exposure listed is the largest that has been published by the research group. This is the exposure that I will try to summarize in later sections.

PHYSICS CAPABILITIES

These detectors are all well designed to observe a number of proton decay modes. Modes that result in only observable particles such as $P \rightarrow e^+\pi^0$ can be strongly constrained by the observations. On the other hand, modes such as $P \rightarrow \nu K^+$ are a challenge to the experimenter because one must first prove that the single observed

TABLE 1. Current Proton Decay Experiments

Detector	Mass (fid.)	Location	Depth (mwe)	Exposure Kt-yr	No. Events
Kolar Gold Fields	60 ton Fe	3°N	7000	0.22	19
Nusex (Mt. Blanc)	100 ton Fe	50°N	5000	0.25	32
IMB	3300 ton H$_2$O	52°N	1570	3.77	401
Kamiokande	880 ton H$_2$O	26°N	2400	0.83	103
Frejus	912 ton Fe	50°N	4800	0.19	13

track is a K^+ and then (by assuming energy and momentum conservation) that the unseen recoiling track is a ν with the correct energy. One is limited by Fermi motion and nuclear rescattering effects to how well one can determine the initial state of the decaying proton.

A number of physics considerations limit the sensitivity of even the well-constrained modes. Many modes result in the production of energetic hadrons in a nucleus. The probability of the hadron interacting before it leaves the parent nucleus is quite high. The problem is accentuated because the momenta involved place the hadrons in the resonance region and they originate in a nucleus, which is a region of high nucleon density. As many as 40% in oxygen and 60% in iron may interact on leaving.

In general, such distortions cannot be compensated, and they are reflected in an efficiency for finding proton decays considerably below the instrumental efficiency. To a great extent, these effects offset any major advances to be gained by making major new detector designs. A better detector will not be able to see a better signal, but it may have better background rejection capabilities.

The best proton decay detector has the conflicting requirements of high mass and fine granularity, while at the same time having low density to reduce particle interactions. In general, all realized detectors are a compromise of these requirements.

The $\Delta B = 2$ process of neutron-antineutron oscillations yields a large number of final states following the annihilation of the antineutron. The general signature is a highly isotropic multitrack event. Because the neutrino background is falling with energy, it is often possible to find a clean selection criterion for the $n \rightarrow \bar{n}$ signal.

Intense interest in magnetic monopoles has led to a number of searches in underground detectors. In general, one looks for a massive (10^{15} GeV) slowly moving object. People have looked for slowly moving objects in scintillation counter arrays and for unusual ionization in gas counters. Proton decay catalysis in the presence of a magnetic monopole provides a direct link of proton decay experiments to magnetic monopole fluxes. Magnetic monopoles in grand unified theories are believed to accelerate proton decay. Cross sections on the order of 100's of millibarns are possible. The signature for a monopole would be multiple interactions in a short interval of time.

If monopoles do cause proton decay, a number of astrophysical limits become relevant. Results on excess neutrinos from the sun or on X-ray emission from neutron stars give stronger limits than terrestrial experiments.

The neutrinos that make up the major background to proton decay can also yield new information. In general, the results from underground experiments are complementary to accelerator or reactor neutrino experiments. The atmospheric neutrinos observed underground are primarily in the 500-MeV to 2-GeV energy range. They originate from all directions and are a mixture of ν_μ, ν_e, $\bar{\nu}_\mu$, and $\bar{\nu}_e$. Because the source is distributed at large distances and over a range of from 8 km to 13,000 km, experiments that depend on long-lived neutrinos can be performed. Neutrino oscillations can be studied with unprecedented sensitivity down to 10^{-5} eV^2. In addition, because the source is a mixture of neutrino types, one can potentially study many components of the 3 × 3 mixing matrix.

While we expect neutrinos to originate from cosmic ray interactions, there are many other potential sources. Many stars that are sources of energetic photons are expected to be sources of energetic neutrinos. The photons and neutrinos are assumed to come from pion decays. One can use the underground data to search for a signal in the face of the atmospheric background. Sensitivity is improved by making directional cuts on the data and using the off-source data to estimate the background.

The energy range over which neutrinos can be studied is limited to <2 GeV. Above this, two factors reduce the rate. Energetic events exit the detector and thus only a lower bound on the energy can be measured. In addition, the flux falls to levels more than a decade below that at lower energies. Both of these problems can be circumvented by studying upward-going muons. These originate from muon neutrino interactions in the rock surrounding the detector. The huge mass of rock gives a much larger target for the smaller flux and the energetic muons produced can travel through the rock and penetrate the detector. Information about the neutrino energy is lost, but the neutrino direction is better known at these energies than at lower values due to smaller scattering angles. In general, only upward muons are chosen so as to be certain of having muons from neutrino interactions and not just penetrating muons from the surface. Some of the detectors have a directional ambiguity and, therefore, they cannot tell upward-moving from downward-moving tracks. They keep events near the horizontal as neutrino interactions.

Surface muons themselves provide a window on multi-TeV cosmic ray primaries. The earth filters out all but the most energetic muons, and these can be studied for very energetic point sources. Several energetic sources are known to be pulsed or periodic and this information can be used to enhance a possible signal and to reduce background. A number of groups have reported signals using this technique.

PHYSICS RESULTS

All experiments in recent times[1-5] that have searched for proton decay have reported signals that are compatible with such decays. The rate of occurrence of such decays is about the same[6] for all experiments. This rate corresponds to about $\tau = 10^{32}$ years for the proton lifetime. TABLE 2 summarizes the data. However, this result is highly ambiguous. At this level, neutrino backgrounds are significant. Quantitative estimates range from the total signal to one-third of the total signal. Many of the candidate events have more than one interpretation and there is no consistent picture of the dominant decay mode or modes.

TABLE 2. Summary of Candidates

Experiment	Exposure	Number of Candidates	τ if Eff. = 100%
KGF	0.22 Kt-yr	4	$(3.0 \pm 1.5) \times 10^{32}$
Nusex	0.25 Kt-yr	3	$(2.0 \pm 1.2) \times 10^{32}$
IMB	3.77 Kt-yr	21	$(0.93 \pm 0.20) \times 10^{32}$
Kamiokande	0.83 Kt-yr	4	$(0.93 \pm 0.20) \times 10^{32}$
Frejus	0.19 Kt-yr	0	$> 0.18 \times 10^{32}$
Total	5.26 Kt-yr	32	$(1.0 \pm 0.18) \times 10^{32}$

In general, results have been reported as lifetime limits (90% C.L.) on a mode-by-mode basis. The major reason for this is to clarify the role of background in the problem. There are very few candidates for the decay mode $e^+\pi^0$ and none in the large detectors. Two factors are significant in interpreting the data. All experiments have an efficiency for finding a specific decay if it has occurred. A number of factors contribute to the efficiency. Actual physics effects such as charge exchange and scattering can distort an event so that it no longer looks like a candidate. Beyond that, Fermi motion and unstable particle width smear the allowed range of momenta for the final state particles.

The other major factor is the background to a specific decay mode as seen in a given detector. While real neutrino interactions can never have the same kinematics as proton decay (because $E = P$), most detectors lose information; therefore, some fraction of the highly inelastic neutrino events will look like proton decays. Because most detectors have poor neutral particle detection, a substantial part of the background rate will be common to all detectors. However, specific details of energy and track resolution can vary the ambiguous region from detector to detector.

Results are usually quoted as 90% confidence limits on the lifetime, with the assumption that all candidates be real events, that is, without background subtraction.

FIGURE 1 shows a comparison of several experiments. The IMB limit[1] is plotted on the horizontal scale. The limit from Kamioka,[2] Nusex,[4] or Kolar Gold Fields[3] is plotted on the vertical scale. If a clear consistent signal were present, the points would lie on the diagonal line. This would also occur if the experiments were background-limited. For the modes shown, it is clear that a longer exposure has produced a larger lifetime limit. Therefore, the experiments as a whole are not background-limited.

The lifetime limit for $e^+\pi^0$ decay is 2.5×10^{32} years from IMB.[1] Combining experiments will not significantly increase the limit. IMB has a limit of 4×10^{31} years for the μ^+K^0 decay. Most experiments have reported candidates for this mode. It is likely that the background is greater for this mode.

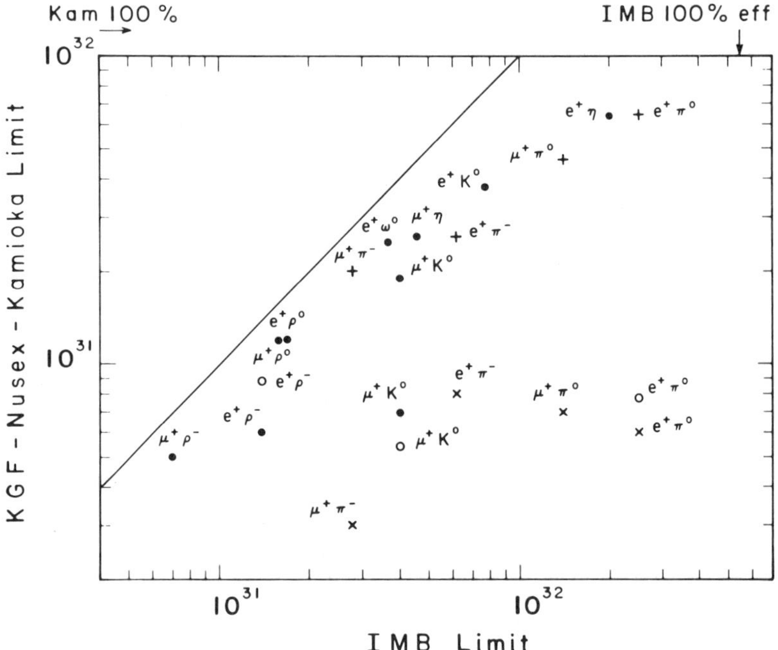

FIGURE 1. Comparison of proton lifetime limits for specific modes from four experiments.

Twenty percent of all protons in water detectors are in the form of hydrogen. These protons give a sample of unbound nucleons free of nuclear effects and Fermi motion. IMB has published limits[1] on free proton decay, with the assumption that all bound decays are suppressed. Due to increased efficiency for detection, the limits are about one-third (not one-fifth) of those for all protons. These results are significant from a theoretical point of view because they are insensitive to decay suppression mechanisms in the nucleus.

The neutron-antineutron oscillation time can be extracted from these experiments. A calculation[7] is needed to convert results in the nucleus to a free oscillation time.

Kamioka[8] quotes a limit of 1.2×10^8 seconds with no candidates. The IMB limit is comparable.

The lowest limit for terrestrial magnetic monopole searches comes from an examination of old mica deposits.[9] A monopole can form a bound state with an aluminum nucleus, and as the state traverses a medium, it should be heavily ionizing. By examining samples with very long exposure times, sensitivity to fluxes as low as $2 \times 10^{-19}/\mathrm{cm}^2/\mathrm{sec}/\mathrm{str}$ at $\beta = 10^{-3}$ can be achieved. Objections are raised because a monopole may already be bound to another nucleus and thus may be unable to penetrate, or to pick up, another nucleus. Such arguments can reduce the sensitivity by two orders of magnitude.

The lowest terrestrial limit for monopole catalysis comes from IMB.[10] A flux limit of $4 \times 10^{-14}/\mathrm{cm}^2/\mathrm{sec}/\mathrm{str}$ is found for an assumed cross section of 100 millibarns.

Kamioka has found limits based on a lack of 35-MeV neutrinos from the sun. Such neutrinos would be produced in a decay chain following monopole catalysis. A limit of $10^{-21}/\mathrm{cm}^2/\mathrm{sec}/\mathrm{str}$ at $\beta = 10^{-3}$ is found. However, if the sun can capture magnetic monopoles, then so can other celestial objects. A limit of $10^{-23}/\mathrm{cm}^2/\mathrm{sec}/\mathrm{str}$ at $\beta = 10^{-3}$ is achieved by considering the effect of monopole catalysis on the X-ray emission from a neutron star.[11]

There has been no evidence for a magnetic monopole flux from any underground experiment. Most proton decay detectors, though, have reported a neutrino flux as measured in their detectors.[4,5,12,13] In general, the agreement with expected fluxes is good. Both the Kamioka detector[8] and the Nusex detector[4] can distinguish ν_e from ν_μ by shower development. They quote a ν_e/ν_μ flux ratio of 0.36 ± 0.08 and 0.28 ± 0.11, respectively. These are lower than the expected value[14] of 0.64. The IMB group has studied the fraction of their contained events resulting in a muon decay.[8] The 26% observed can be converted to a ν_e/ν_μ ratio with a number of assumptions about muon capture in water. If 40% of the ν_μ interactions do not result in a muon decay signal, then the observed value corresponds to $\nu_e/\nu_\mu \approx 1.3$. However, the problem of the ν_e/ν_μ ratio is still under active study.

IMB[13] has used their sample of 401 contained interactions to study atmospheric neutrinos.[13–15] Neutrino oscillations with a baseline of 10^7 meters and a mean neutrino energy of 920 MeV can be studied. These parameters give sensitivities on the order of 10^{-5} eV^2 for maximal mixing (if matter effects[16] are neglected). By varying the zenith angle, neutrino path lengths can run from 10^7 meters to 2.6×10^6 meters. The neutrino energies run from 400 MeV to 2 GeV. Hence, E/L spans a range of 4×10^{-5} MeV/m to 7.7×10^{-4} MeV/m. Sensitivity depends on the statistics available. Because the spectrum is falling with energy, one is less capable of seeing effects of small mixing angles at the larger value of E/L; hence Δm^2. FIGURE 2 summarizes their results for oscillation of ν_e and ν_μ into noninteracting states. The region at maximum mixing, 2.2×10^{-5} eV$^2 < \Delta m^2 < 7.6 \times 10^{-4}$ eV^2, is excluded by this analysis. Results have been obtained for mixing angles $\eta > 22.5°$.

These results may be extended to just ν_μ oscillations. If ν_μ oscillates to ν_τ or to a sterile neutrino state, then matter effects will be irrelevant. If ν_μ oscillates into ν_e, the data rate will not change significantly, but the fraction of events with a muon decay will decrease. No decrease is observed either in the total event rate or in the muon decay fraction as the zenith angle varies. Restricting the analysis to events with a muon decay signature gives the limits in FIGURE 3. These are not quite as low as those for all neutrinos due to the slightly higher energy threshold for ν_μ. The range of 4.2×10^{-5} eV^2 $< \Delta m^2 < 5.4 \times 10^{-4}$ eV^2 is excluded at maximum mixing. Because only 26% of the events can be used in this analysis, the statistical errors are larger than for the sample as a whole.

Searches[17] for point sources of neutrinos with energies below 2 GeV yield no excess in regions of 60°, which is consistent with the angular resolution at these energies. No more than 5.5% of the total observed rate can be attributed to either stellar sources or sources on earth. Less than 3.5% of the total rate can be attributed to a source in the sun.

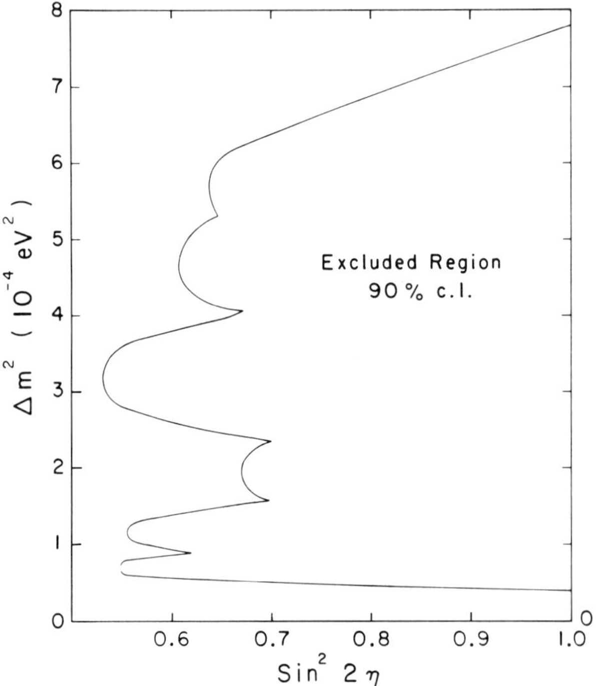

FIGURE 2. Neutrino oscillation limits from IMB for oscillation into sterile neutrinos.

At high energies, results have been reported by three groups.[8,17] Caution is needed in comparing results on neutrino interactions in the rock. Groups are located at different latitudes and have different selection criteria in their local coordinate system. Celestial objects will pass in and out of the local selection region, thus giving rise to reduced sensitivity and perhaps spurious periodic effects. For example,[18] if one requires that the source be below the horizon, then some sources are never visible and some are visible 100% of the time. Comparison should be done with regions at the same declination to avoid any normalization difficulties. TABLE 3 lists limits for sources. Upward muons have also been used to search for neutrino oscillations.[19] This involves

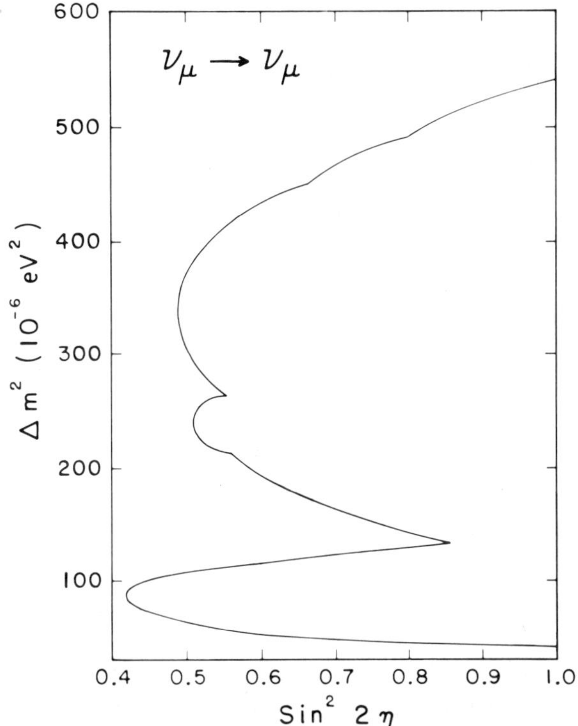

FIGURE 3. Neutrino oscillation limits from IMB for any oscillations of ν_μ.

comparing the observed distribution with an expected one. Because the neutrino energies are higher, the limits are not as low in Δm^2 as for the contained events.

Two groups[20,21] have reported evidence for underground muons associated with the celestial source Cygnus X-3. Both groups have selected muons in the direction of Cygnus X-3 and have analyzed their time structure using the characteristic 4.8-hour period observed in other high energy experiments. A signal emerges at a phase of about 0.7–0.8.

The Soudan group[20] reports a flux of 7×10^{-11} μ/cm^2/sec with a threshold of $E_\mu >$

TABLE 3. Upward Muon Flux Limits (90% C.L.)[a]

Source	IMB Limit	Kamiokande Limit
LMC X-4	3.5×10^{-14}	1.1×10^{-13}
Her X-1	6.9×10^{-14}	6.5×10^{-13}
Cyg X-3	2.1×10^{-13}	4.6×10^{-13}
Crab	8.8×10^{-14}	2.8×10^{-13}
SS 433	5.7×10^{-14}	2.3×10^{-13}
Galactic Center	3.8×10^{-14}	1.6×10^{-13}
Vela X-1	3.5×10^{-14}	1.3×10^{-13}

[a] In units, ν/cm^2/sec.

500 GeV. Nusex[21] reports a flux of 2.5×10^{-12} $\mu/cm^2/sec$ with a threshold of $E_\mu > 2$ TeV. These results have not been confirmed by other groups. Kamioka[22] gives a flux limit of 2.2×10^{-12} $\mu/cm^2/sec$ with a threshold near 600 GeV. IMB has searched,[15] but has not had enough sensitivity to confirm the effect. Frejus[23] has also failed to confirm the effect and quotes a bound of 1.9×10^{-13} $\mu/cm^2/sec$ at a threshold comparable to the Nusex group.

One problem with comparing different experiments is that Cygnus X-3 is known to be a variable source. It is possible that different groups have observed at different times, and in spite of significant overlap, they have failed to get consistent results.

THE FUTURE

In the near future, we can look forward to more data—in particular, the Frejus detector, which is the first of a new generation of massive fine-grain detectors. Background estimates should improve. At present, background subtractions are still statistics limited. I expect the systematic effects associated with the background estimates to decrease at least as fast as the statistical errors on the data.

Kamioka has completed a major upgrade and will attempt to observe a solar neutrino signal via ν_e scattering. IMB will finish an upgrade shortly. The upgrade was meant to increase their sensitivity to low light decay modes by increased light collection. The Soudan II detector—a second-generation tracking calorimeter—is in the construction stage. Completion is anticipated in two years.

There will be an increased interest in astrophysical problems as detectors are improved. Neutrinos and muons are still very good probes of high energy phenomena. These are the source of the only signals in underground detectors.

REFERENCES

1. BLEWITT, G. *et al.* 1985. Phys. Rev. Lett. **55:** 2114; 1985. Search for two prong proton decays. CALT68-1265.
2. ARISAKA, K. *et al.* 1985. J. Phys. Soc. Jpn. **54:** 3213.
3. KRISHNASWAMY, M. R. *et al.* 1985. Proc. of the 19th International Cosmic Ray Conference, La Jolla, California, vol. 8, p. 261–264.
4. BATTISTONI, G. *et al.* 1985. Proc. of the 19th International Cosmic Ray Conference, La Jolla, California, vol. 8, p. 271–274.
5. AACHEN, ORSAY *et al.* COLLABORATION. 1985. Proc. of the 19th International Cosmic Ray Conference, La Jolla, California, vol. 8, p. 257–260.
6. LOSECCO, J. 1985. Comments Nucl. Part. Phys. **XV:** 23
7. DOVER, C. *et al.* 1983. Phys. Rev. **D27:** 1090.
8. KOSHIBA, M. 1986. Talk presented at the 1986 Aspen Winter Conference on Particle Physics.
9. PRICE, P. B. & M. H. SALAMON. 1985. Proc. of the 19th International Cosmic Ray Conference, La Jolla, California, vol. 8, p. 242–245.
10. ERREDE, S. 1983. Experimental limits on monopole catalysis. *In* Monopole '83. J. L. Stone, Ed.: 251. Plenum. New York.
11. FREESE, K., M. TURNER & D. SCHRAMM. 1983. Phys. Rev. Lett. **51:** 1625.
12. KRISHNASWAMY, M. R. *et al.* 1985. Proc. of the 19th International Cosmic Ray Conference, La Jolla, California, vol. 8, p. 4–5.
13. LOSECCO, J. *et al.* 1985. Phys. Rev. Lett. **54:** 2299; Proc. of the 19th International Cosmic Ray Conference, La Jolla, California, vol. 8, p. 116–119.

14. STANEV, T. 1985. Muons and neutrinos. Proc. of the 19th International Cosmic Ray Conference, La Jolla, California, vol. 9, p. 383–398.
15. LoSECCO, J. *et al.* 1986. Neutrino physics with a massive underground detector. Proceedings of the XXI Recontre de Moriond Workshop on Massive Neutrinos (January 1986). O. Fackler & J. Tran Thanh Van, Eds.: 301–312. Editions Frontières. Gif sur Yvette, France.
16. WOLFENSTEIN, L. 1978. Phys. Rev. **D17:** 2369.
17. LoSECCO, J. *et al.* 1986. Contained neutrino interactions in an underground water detector. UND-PDK-86-1.
18. SVOBODA, R. 1985. A search for astrophysical point sources of neutrinos using a large underground Cerenkov detector. Ph.D. thesis, University of Hawaii.
19. BIONTA, R. *et al.* 1984. Proceedings of the DPF Santa Fe Meeting. T. Goldman & M. M. Nieto, Eds.: 255. World Scientific. Philadelphia.
20. MARSHAK, M. L. *et al.* 1985. Phys. Rev. Lett. **54:** 2079; **55:** 1965.
21. BATTISTONI, G. *et al.* 1985. Phys. Lett. **155B:** 263.
22. OYAMA, Y. *et al.* 1986. Phys. Rev. Lett. **56:** 991.
23. CHARDIN, G. 1986. Private communication and Proceedings of the XXI Recontre de Moriond Workshop on Massive Neutrinos (January 1986). O. Fackler & J. Tran Thanh Van, Eds.: 173–180. Editions Frontières. Gif sur Yvette, France.

The Search for Proton Decay:
Where Do We Stand?

MAURICE GOLDHABER

Irvine/Michigan/Brookhaven Collaboration (IMB)[a]
Brookhaven National Laboratory
Upton, New York 11973

We have heard in detail from M. Koshiba and J. LoSecco about progress in the search for proton decay with different detectors. Let me, therefore, confine myself to a few general remarks. While all detectors have recorded proton decay candidates, it seems that at the present time one cannot exclude the possibility that the candidates are due to neutrino background. The question that one might raise is: under what conditions could we ever be sure that some of the candidates could really be ascribed to proton decay? It is my feeling that we can only be sure when we have a sufficient sample of candidates of a particular branch that "fill" the expected phase-space. For instance, if a decay happens in oxygen, we would expect that the candidates would fill the calculated Fermi momentum phase-space. Some of the present candidates have very large net momentum and thus do not correspond to probable proton decay candidates; however, it has usually rather been emphasized that they correspond to improbable neutrino events. Because we know that neutrinos exist, they can also produce "improbable" events, but it may take of the order of ten "probable" examples of a particular proton decay mode before we can compare their distribution with the one expected for that mode. With the improved detectors coming on line, decay branches may become detectable in the next few years if they have a partial lifetime $\tau_B < 10^{33}$ years.

It is now generally accepted that our experiments have ruled out the minimal $SU(5)$ theory—the only one of many GUT approaches that made a testable prediction for the proton lifetime. In the absence of theoretical guidance, experimenters are still keen to extend the presently obtainable limits for the lifetime of the proton.

Future detectors, either of very large water volume as planned in Kamioka, or of very high resolution (liquid argon, and perhaps later methane, because of its free protons) as planned for Gran Sasso, may be able to reach lifetimes of $\sim 10^{34}$ years.

[a]The IMB collaboration consists of: R. M. Bionta, G. Blewitt, C. B. Bratton, D. Casper, A. Ciocio, R. Claus, M. Crouch, S. T. Dye, S. Errede, G. W. Foster, W. Gajewski, K. S. Ganezer, M. Goldhaber, T. J. Haines, T. W. Jones, D. Kielczewska, W. R. Kropp, J. G. Learned, J. M. LoSecco, J. Matthews, H. S. Park, F. Reines, J. Schultz, S. Seidel, E. Shumard, D. Sinclair, H. W. Sobel, J. L. Stone, L. Sulak, R. Svoboda, G. Thornton, J. C. van der Velde, and C. Wuest.

Superstrings—An Overview[a]

JOHN H. SCHWARZ

California Institute of Technology
Pasadena, California 91125

Superstrings originated in a 1971 collaboration with Neveu in which we sought (and found) a dual string theory that would improve upon the bosonic string theory of Veneziano and others.[1] Within the first year, it was shown how one could construct a consistent ghost-free interacting theory of bosonic and fermionic strings.[2] Supersymmetry on the two-dimensional world-sheet was established already in 1971, but it was not until 1976 that the space-time supersymmetry of the theory was understood. In 1972, it was shown that the theory required ten dimensions (nine space and one time) for its consistency. In 1981–82, Green and I demonstrated finiteness of one-loop diagrams, and in 1984, we showed that the cancellation of anomalies requires that the gauge group be $SO(32)$ or $E_8 \times E_8$.[3]

The early development of string theory was motivated by the desire to explain certain features of hadronic physics. Eventually, QCD was developed and widely accepted as the correct theory of the strong interactions. As a result, most workers lost interest in string theory. Thus, after a few years of very intense activity, it suddenly faded into obscurity. About that time, in 1974, Scherk and I noted that two of the most troubling features of string theory for application to hadronic physics could be turned into virtues if the goal was changed.[4] The occurrence of massless spin 1 and spin 2 states is an embarrassment for describing the hadron spectrum, but it is needed for a fundamental theory of elementary particles. Requiring the spin 2 state to interact with Newtonian strength (so that it could be interpreted as the physical graviton) forced us to decrease the fundamental length scale of the string by 20 orders of magnitude from 10^{-13} cm to 10^{-33} cm, the Planck length. Extra dimensions (six in the case of superstrings) are also an embarrassment for hadronic physics, but they could be a virtue in a theory containing gravity in which the space-time geometry is determined dynamically. The extra dimensions provide a natural arena for determining patterns of symmetry breaking and the physical low-mass spectrum. They must be very small, as in the Kaluza-Klein program, although the philosophy is very different. These possibilities, together with the fact that the string loop amplitudes are free from ultraviolet divergences, have sustained my confidence in the importance of this approach to quantum gravity and unification ever since 1974. The idea only received widespread acceptance when it became clear that it could give realistic phenomenology as well.

The standard model based on three families of quarks and leptons in representations of $SU(3) \times SU(2) \times U(1)$, spontaneously broken to an $SU(3) \times U(1)$ subgroup, is enormously successful. However, it cannot be the whole story. It does not contain gravity, and any straightforward attempt to adjoin gravity inevitably gives an

[a]This work was supported in part by the U.S. Department of Energy under Contract No. DEAC 03-81-ER40050.

ill-defined quantum field theory with nonrenormalizable ultraviolet divergences. In addition, it suffers from a number of aesthetic shortcomings that also suggest that a deeper, more fundamental theory should exist. For example, it offers no explanation of why particular Yang-Mills gauge groups and representations should occur. Many parameters must be adjusted arbitrarily to account for mixing matrices, masses, coupling strengths, and so forth. If offers no understanding of why there should be three (or whatever is the correct number) families of quarks and leptons. The description of symmetry breaking by the Higgs mechanism looks ad hoc and introduces additional arbitrary parameters. In superstring theory, we can see the possibility of overcoming all these difficulties.

Five superstring theories are known that seem to be free from divergences and anomalies. The number could increase if other ones are discovered. It could decrease if some are shown to be different versions of the same theory (there are some hints that this may be the case[5]) or to have inconsistencies that have not yet been identified. Ideally, the correct theory of nature would be the only one possible, so we hope that some of the five will succumb to theoretical attack.

Of the five theories, one (called type I) consists of interacting open and closed strings. These strings have no intrinsic orientation and give rise to one $D = 10$ supersymmetry. In other words, the conserved supercharge is a Majorana-Weyl spinor with 16 real components. This theory is only consistent (free from anomalies and infinities) if the gauge symmetry is $SO(32)$. It is unlikely to be realistic, however, which is fortunate because it is also the most difficult of the five theories to analyze.

The other four superstring theories contain oriented closed strings only. Two of them, called type II theories, have two $D = 10$ supersymmetries. In one case (type IIA), the two supercharges have opposite chirality and the theory is left–right symmetric. This feature rules it out as a candidate for a realistic theory. In the other case (type IIB), the two supercharges have the same chirality, so this objection does not apply. However, the type II theories do not possess a gauge symmetry group in ten dimensions, and it appears very unlikely that realistic symmetries and patterns of particles could arise from compactification of six dimensions.

The remaining two superstring theories, also based on oriented closed strings only, are called "heterotic" theories.[6] This name was introduced to refer to the advantages of hybridizing bosonic strings and superstrings. Modes that propagate on the string in one direction are described by the same mathematics as in the type II theories, whereas the ones that propagate in the other direction are described by the mathematics of the $D = 26$ bosonic string. The extra $26 - 10 = 16$ left-moving modes describe group-theory degrees of freedom. The only consistent choices give the gauge groups, $SO(32)$ (a second theory with this group) and $E_8 \times E_8$. The latter theory currently appears the most promising for realistic phenomenology.

The four theories with only oriented closed strings (type II and heterotic) have a remarkably simple perturbation expansion. The motion of a string sweeps out a two-dimensional surface in space-time, which is called the world-sheet. Thus, whereas Feynman diagrams in theories of point particles consist of lines connected up in various different ways, the Feynman diagrams of string theory are two-dimensional surfaces. As a technical matter, it is convenient to do a continuation from Minkowski space to Euclidean space. Then, the topological classification of two-dimensional surfaces, which characterizes the different possibilities for Feynman diagrams, is quite easy. For the four theories based on oriented closed strings, the relevant diagrams are closed oriented two-dimensional surfaces. The topology of such surfaces is completely specified by the "genus" g (the number of handles). The sphere has $g = 0$, the torus has $g = 1$, and so forth. The number g also corresponds to the number of loops in the string perturbation expansion. Thus, there is only one diagram at each order of the expansion. It is even conceivable that the series could be summed by a linear integral equation. (The kernel would be the "handle adder.")

Explicit evaluation of multiloop ($g > 1$) amplitudes involves state-of-the-art methods in the theory of Riemann surfaces. The subject is being developed very rapidly and the complete formulas should be available soon. There are a number of handwaving arguments that suggest that each of the diagrams should be finite, aside from divergences that are well understood and cause no trouble. (I find them compelling.) In any case, once the formulas are known, it should not be very hard to check this explicitly.

How can it be that string theory has such a simple perturbation expansion, whereas Einstein gravity, which corresponds to a low-energy approximation, has such a complicated one? The reason, basically, is that the infinite sequence of contact terms that arise from an expansion of the Einstein-Hilbert action are all replaced by exchanges of massive string modes. This results in simplified mathematics and the automatic incorporation of a scale (the string length or tension parameter) that serves as a cutoff for ultraviolet divergences. This is analogous to the replacement of the four-Fermi theory of the weak interactions by the standard $SU(2) \times U(1)$ model. In that case, a four-Fermi contact term is replaced by exchanges of massive W, Z, and Higgs particles, and nonrenormalizable divergences are replaced by renormalizable ones.

Each of the superstring theories is free from adjustable dimensionless parameters in its fundamental equations. It is therefore conceivable that all the parameters of the standard model may be calculable. There is also no freedom in choosing groups and representations, so these features might be understood as well. The main caveat here is that even though the equations are unique, they may have many different solutions ("vacua") that would be characterized by a number of parameters. We hope, of course, that this will not turn out to be the case, but it is a disturbing possibility. Even in this case, the correct solution (if found) should have considerable predictive power. A priori, the parameters of the heterotic theories (to give a specific example) are $\hbar$, c, the string tension T, the gauge coupling constant g, and the gravitational coupling constant κ. However, there is a relation of the form $Tg^2 \sim \kappa^2$. Moreover, one parameter, κ say, can be absorbed in the normalization of a scalar field ϕ (called the "dilaton"). Its value is then determined dynamically by the vacuum expected value $\langle \phi \rangle_0$. The remaining parameters simply determine units of mass, length, and time.

When classical field theories have conserved charges, it can happen that the corresponding quantum theories have a breakdown of the conservation law due to one-loop effects known as anomalies. If the conserved charges are associated with fundamental gauge fields of the theory, such an anomaly would imply that unphysical longitudinal modes, decoupled in the classical theory, can no longer be consistently eliminated. In this case, there is a breakdown of gauge invariance, and hence of unitarity and causality. In short, gauge-current anomalies are a bad thing that must be absent from a sensible quantum theory.

Anomalies were first studied in four-dimensional gauge theories by Adler, Bell, Jackiw, and others. They can arise from the contributions of chiral fermion loops in certain triangle diagrams. In fact, anomalies only occur in "chiral theories" in which the left-handed and right-handed fermions enter differently. One of the successes of the standard model is that the quarks and leptons give canceling contributions to anomalies so that the model is altogether anomaly-free. In ten dimensions, the simplest diagrams that can give anomalies are hexagons. Again, anomalies arise from contributions of chiral fermions (and bosons, too). In the IIA theory, there is left-right symmetry and trivial anomaly cancellation. In the other four cases, the theories are chiral and the cancellation of anomalies is highly nontrivial.

In ten dimensions, there can be anomalies not only in gauge currents, but in the energy-momentum tensor (the gravitational current) as well.[7] These are just as fatal for a theory. Such anomalies are referred to as "gravitational anomalies." In $D = 10$, there are massless point-particle theories that correspond to low-energy approximations to string theories. The type IIA supergravity theory is trivially anomaly-free (just as the string theory is) because it is nonchiral. In the case of type IIB supergravity, there are three nontrivial cancellations that eliminate the gravitational anomalies.[7] The remaining point-particle theories consist of $N = 1$ supergravity coupled to super Yang-Mills multiplets with a gauge group G. This coupled system turns out to have both gauge and gravitational anomalies for every possible choice of group G. For this reason, many people assumed the same must be true for superstring theories that reduce to these point-particle theories at low energy. It turns out, however, that string effects can result in cancellations for two choices of the group G.[3]

I will not go into the details of the anomaly analysis here because it has been reviewed elsewhere.[8] Let me simply review the result of the analysis. First, cancellation of pure gravitational anomalies requires that the group G have 496 generators because the coefficient of one potentially fatal term is proportional to $n - 496$. The -496 is provided by loops of chiral fermions from the supergravity multiplet, whereas the n comes from loops of chiral fermions from the Yang-Mills multiplet. Given that $n = 496$, the second condition required for the cancellation of all other gauge as well as mixed gauge-gravitational anomalies is

$$\mathrm{tr}\, F^6 = \frac{1}{48}\, \mathrm{tr}\, F^4\, \mathrm{tr}\, F^2 - \frac{1}{14{,}400}\, (\mathrm{tr}\, F^2)^3.$$

This formula is required to hold for an arbitrary 496×496 matrix F representing a generator of the group G. Remarkably, this condition is satisfied, with exactly the coefficients given, in just two cases: $SO(32)$ and $E_8 \times E_8$. Each of these groups has 496 generators. (Recall that $\dim E_8 = 248$.) As we explained earlier, two superstring theories have been constructed for the $SO(32)$ case. There is just one—the heterotic string—for the $E_8 \times E_8$ case. This group looks much more appealing from a phenomenological point of view.

As soon as it was known that superstring unification determines that the fundamental gauge group is $SO(32)$ or $E_8 \times E_8$, phenomenology could begin in earnest. Six of the spatial dimensions need to be rolled up into a tiny compact space K that is sufficiently small not to have been observed, while the others give four-dimensional Minkowski space. Thus, altogether, the ten-dimensional space-time should be given by $M_4 \times K$. In principle, this compactification should occur "spontaneously", meaning that it is a consistent ground state solution of the quantum mechanical equations of motion. So far, only classical solutions have been investigated and it appears that there are very many of them. Perhaps stability considerations or some subtle quantum mechanical effects will single out one of them, but so far this is only wishful thinking.

The first scheme for realistic compactification to be considered was one in which the space K is a Ricci flat (i.e., no Ricci curvature) Kähler manifold of $SU(3)$ holonomy.[9] This means that the curvature tensor takes values only in an $SU(3)$ subgroup of the $SO(6)$ group of rotations of the tangent space of K. It results in an effective $D = 4$ theory with $N = 1$ supersymmetry, which is required to solve the hierarchy problem. The manifold K in this case is called a Calabi-Yau space because it is a type whose existence was conjectured by Calabi and proved by Yau. In order to satisfy the classical superstring field equations, it is necessary to give nonzero vacuum values to some of the gauge fields. Specifically, an $SU(3)$ subgroup of gauge potentials should be equated with the spin connection of K. This can only be done in the case of the $E_8 \times E_8$ theory if the $SU(3)$ is embedded entirely in one E_8 factor according to the embedding $E_8 \supset E_6 \times SU(3)$, under which the adjoint breaks down by the branching rule,

$$248 = (78, 1) + (27, 2) + (\overline{27}, \overline{3}) + (1, 8).$$

When the symmetry breaking described above takes place, one is left with an E_6 gauge theory containing chiral supermultiplets in the $\underline{27}$ representation. This is the multiplet that one ordinarily uses for quarks and leptons in E_6 grand unification. The number of generations is determined by an elegant topological invariant of K, namely, its Euler characteristic. Specifically, one has $N_g = \frac{1}{2}|\chi(K)|$.

It is necessary to break E_6 down to $SU(3) \times SU(2) \times U(1)$ or perhaps to a somewhat larger group. There is a natural mechanism for doing this if K is not simply connected. In this case, there exist curves in K that are not contractible to a point. It is then possible for a particular E_6 gauge potential to have a nonzero line integral around such curves, even though the associated field strength vanishes. This is analogous to the familiar Bohm-Aharonov effect that also exhibits physical consequences associated with nonzero potentials having vanishing field strengths. These "Wilson loops" can be used to break E_6 to various possible subgroups. It is not possible to reach exactly $SU(3) \times SU(2) \times U(1)$ in this way. There must be some extension, such as an additional $U(1)$, that would imply a second Z boson.

The E_6 scheme described above may not be entirely satisfactory. The 27s of E_6 are supposed to describe Higgs supermultiplets H, as well as the quark/lepton ones, Q. Ideally, the only dimension-four couplings of supermultiplets would be Q^2H and H^3. However, in the E_6 context, Q^3 and QH^2 occur as well. This implies, in particular, very rapid proton decay. Another problem with E_6 is that a 27 contains 12 unknown states that must somehow be hidden from view. Also, there is no evidence for a second Z boson, although it is not yet ruled out.

A scheme that gives $SO(10)$ instead of E_6 would probably be preferable because one could use the $\underline{16}$ for the Q multiplets and the $\underline{10}$ for the H multiplets. Then, Q^3 and QH^2 couplings would be ruled out by group theory. Recent work by Witten, Strominger, and Hull suggests that there exist $SO(10)$ solutions with $N = 1$ supersymmetry at low energies. When the spin connection of K is projected onto the string world-sheet, it has two components, ω_+^{rs} and ω_-^{rs}, associated with right- and left-moving modes of the string. In the heterotic theory, the right-moving modes are responsible for supersymmetry, whereas the left-moving ones have to do with the group theory. Therefore, if ω_+ takes values in $SU(3)$, one obtains $N = 1$ supersymmetry in $D = 4$ as before. However, if ω_- takes values in $SO(6)$, then this is the group that must be embedded in E_8, which breaks it to $SO(10)$ according to the rule $E_8 \supset SO(10) \times SO(6)$, with

$$248 = (45, 1) + (16, 4) + (\overline{16}, \overline{4}) + (10, 6) + (1, 15).$$

The number of 4 and $\overline{4}$ zero modes on K determines the number of Q supermultiplets, whereas the number of 6 zero modes determines the number of H supermultiplets. The asymmetry between ω_+ and ω_- in this construction depends on the existence of torsion in the solution. This type of scheme is just beginning to be analyzed, and it is not yet clear how realistic it can be made. However, it does seem to represent an improvement on the original E_6 approach.

In addition to obtaining quark/lepton and Higgs supermultiplets with realistic quantum numbers and multiplicities, we would like to know all their dimension-four couplings so that we could deduce all masses, mixing angles, lifetimes, etc. If this required explicit knowledge of the metric of K and the precise values of all other background fields, it would probably be hopelessly difficult. There are encouraging indications, however, that many (and maybe even all) of the couplings of the low-energy effective theory can be deduced from topological properties of K.[10] This is a very exciting possibility because these can probably be calculated analytically (using fancy methods of algebraic geometry.) The numbers obtained in this way would be exact in the tree approximation of the string theory. Which of them would survive quantum corrections is not yet known. This is important to sort out because there are indications that radiative corrections could be quite large.[11]

When K is not simply connected (as we need for symmetry breaking), strings can wrap around noncontractible curves, thus describing new "soliton" states of the theory. Wen and Witten have shown that such states describe particles with fractional electric charge e/n.[12] The integer n depends on detailed properties of the fundamental group of K, the noncontractible curve in question, and its associated Wilson loop. These particles are color singlets with a mass of order 10^{18} GeV. Thus, they are unconfined and have nothing whatsoever to do with quarks. The same theories also admit magnetic monopole solutions analogous to those of standard GUT theories. The details are different, however, because it is necessary to investigate whether a gauge field configuration can be unwrapped in ten dimensions.[12] The minimum magnetic charge turns out to be n times the Dirac unit, which is required by the Dirac quantization condition for a theory containing electric charge e/n.

Superstring theory automatically gives rise to axions with the properties required to account for CP conservation in QCD. Whether the decay constant is consistent with cosmological bounds is not yet clear. CP violation in the electroweak interactions can be achieved if the internal manifold K has no "orientation reversing isometry", that is, no reflection symmetry.[10]

One of the problems in ordinary GUTS is the so-called "fine-tuning" problem. The parameters of Higgs potentials need to be adjusted to high precision in order to produce Higgs doublets at the electroweak scale, while keeping associated triplets at ultrahigh scales. The Wilson-loop symmetry-breaking mechanism of superstring theory involves operators whose possible values are discrete. For certain choices, the desired hierarchy can be achieved without dialing parameters.

An especially intriguing feature of the $E_8 \times E_8$ theory concerns the role of the second E_8. All observed gauge forces presumably arise from one E_8 factor, whereas the second one gives gauge fields that do not couple to matter that we are familiar with. There should exist "shadow matter" carrying quantum numbers of the second E_8, which would only be observable through effects of gravitational strength. It is plausible that shadow matter could account for about half of the mass of the universe, thus being an important ingredient in the solution of the dark-matter problem. In most schemes that have been considered so far, the symmetry-breaking patterns of the two E_8 factors are different. If the physics were symmetrical, this would have some amusing science fiction–like implications, but that seems unlikely to be the case.

A great deal of work remains to be done in developing the theoretical machinery of superstring theory. At the moment, we find ourselves in the peculiar position of knowing the theories well enough to calculate Feynman diagrams. However, because we are dealing with a unified theory of gravity and all other forces, it is clear that there should be a profound underlying geometric principle. So far, we have not succeeded in understanding this principle. The suspicion is that a new branch of geometry, which might be called "string geometry", is required. It would presumably represent a string modification of Riemannian geometry at short distances. The historical evolution is backwards from that of general relativity. Einstein began by formulating the equivalence principle, and then developed the mathematical machinery for building a theory that incorporates it. We know the superstring equations, but not the underlying principle.

A field theory description of interacting open strings has been developed.[13] It is a beautiful gauge theory with an incredible amount of symmetry. However, much work still remains to be done because gravity is associated with closed strings. Therefore, the truly profound geometric issues are only expected to arise in the construction of a gauge-field theory of closed strings. Once this is achieved, we should be in a better position to discuss compactification. Until now, the compact space K has been described using the language of Riemannian geometry. If, as we expect, the size of K is comparable to the characteristic size of strings, this could be quite misleading. String geometry should be a much better tool.

At present, we know a theory—the $E_8 \times E_8$ heterotic string—that has no adjustable dimensionless parameters and that could give a correct description of all fundamental physics. This is certainly a remarkable state of affairs, but we still want much more. As far as we can tell, there are a large number of candidate vacuum solutions that solve the equations. This forces us to make a phenomenologically-based choice among them. It would be far preferable if a single solution were singled out by theoretical criteria. This would be tantamount to deducing the initial wave function of the universe, as well as the Schrödinger equation. Then, we could calculate (at least in principle) the complete wave function of the universe. The problem thus would be to understand how probability amplitudes get converted into our particular reality.

REFERENCES

1. RAMOND, P. 1971. Phys. Rev. **D3:** 2415; NEVEU, A. & J. H. SCHWARZ. 1971. Nucl. Phys. **B31:** 86.

2. For a collections of reprints, see: SCHWARZ, J. H. 1985. Superstrings. World Scientific. Singapore.
3. GREEN, M. B. & J. H. SCHWARZ. 1984. Phys. Lett. **149B:** 117.
4. SCHERK, J. & J. H. SCHWARZ. 1974. Nucl. Phys. **B81:** 118.
5. FREUND, P. 1985. Phys. Lett. **151B:** 387; CASHER, A., F. ENGLERT, H. NICOLAI & A. TAORMINI. 1985. Phys. Lett. **162B:** 121; HARVEY J., Unpublished.
6. GROSS, D., J. HARVEY, E. MARTINEC & R. ROHM. 1985. Phys. Rev. Lett. **53:** 502.
7. ALVAREZ-GAUMÉ, L. & E. WITTEN. 1983. Nucl. Phys. **B233:** 269.
8. SCHWARZ, J. H. 1986. Superstrings and Supergravity. A. T. Davies & D. G. Sutherland, Eds.: 301. Proceedings of the 1985 Scottish Univ. Summer School in Physics.
9. CANDELAS, P., G. HOROWITZ, A. STROMINGER & E. WITTEN. 1985. Nucl. Phys. **B258:** 46.
10. STROMINGER, A. 1985. Unified String Theories. M. Green & D. Gross, Eds.: 654. World Scientific. Singapore; WITTEN, E. 1985. Nucl. Phys. **B258:** 75.
11. DINE, M. & N. SEIBERG. 1985. Phys. Rev. Lett. **55:** 366; Phys. Lett. **162B:** 299; KAPLUNOVSKY, V. 1985. Phys. Rev. Lett. **55:** 1036.
12. WEN, X. G. & E. WITTEN. Princeton Univ. preprint.
13. There is a vast literature by now including: SIEGEL, W. 1985. Phys. Lett. **151B:** 391, 396; NEVEU, A. & P. C. WEST. CERN preprints; BANKS, T. & M. PESKIN. 1986. Nucl. Phys. **B264:** 513; WITTEN, E. 1986. Nucl. Phys. **B268:** 253.

An Introduction to String Field Theory

A Pedestrian Approach to the Covariant Formulation[a]

GEOFFREY B. WEST

Los Alamos National Laboratory
Los Alamos, New Mexico 87545

INTRODUCTION

One of the most appealing aspects of the field theoretic formulation of elementary particle interactions is that it very naturally incorporates general symmetry properties such as Lorentz and gauge invariance. Furthermore, nonperturbative effects arising from spontaneous symmetry breaking or from the underlying topological structure of the theory are most easily accommodated within this framework. Indeed, the usual formulation and understanding of the standard model is almost unthinkable outside of field theory. Therefore, it is quite natural to attempt a similar formulation for string theories because with the recent discovery of anomaly cancellations,[1] these have emerged as serious candidates for a completely unified description of all the interactions including gravitation.[2] In the string case, there is perhaps even stronger motivation for a gauge-invariant field theoretic description because one of its more remarkable properties is the automatic appearance of massless gauge particles (in particular, the graviton) without the explicit introduction of some fundamental principle akin to Einstein's general coordinate invariance or the principle of equivalence. Presumably, part of this mystery is due to the fact that these theories were formulated in the physical light-cone gauge, thereby masking any general underlying gauge symmetry. The formulation of a completely gauge-invariant action for the interacting string would hopefully lead to a deeper understanding of how and why these theories work and, ultimately, what the underlying principle actually is. Furthermore, without such a formulation, it would seem unlikely that the question of compactification to a four-dimensional "real" world could be dynamically understood because this requires an understanding of the vacuum structure of the theory.

Ordinary field theory is based upon the Newtonian concept of point particles whose motion is described by one-dimensional world lines, $x_\mu(\tau)$, parametrized by the evolution parameter (time) τ (see FIGURE 1). In principle, one could set up particle interactions in terms of these first quantized world lines without explicit reference to a field and thereby calculate all S-matrix elements. Typically, this is not what is done; instead, one introduces a local field $\phi(x)$ whose Fourier components represent the creation or destruction of particles of a given momentum. In this formalism, the scattering of particles is ascribed to local interactions among the various fields. As the simplest possible example of this duality between a world-line and field formulation, I shall examine (below) the case of a free relativistic point particle and show how the

[a]This work was supported by the U.S. Department of Energy under Contract No. W-7405-ENG-36.

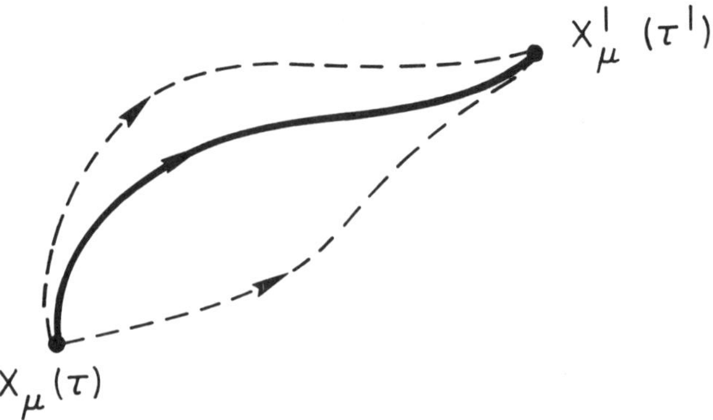

FIGURE 1. World-line trajectories for particle propagation. The solid line represents the classical trajectory, whereas the broken line represents quantum fluctuations (see equation 6).

Feynman propagator, usually derived from field theory, can be calculated from a world-line approach. This is instructive not only because it can, in principle, be generalized to the string, but also because it exposes at a very basic level the role of gauge invariance and its relationship to reparametrization invariance. The generalization to the string involves replacing the evolution of world lines, $x_\mu(\tau)$, by the evolution of world sheets, $X_\mu(\sigma, \tau)$, which are swept out by the motion of the string[3] [σ is the string coordinate that can itself be parameterized by τ: $\sigma(\tau)$; see FIGURE 2]. Notice that because X_μ in general depends on a two-dimensional parameter $\sigma_a \equiv (\sigma, \tau)$ rather than simply on a one-dimensional parameter τ, as in $x_\mu(\tau)$, it behaves like a 2-D field. Thus, if the dynamics of the string are to be formulated in terms of a string field[4] $\Phi[X_\mu(\sigma)]$ that depends locally on the coordinate field $X_\mu(\sigma, \tau)$, then mathematically it must behave like a functional of X_μ. This is in contrast to ordinary field theory where

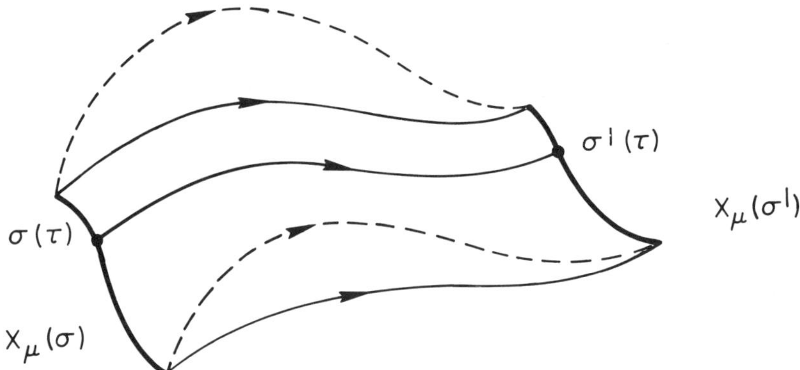

FIGURE 2. World sheets representing evolution of an open string. A point on the string can be parametrized by $\sigma(\tau)$. The solid lines represent classical paths, whereas the broken ones are quantum fluctuations (see equation 18).

the field of a particle $\phi(x_\mu)$ behaves like an ordinary function of $x_\mu(\tau)$, and its action, for example, $S[\phi(x)]$, behaves like a functional. We can therefore make the following correspondence:

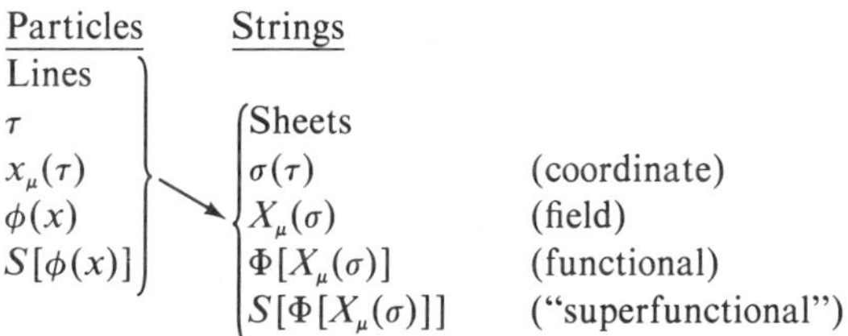

The string concept was originally introduced as a way of understanding the linearly rising Regge trajectories of the hadronic spectrum. Because the string functional incorporates all excitations of the string, it is not surprising that it can be decomposed into towers of particle states with ever-increasing mass and spin. This suggests that the construction of a covariant gauge-invariant action for Φ can be expected to be closely related to the analogous construction for ordinary massive fields of arbitrary spin. I shall discuss this analogy for the low-spin components of Φ in some detail below and use it to show how to construct the free gauge-invariant equations for Φ. The technique is simple and quite unsophisticated, particularly when compared to other methods in the literature.[5] It is not clear however that this simple approach is generalizable to the serious problem of including interactions;[6] on the other hand, it does, I think, allow one to see reasonably clearly what is going on. As an example of this simplicity, I shall show how to construct a free action outside of the critical dimension (26 for the bosonic string).[7] Whether this is an advantage or has any deep significance, I will not argue because interactions certainly need to be included.

Interactions introduce some highly nontrivial technical questions (including some questions of principle) that I shall not have time to discuss here.[6] Let me only remark that at the first quantized level (i.e., in terms of X_μ), interactions arise in a conceptually very simple manner, namely, from the breaking and joining of strings (as illustrated in FIGURE 3). Thus, if string 3 breaks into two pieces—string 1 and string 2, say—then symbolically $X_\mu(\sigma_3) \sim X_\mu(\sigma_2) + X_\mu(\sigma_1)$. Translating this into a field theoretic statement suggests an interaction term of the form $\Phi[X_1]\Phi[X_2]\Phi[X_3]\,\delta\,[X_1 + X_2 - X_3]$, which can be thought of as the analogue of a $\phi^3(x)$ interaction in ordinary field theory. However, unlike the situation there, the string interaction is nonlocal in X_μ, thus violating our primitive prejudices concerning fundamental interactions. A further question is what sort of gauge transformation and what sort of fundamental symmetry could possibly generate such kinds of interactions? These are the sorts of questions that are beginning to be asked, and some partial answers are beginning to emerge.[6] The present thrust would appear to involve new mathematics, and it may well be that to ultimately understand the string, we are all going to have to reeducate ourselves in erudite areas of differential geometry, group manifolds, and the like. Meanwhile, I am going to stay at a relatively pedestrian level and only use concepts and mathematics that we are all used to from ordinary common-or-garden field theory. I apologize to my more sophisticated colleagues and hope that the noncognoscenti can get a flavor of what string field theory involves from my remarks.

WORLD-LINE PARTICLE TRAJECTORIES AND
ORDINARY FIELD THEORY

I want first to briefly explore the relationship of particle field theory to a more physical description in terms of world-line trajectories. For simplicity, only the noninteracting case will be considered here.

The classical action for a free relativistic point particle is given by[8]

$$S = m \int ds, \tag{1}$$

where $ds = (-dx^\mu \, dx_\mu)^{1/2}$ is the world-line element for the particle's trajectory. This

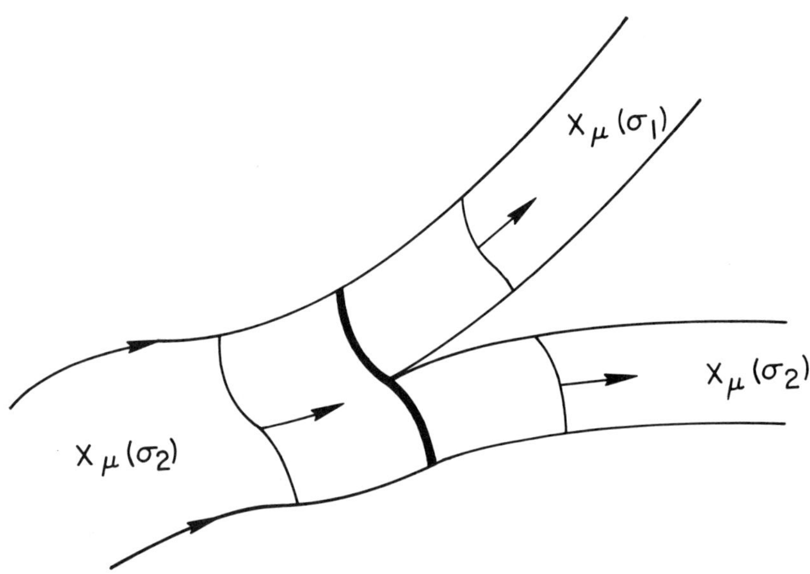

FIGURE 3. The breaking of a string into two separate strings, thus representing a "cubic" string field functional interaction.

can be reexpressed as

$$S = m \int \left[-\frac{dx^\mu}{d\tau} \frac{dx_\mu}{d\tau} \right]^{1/2} d\tau, \tag{2}$$

thereby defining the Lagrangian to be

$$L(x_\mu, \dot{x}_\mu) = [-\dot{x}^\mu \, \dot{x}_\mu]^{1/2} \tag{3}$$

(where $\dot{x}_\mu \equiv dx_\mu/d\tau$). We are interested in the quantum mechanical propagation of the particle and so need to calculate the matrix element $\langle x'(\tau') | x(\tau) \rangle$ (see FIGURE 1). In a "naive" path integral representation, this is governed by the action of equation 2.

At first sight, this seems problematical because it is nonlinear in x_μ, thus making an evaluation of the path integral intractable. There is also a small question of principle because there appear to be two "times" in the problem: τ and x_0. It is quite instructive to see how the formalism actually circumvents these two problems.

First, let us calculate the canonical momentum:

$$p_\mu \equiv \frac{\partial L}{\partial \dot{x}^\mu} = \frac{m\dot{x}_\mu}{[-\dot{x}_\mu^2]^{1/2}}. \tag{4}$$

From this, the Hamiltonian $H \equiv p^\mu \dot{x}_\mu - L$ is seen to vanish, and therefore it cannot be the generator of the time (i.e., τ) evolution of the particle. A further consequence of equation 4 is the τ-independent constraint,

$$p^2 + m^2 = 0, \tag{5}$$

representing the classical mass-shell condition. Because this must remain valid as the system evolves, not all of the canonical coordinates are independent and care must be taken when quantizing them. The general problem of quantizing such constrained systems within the canonical formalism was solved many years ago in an elegant fashion by Dirac.[9] In more recent times, his ideas have been applied to the path integral formulation by Fade'ev,[10] whose work forms the foundation for the quantization of gauge field theories and the subsequent appearance of Fade'ev-Popov ghosts. Fade'ev's general expression is given by

$$\langle x'_\mu(\tau)|x_\mu(\tau)\rangle = \int Dp_\mu \int Dx_\mu \, e^{i\int_\tau^{\tau'} [p^\mu \dot{x}_\mu - H]d\tau} \, \delta[f(x_\mu, p_\mu)] \, \delta[g(x_\mu, p_\mu)] \det\{f, g\}, \tag{6}$$

where $f(x_\mu, p_\mu) = 0$ is the "primary" constraint (here, equation 5) and $g(x_\mu, p_\mu) = 0$ is a "gauge choice" that ensures that only the correct number of independent degrees of freedom get quantized. The determinant of the Poisson bracket $\{f, g\}$ ensures that the matrix element is "gauge invariant," that is, that it does not depend on the choice of g. For the case considered here, $H = 0$ and $f = p^2 + m^2$, so $\{f, g\} = 2p_\mu \partial^\mu g$. Thus, the nonlinear nature of L does not actually appear in the path integral, and so it causes no explicit technical problem. Furthermore, the question of the "two times" is "solved" by the gauge freedom implicit in the choice of g. For example, the "natural gauge" is to identify $x_0 = \tau$ by choosing $g(x_\mu, p_\mu) = (x_0 - \tau)/2p_0$. This has the added advantage that $\det\{f, g\} = 1$ so that there are no "Fade'ev-Popov ghosts." In this gauge,

$$\langle x'_\mu(\tau')|x_\mu(\tau)\rangle = \int Dp \int Dx \, e^{-i\int_\tau^{\tau'} [p\cdot\dot{x} - \sqrt{p^2+m^2}]d\tau}, \tag{7}$$

allowing one to identify $\sqrt{p^2 + m^2}$ as an effective Hamiltonian. It is straightforward to carry out the integral over x and show that equation 7 reduces to the standard Feynman propagator:[11]

$$\Delta_F(x' - x) = \int \frac{d^4p \, e^{ip\cdot x}}{p^2 + m^2 - i\epsilon}. \tag{8}$$

It is worth noting that although the "natural" gauge choice is not covariant (because it preferentially sorts out x_0), the final result is. Covariant gauge choices can of course be made; for example, $g \sim x \cdot p - h(\tau)$, which contains an arbitrary function of τ. The possibility of such arbitrariness in the gauge choice is a reflection of reparametrization invariance—$\tau \rightarrow \tau + \xi(\tau)$, say—which clearly cannot affect the physics.

In the field theory description, the propagator is derived from the time-ordered product[10]

$$\langle T[\phi(x')\phi(x)] \rangle \equiv \int D\phi \, e^{iS[\phi(x)]}\phi(x')\phi(x), \tag{9}$$

where S is the field action functional. Notice that no reference to a constrained Hamiltonian or gauge choice occurs in this description. In addition, when writing $\phi(x)$, x_0 is generally understood to be the time and no explicit reference to τ is made. In the language of the first-quantized formalism, this corresponds to working in the natural gauge $x_0 = \tau$. The field representation (equation 9) is presumably to be identified with equation 6 and it subsequently must give the Feynman result (equation 8). Suppose we express S in bilinear form:

$$S[\phi(x)] = \tfrac{1}{2}\int d^4x \, \phi(x) \, K\phi(x); \tag{10}$$

then in order to reproduce equation 8, a standard Gaussian integration must reveal that $K = (\partial^2 - m^2)$. The consequent equation of motion for $\phi(x)$ is the standard Klein-Gordon one,

$$(\partial^2 - m^2) \, \phi(x) = 0, \tag{11a}$$

which, with the de Broglie substitution of $p_\mu \rightarrow - i\partial_\mu$, can be expressed as

$$(p^2 + m^2) \, \phi(x) = 0. \tag{11b}$$

The above rather trivial sequence of steps (which reverses the usual textbook ones) can be viewed as a "derivation" of the covariant Klein-Gordon operator appropriate to free field theory starting from the particle world-line trajectory viewpoint. This suggests that we can similarly "derive" the analogous covariant free action for the string field functional from a first-quantized formalism expressed in terms of X_μ. Unfortunately, the problem is not quite so trivial as for the particle, as we shall now discuss.

WORLD-SHEET STRING TRAJECTORIES AND STRING FIELD THEORY

We are now going to mimic the above discussion and apply it to the string with the view to obtaining the covariant action for its field functional. The classical action analogous to equation 1 for a free string is simply obtained by replacing the invariant line element ds by the invariant area element dA:[3]

$$S = \int dA. \tag{12}$$

This can be expressed in the form

$$S = \int_{-\infty}^{\infty} d\tau \int_0^\pi d\sigma \, \sqrt{-g}, \tag{13}$$

where the length of the string has been rescaled to be π.

The integrand can be interpreted as a Lagrangian density for the 2-D field

$X_\mu(\sigma, \tau)$: $L[X_\mu, \partial_a X_\mu] = \sqrt{-g}$, with $g = \det g_{ab}$ and $g_{ab} = (\partial_a X^\mu)(\partial_b X_\mu)$ as the metric of the world sheet. This can be reexpressed in a form analogous to equation 3:

$$L[X_\mu, \partial_a X_\mu] = [\dot{X}^2 X'^2 - (\dot{X}^\mu X'_\mu)^2]^{1/2}, \tag{14}$$

where $\dot{X} \equiv \partial_\tau X$ and $X' \equiv \partial_\sigma X$. The canonical momentum conjugate to X_μ is

$$\Pi_a^\mu = \partial L / \partial X_\mu^a = \sqrt{-g}\,\partial_a X^\mu. \tag{15}$$

This is easily seen to satisfy the constraint

$$\Pi_a^\mu \Pi_\mu^b + g\delta_a^b = 0, \tag{16}$$

which is the string analogue to $p^2 + m^2 = 0$. It is only the τ-independent components of these constraints that must be made explicit in the path integral. These are

$$f_1(\Pi_\mu^\tau, X_\mu) \equiv \Pi_\tau^2 + X'^2 = 0 \tag{17a}$$

and

$$f_2(\Pi_\mu^\tau, X_\mu) \equiv \Pi_\tau^\mu X'_\mu = 0. \tag{17b}$$

In contrast to the particle case, notice that these explicitly break (σ, τ) covariance; this is a little like the case of Coulomb's law ($\nabla \cdot E = 4\pi\rho$) in QED, which is a time-independent constraint that destroys the explicit Lorentz covariance.

As in the particle case, the Hamiltonian $H \equiv \Pi_\tau^\mu \dot{X}_\mu - L$ vanishes here as well, thus taking with it the obnoxious nonlinear dependence on X_μ residing in the original classical action (equation 14). We can now write the analogue to equation 6 for the string propagator (see FIGURE 2):

$$\langle X'_\mu(\sigma', \tau') | X_\nu(\sigma, \tau) \rangle = \int DX_\mu \int D\Pi_\mu^\tau\, e^{i \int d\tau \int d\sigma \Pi_\tau^\mu \dot{X}_\mu}$$

$$\times\, \delta[\Pi_\tau^2 + X'^2]\, \delta[\Pi_\tau^\mu X'_\mu]\, \delta[g_1(X_\mu, \Pi_\mu^\tau)]$$

$$\times\, \delta[g_2(X_\mu, \Pi_\mu^\tau)]\, \det\{f_1, g_1\}\, \det\{f_2, g_2\}. \tag{18}$$

This involves two gauge choices, g_1 and g_2, corresponding to the two contraints, equations 17a and 17b, respectively. These can clearly be associated with the invariance of the physics to reparametrizations

$$\sigma \rightarrow \sigma'(\tau, \sigma) \tag{19a}$$

and

$$\tau \rightarrow \tau'(\tau, \sigma). \tag{19b}$$

A popular and technically very useful gauge is the "light-cone" gauge[3] in which the two choices are $X_0 + X_1 \sim \tau$ and $\Pi_\tau^\mu \sim \dot{X}_\mu$. This gauge can be thought of as analogous to the Coulomb gauge of electrodynamics because only "physical" transverse modes survive. In addition, at the classical level, this gauge has the nice property that the 2-D metric is flat and the Langrangian is linear.

We would like to identify equation 18 with a field functional representation (much as equation 9 was identified with equation 6) and thereby determine the corresponding field functional action:

$$\langle T\Phi[X'_\mu]\,\Phi[X_\nu]\rangle = \int D\Phi\, e^{iS[\Phi]}\,\Phi[X'_\mu]\,\Phi[X_\nu]. \tag{20}$$

As in equation 10, one expects S to be expressible in bilinear form, for example,

$$S[\Phi] = \int DX_\mu\,\Phi[X_\mu]\,K\,\Phi[X_\mu]. \tag{21}$$

(Notice, incidentally, that S itself must be a functional integral because Φ is a functional.) Just as in the particle case, one can now imagine determining K by identifying equation 20 with equation 18. Although carrying this out is apparently "just" a technical problem, unfortunately no simple gauge invariant expression for equation 18 analogous to the Feynman propagator exists. In general, there is also no guarantee that the result is expressible as a simple path integral of the form of equation 20 as in the particle case. It is not inconceivable that other auxiliary fields are required to "absorb" the extra gauge invariance associated with the string. That this may indeed be the case is hinted at by using an *ansätze* suggested by the particle case. There, the resulting equations of motion were essentially tantamount to enforcing the constraint (equation 5) as an operator on the field $\phi(x)$ (using the de Broglie replacement $p_\mu \rightarrow -i\partial_\mu$ as in equations 11). In the string case, the analogous procedure would be to replace Π_μ by $-i\delta/\delta X^\mu$ in equations 17, thereby leading to the functional field equations,

$$\left[\frac{\delta^2}{\delta X_\mu^2} - X'^2\right]\Phi\,[X_\mu] = 0 \tag{22}$$

and

$$X'_\mu\frac{\delta}{\delta X_\mu}\,\Phi\,[X_\mu] = 0. \tag{23}$$

It was suggested by Siegel[12] that these be viewed as an equation of motion, namely, equation 22, in a "gauge" defined by equation 23, with gauge transformations being defined as reparametrizations of the string. This is reminiscent of situations that one encounters in ordinary field theory: for example, a massive spin-one particle is described by the equation of motion

$$(\partial^2 - m^2)\,A_\mu(x) = 0 \tag{24a}$$

subject to

$$\partial^\mu A_\mu(x) = 0. \tag{24b}$$

This latter constraint can be viewed as a gauge choice (the usual Lorentz gauge of QED) even though these equations represent a massive vector particle. It is clearly natural to see whether one can carry this gauge idea further and construct an action for the massive field that is invariant to a standard gauge transformation of the kind

$$\delta A_\mu = \delta_\mu \lambda. \tag{24c}$$

Such an action was indeed constructed many years ago by Stueckelberg,[13] and below I shall indicate how one can systematically extend his result to massive fields of arbitrary spin.[7] The important point here, however, is that in order to carry out such a construction, it turns out that an additional auxiliary field must be added to the action. If one were to insist on using only the original spin field, then the price paid is the loss of a local action. Thus, one might guess that the identification of equation 18 with equations 20 and 21 would lead to a nonlocal expression for K.

At this stage, then, we shall abandon the approach based explicitly on equation 18 and explore rather how an invariant action, analogous to Stueckelberg's action, can be constructed starting from the *ansätze* (equations 22 and 23). Before doing this, however, I want to spend a little time examining the meaning and content of these equations in order to get a better feel for what Φ actually represents and in order to show the intimate connection to massive spin fields.[14]

COMPONENT EQUATIONS FOR Φ

Because the σ dependence of $X_\mu(\sigma)$ represents a finite string, it is natural to make a Fourier decomposition into its normal modes. To keep things simple, throughout this section,[14] I shall only deal with the open string that satisfies the boundary conditions $X'(0) = X'(\pi) = 0$. Thus,

$$X_\mu(\sigma) = x_\mu + 2 \sum_{n=1}^{\infty} x_\mu^n \cos n\sigma, \tag{25}$$

with the zero mode x_μ interpreted as the "ordinary" d-dimensional space-time coordinate. It will prove convenient to introduce $P_\mu \equiv \Pi_\mu^\tau + X_\mu \rightarrow -i\delta/\delta X^\mu + X_\mu$, whose Fourier expansion is

$$P_\mu(\sigma) = \frac{1}{\sqrt{\pi}} \sum_{n=1}^{\infty} a_\mu^n \, e^{in\sigma}. \tag{26}$$

It is not difficult to check that

$$a_n^\mu = \frac{i}{2\sqrt{\pi}} \left[\frac{\delta}{\delta x_\mu^n} + 2\pi x_n^\mu \right], \qquad n \neq 0, \tag{27a}$$

$$a_0^\mu = \frac{i}{\sqrt{\pi}} \partial^\mu \text{ (representing the "ordinary" momentum operator)}, \tag{27b}$$

and that the a_n^μ satisfy the following:

$$a_n^\mu = a_{-n}^{\mu +} \tag{28a}$$

and

$$[a_n^\mu, a_m^{\nu +}] = n \, \delta_{mn} \, g^{\mu\nu}. \tag{28b}$$

The field functional $\Phi[X_\mu(\sigma)]$ can now be expanded in terms of these modes. To expose their physical content, it is advantageous to separate out the zero mode because this is going to be interpreted as "ordinary" space-time:

$$\Phi[X_\mu(\sigma)] = \sum_{\substack{mn... \\ \mu\nu...}} \phi^{\mu\nu...}_{mn...}(x) \, [a^{m+}_\mu \, a^{n+}_\nu \, ...] \, \Phi_0. \tag{29}$$

The ground state Φ_0 is defined by $a^n_\mu \Phi_0 = 0$ and represents the ground state of the infinite number of ordinary harmonic oscillators of the vibrating string: in an explicit Schrödinger basis, Φ_0 is simply an infinite product of Gaussians in x^n_μ. The creation operators produce all possible states of the string with weights $\phi^{\mu\nu...}_{mn...}(x)$. These functions of ordinary space-time are the component fields of the string and they represent particles of ever-increasing mass and spin. The fact that Φ can be decomposed into these infinite towers of ordinary harmonic oscillators with a linear spectrum is the reason that the string was invented to describe the "old" hadronic spectrum with its linearly rising Regge trajectories. More explicitly, equation 29 can be written as

$$\Phi[X_\mu(\sigma)] = [\phi(x) + A^\mu_n(x) \, a^{n+}_\mu + \theta^{\mu\nu}_{mn}(x) \, a^{m+}_\mu \, a^{n+}_\mu \, ...]\Phi_0, \tag{30}$$

where $\phi(x)$ is a single scalar, $A^\mu_n(x)$ is an infinite tower of vectors, etc.

To incorporate the constraints (equations 17) and subsequent equations of motion (equations 22 and 23), we shall need the expansion of $P^2_\mu(\sigma)$:

$$P^2_\mu(\sigma) \equiv 1/\sqrt{\pi} \sum_{n=1}^{\infty} L_n \, e^{in\sigma}. \tag{31}$$

This defines the operators

$$L_n = L^+_n = \int_{-\pi}^{\pi} d\sigma \, e^{in\sigma} \, P^2_\mu(\sigma) = \frac{1}{2} \sum_{k=-\infty}^{\infty} \cdot a^+_k a_{k+n}; \tag{32}$$

which generate the Virasoro algebra[3]

$$[L_m, L_n] = (m - n) \, L_{m+n} + d/12 \, m(m^2 - 1) \, \delta_{m,-n}. \tag{33}$$

The normal ordering in equation 32 is included in order to define away the infinite zero-point energy associated with each harmonic oscillator. Notice that it is only relevant for the definition of L_0, along with being the origin of the c-number term in the algebra (equation 33). This has an important consequence because the proposed equations of motion (equations 22 and 23) in this basis read $L_n\Phi = 0$, which by virtue of the c-number term is inconsistent with the algebra. To circumvent this difficulty, Siegel[12] suggested that the appropriate equations be

$$(L_0 - 1) \, \Phi = 0 \tag{34a}$$

and

$$L_n\Phi = 0, \qquad n > 0. \tag{34b}$$

Now, from equation 32, one can see that $L_0 = -(\partial^2/2\pi - N)$, where $N = \Sigma^\infty_{k=1} a^+_k a_k$ is the number operator and thus acts like a Klein-Gordon operator. Furthermore, the operators $L_n \sim i/\sqrt{\pi}\, a_n^\mu \partial_\mu + \cdots$ $(n > 0)$ act like a divergence operator. Thus, at least superficially, equations 34 are quite analogous to equations 24 for a massive vector field. This analogy can be made precise by breaking equations 34 down into component form using equations 30 and 32, together with equation 28:

$$
\begin{array}{ll}
\underline{(L_0 - 1)\,\Phi = 0} & \underline{L_n \Phi = 0} \\[1em]
(\partial^2 + 2\pi)\phi(x) = 0 & \\[0.8em]
\partial^2 A_\mu^1(x) = 0 & \partial^\mu A_\mu^1(x) = 0 \\[0.8em]
(\partial^2 - 2\pi)\,A_\mu^2(x) = 0 & A_\mu^2(x) + 1/\sqrt{\pi}\,\partial^\nu \theta_{\mu\nu}^{11}(x) = 0 \\[0.8em]
(\partial^2 - 2\pi)\,\theta_{\mu\nu}^{11}(x) = 0 & 1/\sqrt{\pi}\,\partial^\mu A_\mu^2(x) + \tfrac{1}{2}\theta_{\;\mu}^{11\mu}(x) = 0 \\[0.8em]
\qquad\text{etc.} & \qquad\text{etc.}
\end{array}
\tag{35}
$$

Notice the presence of a scalar tachyon and a massless vector; in the closed string, the latter becomes a massless tensor (the "graviton"), whereas in the supersymmetric version, the former disappears. The most natural interpretation of these equations is that of equations of motion of gauge fields in a generalization of the Lorentz gauge. This immediately brings up the question: what is the gauge transformation for Φ (analogous to $\delta A_\mu = \partial_\mu \lambda$; equation 24c) and what is the gauge invariant form for its action (analogous to $\tfrac{1}{4}\,F_{\mu\nu}^2$)? The most appealing answer to this is that gauge invariance is simply a reflection of reparametrization invariance (equations 19). Indeed, under a reparametrization of the string $\sigma \to \sigma + \xi(\sigma)$, $\delta\Phi = i\Sigma_{n=-\infty}^{\infty} L_{-n}(\xi_n \Phi)$, where ξ_n is the Fourier component of $\xi(\sigma)$. Siegel therefore suggested that the correct gauge transformation for Φ be

$$
\delta\Phi[X_\mu(\sigma)] = \sum_{n=1}^{\infty} L_{-n}\Lambda_n[x_\mu(\sigma)],
\tag{36}
$$

where Λ_n is an arbitrary field functional with an expansion analogous to equation 29 or 30:

$$
\Lambda_n[X_\mu(\sigma)] = [\lambda_n(x) + \lambda_{mn}^\mu(x)\,a_\mu^{m+} + \ldots]\,\Phi_0.
\tag{37}
$$

Notice that the index n on Λ_n becomes an added mode index on each component field. It should be emphasized that equation 26 is no more a proven consequence of reparametrization invariance than equations 34 are a proven consequence of the original primary constraints on the dynamics of the world sheet. They should both be viewed simply as suggestive *ansätze* for going from a world-sheet description to a field-functional description, thereby shortcutting the technicalities discussed at the close of the last section. Just as equations 34 apparently lead to sensible-looking equations, namely, equation 35, so here equation 36 leads to sensible-looking gauge transformations, as can be seen by making a component decomposition:

$$
\delta A_\mu^1(x) = \partial_\mu \lambda^1(X),
$$

$$
\partial A_\mu^2(x) = \lambda_\mu^{11}(x) + 1/\sqrt{\pi}\,\partial_\mu \lambda^2(x),
$$

$$
\delta\theta_{\mu\nu}^{11}(x) = \tfrac{1}{2}\sqrt{\pi}\,[\partial_\mu \lambda_\nu^{11}(x) + \partial_\nu \lambda_\mu^{11}(x)] + \tfrac{1}{2}g_{\mu\nu}\lambda^2(x), \text{ etc.}
\tag{38}
$$

Before discussing the gauge invariant formulation of these equations, I would first like to digress and briefly discuss the question as to whether they really are a gauge-fixed version of the correct equations. I shall only investigate this in the limited sense that they must at least contain the correct number of physical degrees of freedom as expressed, for example, by a count in the light-cone gauge.

COUNTING DEGREES OF FREEDOM (AND $d = 26$)

Let us denote a "level" by N so that, for instance, the $N = 0$ level contains the scalar tachyon, the $N = 1$ level contains the massless vector, and so on. Then, let us introduce a generating function, $M_N(d)$, defined by the following algebraic relationship:[14]

$$\sum_{k=1}^{\infty} (1 - x^k)^{-d} \equiv \sum_{N=0}^{\infty} M_N(d) x^N. \tag{39}$$

If we do this, then $M_N(d)$ is the number of degrees of freedom (df) at the N-th level contained in Φ: furthermore, $M_N(d - 2)$ is the number of physical (i.e., light-cone gauge) df at that same level:

	Scalar Tachyon	Massless Vector	Massive Vector and Tensor	$\phi_{\mu\nu\lambda}^{111}, \theta_{\mu\nu}^{12}, A_{\mu}^{3} \ldots$
	ϕ	A_{μ}^{1}	$A_{\mu}^{2}, \theta_{\mu\nu}^{11}$	
N (Level)	0	1	2	3
$M_N(d)$ (df in ϕ)	1	d	$1/2d(d + 3)$	$1/6d(d + 1)(d + 8)$
$M_N(d - 2)$ (Physical df)	1	$d - 2$	$1/2(d - 2)(d + 1)$	$1/6(d - 2)(d - 1)(d + 6)$

The above table gives the explicit expression for these numbers up to the third level. As a simple illustration, consider $N = 1$ where Φ contains only the massless vector field $A_{\mu}^{1}(x)$. Naively, this has d df, not all of which, however, are physical. The gauge condition $\partial_{\mu}A^{\mu} = 0$ removes one and the residual gauge freedom $\delta A_{\mu} = \partial_{\mu}\lambda$ (with $\partial^2\lambda = 0$) removes another. There are thus $(d - 2)$ physical df at this level, which is in agreement with the table. In general, then, the number of physical df at a particular level is obtained by reducing the number of df in Φ by (a) the number of independent constraints implied by the gauge condition $L_n\Phi = 0$ and (b) the number of independent gauge transformations (or parameters) in Λ_m. Although the former is fairly straightforward to impose, the latter is not; this is because it is possible to make transformations on the Λ_m that leave Φ invariant. As an example of this from ordinary field theory, consider a gauge transformation on some field $\phi_{\mu\nu\lambda} \ldots$ of the form[7]

$$\delta\phi_{\mu\nu\lambda} \ldots = \partial_{\mu} \lambda_{\nu\lambda} \ldots + g_{\mu\nu} \eta_{\lambda} \ldots, \tag{40}$$

where symmetrization on all indices is to be understood. Clearly, the arbitrary gauge functions λ and η are undefined up to the residual gauge transformations of the forms,

$$\delta\lambda_{\nu\lambda\rho}\ldots = g_{\nu\lambda}\,\chi_\rho\ldots,$$
$$\delta\eta_{\lambda\rho}\ldots = -\partial_\lambda\chi_\rho\ldots, \tag{41}$$

because this joint transformation leaves ϕ unchanged. Therefore, counting the number of independent gauge parameters is a nontrivial exercise. Let us explore this a little further in the context of the string field.

The residual gauge invariance of equations 34 is obtained by examining their response to the transformations in equation 36. From equation 34a, we thereby obtain the generalization of $\partial^2\lambda = 0$ for the Maxwell field, viz.,

$$(L_0 + n - 1)\Lambda_n = 0, \tag{42}$$

whereas from equation 34b, we obtain two equations,

$$L_n\Lambda_m + (2n + m)\,\Lambda_{m+n} = 0 \tag{43}$$

and

$$\left(L_0 - \frac{(n-1)}{24}\,\{n(d-26) + (d-2)\}\right)\Lambda_n = 0. \tag{44}$$

The consistency of equations 42 and 44 requires the well-known remarkable constraint that $d = 26$; later, I shall return to this question and discuss how one can avoid this, at least within free field theory. The remaining equation (equation 43) has the consequence that all Λ_n can be expressed in terms of Λ_1 alone,

$$\Lambda_{n+1} = -\frac{L_{-n}}{2n + 1}\,\Lambda_1, \tag{45}$$

and that Λ_1 itself is constrained by the series of equations

$$L_n\Lambda_1 = (-1)^{n+1}\frac{2(2n + 1)}{(n + 2)!}\,L_1^n\Lambda_1. \tag{46}$$

Thus, when counting the independent physical df in Φ, only Λ_1 need be considered. In fact, not all of its component fields are independent because it is constrained by equation 46. Furthermore, notice that it satisfies the Klein-Gordon equation $L_0\Lambda_1 = 0$, which, together with equation 46, gives a system of equations similar in structure to the original set for Φ, namely, equations 34. It is therefore not surprising that this system of equations for Λ_1 will be invariant to gauge transformations analogous to equation 36, namely,

$$\delta\Lambda_1 = L_{-m}\chi_m, \tag{47}$$

with

$$(L_0 + m)\,\chi_m = 0 \qquad (\chi_1 = 0) \tag{48}$$

and

$$L_n\chi_m + (2n + m)\,\chi_{m+n} = 0. \tag{49}$$

However, these equations for χ_m are identical in structure to equations 42 and 43 for Λ_m and they therefore lead to a further elimination of gauge parameters. Clearly, this procedure continues ad infinitum, thus making a general counting argument rather complicated. I shall not go into further details of how such an argument proceeds, but I shall simply remark that such an argument has, in fact, been constructed in a slightly different context by Banks and Peskin.[5] The important points I want to stress here are (a) the somewhat unexpected appearance of an infinite sequence of residual gauge symmetries and (b) the apparently special role of $d = 26$.

I shall now return to the discussion of how one constructs gauge invariant actions given the gauge-fixed equations of motion.

CONSTRUCTION OF GAUGE-INVARIANT ACTIONS FOR MASSIVE BOSONIC FIELDS

To set the stage, I shall begin with the relatively simple example of the massless Maxwell field in Lorentz gauge whose equation of motion is

$$\partial^2 A_\mu = 0 \tag{50}$$

subject to

$$\partial^\mu A_\mu = 0. \tag{51}$$

The problem is to find an unconstrained gauge-invariant action whose equation of motion reduces to equation 50 when the Lorentz gauge is chosen. The constraint can be incorporated into the action by the introduction of a conventional Lagrange multiplier χ:

$$S_0 = \int d^4x \, [\tfrac{1}{2} A^\mu \, \partial^2 A_\mu + \chi \, \partial^\mu A_\mu]. \tag{52}$$

Now, under a gauge variation $\delta A_\mu = \partial_\mu \lambda$,

$$\delta S_0 = \int d^4x \, [\partial^\mu A_\mu \, (\delta\chi - \partial^2\lambda) + \chi\partial^2\lambda]. \tag{53}$$

If χ is kept fixed (i.e., $\delta\chi = 0$), then δS_0 vanishes if λ is harmonic, that is, $\partial^2\lambda = 0$ (as appropriate to the Lorentz gauge). The structure of equation 53, however, suggests an alternative way of obtaining an invariant action that, at the same time, circumvents this constraint on λ: namely, if we choose χ such that its variation $\delta\chi = \partial^2\lambda$, then equation 53 becomes

$$\delta S_0 = \int d^4x \, \tfrac{1}{2}\delta\chi^2. \tag{54}$$

Therefore, if one adds $-\tfrac{1}{2}\chi^2$ to S_0, then $\delta S = 0$ and an invariant action results. Thus, the new action

$$S \equiv \int d^4x \, [\tfrac{1}{2} A^\mu \partial^2 A_\mu + \chi\partial^\mu A_\mu - \tfrac{1}{2}\chi^2] \tag{55}$$

is invariant to the joint variations, $\delta A_\mu = \partial_\mu \lambda$ and $\delta \chi = \partial^2 \lambda$. Clearly, χ is not a dynamical field, and it can be trivially integrated out by making the shift $\chi \to \chi + \partial^\mu A_\mu$, which gives the standard gauge invariant action

$$S = \int d^4 x \ [\tfrac{1}{2} A^\mu \partial^2 A_\mu + \tfrac{1}{2}(\partial_\mu A^\mu)^2] \tag{56}$$

$$= \tfrac{1}{4} \int d^4 x \ F^2_{\mu\nu}. \tag{57}$$

It is very instructive to see how this can be extended to the massive case whose equations are given in equation 24. The initial form of the action (S_0) incorporating the gauge condition can be written as in equation 52, but with $\partial^2 \to (\partial^2 - m^2)$. Although this has no explicit gauge symmetry, we can investigate its response to the gauge transformation $\delta A_\mu = \partial_\mu \lambda$ with the view to generating a new action that does manifest such a symmetry. Under the transformation, clearly

$$\delta S_0 = \int d^4 x \ [\partial_\mu A^\mu \{\delta \chi - (\partial^2 - m^2)\lambda\} + \chi \partial^2 \lambda]. \tag{58}$$

Thus, if χ is chosen such that $\delta \chi = (\partial^2 - m^2)\lambda$, then we have, in place of equation 54,

$$\delta S_0 = \int d^4 x \ \tfrac{1}{2}\delta \left[\chi \frac{\delta^2}{(\partial^2 - m^2)} \chi \right]. \tag{59}$$

A gauge-invariant action will thus result from adding a term $-\tfrac{1}{2}\chi \partial^2 (\partial^2 - m^2)^{-1} \chi$ to S_0, which obviously leads to a nonlocal and therefore unsatisfactory action. In order to avoid precisely this situation, additional auxiliary fields need to be added.

To proceed, we still choose $\delta \chi = (\partial^2 - m^2)\lambda$, but we rewrite equation 58 as

$$\delta S_0 = \int d^4 x \ [\chi(\partial^2 - m^2)\lambda + m^2 \chi \lambda] \tag{60}$$

$$= \int d^4 x \ [\tfrac{1}{2}\delta \chi^2 + m^2 \chi \lambda]. \tag{61}$$

The trick is to add the term $(-\tfrac{1}{2}\chi^2 - m\chi\phi)$ to S_0 where under the gauge transformation, $\delta\phi = m\lambda$. We do this because then the remaining variation of the action becomes

$$\delta S_1 = -\int d^4 x \ m \ \phi \ \delta \chi \tag{62}$$

$$= \int d^4 x \ \delta[\tfrac{1}{2}\phi \ (\partial^2 - m^2)\phi], \tag{63}$$

which is a total derivative. Thus, a ghostlike kinetic energy term, $\tfrac{1}{2}\phi(\partial^2 - m^2)\phi$, added to the action will guarantee invariance. Putting these pieces together, we see that an action

$$S = \int d^4 x \ [\tfrac{1}{2} A (\partial^2 - m^2) A_\mu + \chi \partial^\mu A_\mu - \tfrac{1}{2}\chi^2 - m\chi\phi + \tfrac{1}{2}\phi(\partial^2 - m^2)\phi] \tag{64}$$

has been generated that is invariant to the joint gauge transformations,

$$\delta A_\mu = \partial_\mu \lambda, \qquad \delta \chi = (\partial^2 - m^2)\lambda, \qquad \text{and} \qquad \delta\phi = m\lambda. \tag{65}$$

As before, the auxiliary Lagrange multiplier field χ has no kinetic energy and can be eliminated to give

$$S = \int [\tfrac{1}{2}A_\mu(\partial^2 - m^2)A^\mu + \tfrac{1}{2}(\partial_\mu A - m\phi)^2 + \tfrac{1}{2}\phi(\partial^2 - m^2)\phi]d^4x, \tag{66}$$

which is essentially Stueckelberg's result.[13]

Notice that if one uses the gauge function λ to set $\partial^\mu A_\mu = 0$, then A_μ and ϕ decouple and the resulting equations of motion reproduce equation 24a. On the other hand, one can gauge away ϕ to obtain the standard Proca Lagrangian, $\tfrac{1}{4}F^2_{\mu\nu} - \tfrac{1}{2}m^2A^2_\mu$, which has the property that its equations of motion reproduce both equations 24.

Without much difficulty, this procedure can be extended to fields of arbitrary spin that are represented by a symmetric divergenceless and traceless tensor $\phi_{\mu\nu\lambda}\ldots$ that satisfies a Klein-Gordon equation.[7] As in the above, these constraints (or gauge conditions) can be incorporated into the action via Lagrange multipliers and the response to the general gauge transformation (equation 40) can be balanced by a suitable judicious choice of auxiliary fields.[15] Upon integrating out the Lagrange multipliers, a gauge-invariant action analogous in structure to equation 66 results.

CONSTRUCTION OF A GAUGE-INVARIANT STRING FIELD ACTION

I will now show how the above procedure can straightforwardly be applied to the string field functional equations. We want to construct a local covariant action that is invariant to the gauge transformation

$$\delta\Phi = L_{-k}\Lambda_k \tag{36}$$

and that in the gauge where

$$L_n\Phi = 0 \tag{34b}$$

reproduces the equation of motion

$$(L_0 - 1)\Phi = 0. \tag{34a}$$

Following the procedure outlined above, we first incorporate the gauge fixing into the action using a Lagrange multiplier string functional $\chi_n[X_\mu]$:

$$S_0 = \int DX_\mu \,[\tfrac{1}{2}\Phi\,(L_0 - 1)\Phi + \chi_n L_n\Phi]. \tag{67}$$

For simplicity, we shall now suppress the path integral symbols (and so work with the Lagrangian).

Under the variation of equation 36,

$$\delta S = \left[L_m\Phi(L_0 + m - 1) - 2m\chi_m\left\{L_0 + \left(\frac{m-1}{24}\right)([d-26]m + d - 2)\right\}\right]\Lambda_m$$

$$- \chi_{n;m}\Lambda_{m;n} + (L_m\Phi)\delta\chi_m, \tag{68}$$

where $\chi_{m;n} \equiv L_m\chi_n + (2m + n)\chi_{m+n}$, etc.

In deriving this, extensive use has been made of the Virasoro algebra (equation 33). One possible procedure for obtaining an invariant action is to write $\chi_m = O_m\phi$ and determine the operator O_m by setting $\delta S_0 = 0$. It is clear (even with $d = 26$) that O_m will contain nonlocal operators and will therefore be unsatisfactory.[16] Instead of eliminating χ_m, we can follow the procedure used for ordinary spin fields and choose its variation so that

$$\delta \chi_m = -(L_0 + m - 1)\Lambda_m, \tag{69}$$

which explicitly eliminates Φ from δS_0. If we now add a term $-m\chi_m^2$ to S_0, then (with $d = 26$) the remaining variation in S is simply

$$\delta S_1 = -\chi_{n;m}\Lambda_{m;n}. \tag{70}$$

Thus, if a further term, $\chi_{n;m}\Psi_{mn}$, is added with

$$\delta \Psi_{mn} = \Lambda_{m;n}, \tag{71}$$

then

$$\delta S = \Psi_{mn}\delta\chi_{n;m}$$

$$= \Psi_{mn}(L_0 + m + n - 1)\Lambda_{n;m} \quad \text{(from equation 69)}$$

$$= \tfrac{1}{2}\delta[\Psi_{mn}(L_0 + m + n - 1)\Psi_{nm}] \quad \text{(from equation 71)}.$$

An invariant action therefore results if one finally adds the ghostlike kinetic term $-\tfrac{1}{2}\Psi_{mn}(L_0 + m + n - 1)\Psi_{nm}$ to the action. Collecting these pieces together, we see that we have constructed an action

$$S = \tfrac{1}{2}\Phi(L_0 - 1)\Phi + \chi_n L_n\Phi - n\chi_n^2 + \chi_{n;m}\Psi_{mn} - \tfrac{1}{2}\Psi_{mn}(L_0 + m + n - 1)\Psi_{nm} \tag{72a}$$

that is invariant to the joint gauge transformations,

$$\delta\Phi = L_{-m}\Lambda_m, \qquad \Delta\chi_n = (L_0 + m - 1)\Lambda_m, \qquad \text{and} \qquad \delta\Psi_{mn} = \Lambda_{m;n}. \tag{72b}$$

Because χ_n is not dynamical, it can clearly be eliminated to give

$$S = \tfrac{1}{2}\Phi(L_0 - 1)\Phi + \tfrac{1}{4}n[L_n\Phi - L_{-m}\Psi_{nm} + (m + n)\Psi_{n-m,m}]^2$$

$$\cdot \tfrac{1}{2}\Psi_{mn}(L_0 + m + n - 1)\Psi_{nm}. \tag{72c}$$

In deriving this result, notice that both the number of dimensions ($d = 26$) and the value of the "Regge intercept" (the "1" in $L_0 - 1$) appear crucial, as evidenced in equation 68. Below, we shall discuss how this can, in principle, be circumvented by adding further auxiliary fields.

As anticipated above, equations 72 are quite similar in structure to the gauge-invariant action for a massive spin field, for example, equation 66. It can, of course, be expanded in component form by using the expansions for Φ (equation 30), Λ_m (equation 37), and the analogous expansion for Ψ_{MN}:

$$\Psi_{MN}[X] = [\Psi_{MN}(x) + i\Psi_{MNm}(x)a_\mu^{m+} + \Psi_{MNmn}^{\mu\nu}(x)a_\mu^{m+}a_\nu^{n+} + \cdots]\Phi_0. \tag{73}$$

The action S naturally splits into levels: $\Sigma_{N=0}^{\infty} S_N$, with each S_N being a complicated expression of ordinary spin fields. Even for the lower levels ($N = 2, 3, \ldots$), it is quite tedious to verify that these can be transformed into the corresponding expressions for a given spin; this involves redefinitions of fields and gauge parameters and, in some cases, a specific gauge choice. I shall not discuss this here nor the specific properties of equation 72, except to remark that one can gauge away Ψ_{mn} to give an action[14] that is the analogue of the Proca action for spin-1 ($\frac{1}{4}F_{\mu\nu}^2 - m^2 A_\mu^2$),

$$S = \frac{1}{2}\Phi(L_0 - 1)\Phi + \frac{1}{4}n(L_n\Phi)^2, \tag{74}$$

which allows the symbolic identification

$$[\frac{1}{2}\Phi L_0\Phi + \frac{1}{4}n(L_n\Phi)^2] \sim [\frac{1}{2}A_\mu^2\partial^\mu A - \frac{1}{2}(\partial_\mu^\mu A^2)] \sim \frac{1}{4}F_{\mu\nu}^2.$$

BEYOND $d = 26$

The covariant action (equation 72) was derived by specializing to 26 dimensions [and using a Regge intercept of 1; this is the "1" in $(L_0 - 1)$]. The naive reason for this is apparent from equation 68, which gives the response of the action to a gauge transformation. When $d = 26$, the expression in curly braces coming from the Lagrange multiplier term reduces to $(L_0 + m - 1)$, which is the same operator that arises from the kinetic energy term. This consequently allows the variation from the Lagrange multiplier term to be expressed as a total derivative, $m\delta\chi_m^2$, which ultimately leads to equation 72. Not surprisingly, a similar situation occurred when counting the physical degrees of freedom in Siegel's gauge (see equation 44). Naively, without the simplification that occurs when $d = 26$, it would appear that the procedure could not in general work and that this would prohibit the construction of a local gauge invariant action. We shall now show that this is not, in fact, the case. To illustrate how a local action can be derived even if $d \neq 26$, we return to equation 68, which gives the response of the action (equation 67) to a gauge transformation. As before, we can again choose $\delta\chi_m = -(L_0 + m - 1)\Lambda_m$ (equation 69) to obtain

$$\delta S = m\delta\chi_m^2 + \frac{1}{12}\epsilon m(m^2 - 1)\chi_m\Lambda_m - \chi_{n;m}\Lambda_{m;n}, \tag{75}$$

where $\epsilon \equiv 26 - d$. Working as before, we can add to S the terms

$$-m\chi_m^2 - \frac{1}{12}\epsilon m(m^2 - 1)\chi_m\Delta_m + \chi_{n;m}\Psi_{mn},$$

where the gauge variations of Δ_m and Ψ_{mn} are given by

$$\delta\Delta_m = \Lambda_m \quad \text{and} \quad \delta\Psi_{mn} = \Lambda_{m;n}. \tag{76}$$

Then

$$\delta S = \frac{1}{24}\epsilon m(m^2 - 1)\delta[\Delta_m(L_0 + m - 1)\Delta_m] - \frac{1}{2}\delta[\Psi_{mn}(L_0 + m + n - 1)\Psi_{nm}]. \tag{77}$$

Adding these kinetic terms for Δ_m and Ψ_{mn} to S and then integrating out the auxiliary field χ_n results in an action

$$S \equiv \frac{1}{2}\Phi(L_0 - 1)\Phi - \frac{1}{4n}[L_n\Phi - \frac{1}{12}\epsilon n(n^2 - 1)\Delta_n + L_{-m}\Psi_{nm} + (m + n)\Psi_{n-m,m}]^2$$

$$+ \frac{1}{24}\epsilon n(n^2 - 1)\Delta_n(L_0 + n - 1)\Delta_n - \frac{1}{2}\Psi_{mn}(L_0 + m + n - 1)\Psi_{nm}, \tag{78}$$

which is invariant to the joint gauge transformations,

$$\delta\Phi = L_{-m}\Lambda_m, \qquad \delta\Lambda_m = \Lambda_m, \qquad \text{and} \qquad \delta\Psi_{mn} = \Lambda_{m;n}, \tag{79}$$

even if $d \neq 26$. Thus, a new set of Stueckelberg fields must be introduced to accommodate the $d \neq 26$ situation. Although we have not examined the consequences of this in any detail, these extra fields are presumably associated with longitudinal modes of the string that decouple in 26 dimensions.[17] It should be noted that the massless mode A_μ^1 is unaffected by the presence of the new fields in Δ_n.

The above procedure can be generalized to a situation where the Regge intercept differs from unity. For example, suppose that instead of equation 34a, we started with

$$(L_0 - 1 + \delta)\Phi = 0, \tag{80}$$

but which is still subject to

$$L_n\Phi = 0. \tag{34b}$$

It is not clear that this can necessarily be associated with a consistent free first-quantized string; the question of the roles of the longitudinal modes and the normal ordering implicit in L_0 would require some detailed examination. Nevertheless, a priori, there is no reason that it should not describe a respectable string field functional. Repeating the previous derivation, we can easily see that the addition of $\delta \neq 0$ requires no new additional fields. We can also see that the local covariant action is still given by equation 78, but with the kinetic energy operator replaced everywhere by $(L_0 - 1 + \delta)$ and the parameter, $(11/24)\epsilon(n^2 - 1)$, replaced by $(11/24)\epsilon(n^2 - 1) + \delta$.

CONCLUSIONS

In conclusion, I have tried in this discussion to give a relatively elementary account of what a string field represents and what is involved in the construction of its covariant action. The emphasis has been on drawing a correspondence with similar problems in ordinary field theory by using, in particular, the language and mathematics with which most of us are quite familiar. Only the free string was discussed; as pointed out in my introduction, the inclusion of interactions is technically a more complicated problem. Indeed, the present trend seems to be moving in a direction that requires new mathematical techniques beyond the plodding pedestrianism expressed in this paper. Whether the ultimate formulation will actually require such sophistication is not yet clear. Clearly, the test will be whether it is concise, aesthetic, and physically appealing, and above all, whether it exposes the deep principles that many believe lay hidden in the depths of the string.

NOTES AND REFERENCES

1. GREEN, M. B. & J. H. SCHWARZ. 1984. Phys. Lett. **149B:** 117.
2. GROSS, D. J. *et al.* 1985. Phys. Rev. Lett. **54:** 502.
3. For a review of early work on the string, see, for example: SCHERK, J. 1975. Rev. Mod. Phys. **47:** 123.
4. The original idea of a string field is due to: KAKU, M. & K. KIKKAWA. 1974. Phys. Rev. **D10:** 1110, who confined themselves to the light-cone gauge.

5. Some recent papers include: SIEGEL, W. & B. ZWIEBACH. 1986. Nucl. Phys. **B263:** 105; BANKS, T. & M. E. PESKIN. Nucl. Phys. **B276:** 513; NEVEU, A., H. NICOLAI & P. WEST. Ibid, p. 573.
6. See, for example: WITTEN, E. 1986. Nucl. Phys. **B268:** 253; LYKKEN, J. & S. RABY. 1986. Nucl. Phys. **B278:** 256.
7. WEST, G. B. 1986. Nucl. Phys. **B277:** 125.
8. LANDAU, L. D. & E. M. LIFSHITZ. 1971. The Classical Theory of Fields. Third edition, p. 24. Addison–Wesley. Reading, Massachusetts.
9. DIRAC, P. A. M. 1958. Proc. R. Soc. **A246:** 326.
10. FADE'EV, L. D. 1969. Theor. Math. Phys. **1:** 1; see also: ITZYKSON, C. & J-B. ZUBER. 1980. Quantum Field Theory. McGraw–Hill. New York.
11. To my knowledge, this derivation of the Feynman propagator from a world-line approach has not appeared in the literature.
12. SIEGEL, W. 1985. Phys. Lett. **151B:** 391, 396.
13. STUECKELBERG, E. C. G. & D. RIVIER. 1949. Helv. Phys. Acta **22:** 215.
14. Most of this and the next section are based on work done in collaboration with S. Raby and R. C. Slansky; see the Los Alamos preprint "Towards a covariant string theory" (LA-UR-85-3794). See also: SLANSKY, R. C. 1985. Los Alamos preprint no. LA-UR-85-2768.
15. A similar procedure based on equations of motion rather than on the action has been devised by: PFEFFER, D., P. RAMOND & V. G. J. RODGERS. 1985. Univ. of Florida preprint no. UFTP-85-19.
16. Such an action has indeed been constructed by different means by: FRIEDAN, D. 1985. Univ. of Chicago preprint no. EFI-85-27. See also: KAKU, M. & J. LYKKEN. 1985. CCNY preprint no. Print-85-412.
17. PATRASCIOU, A. 1974. Nucl. Phys. **B81:** 525; BARDEEN, W. *et al.* 1976. Phys. Rev. **D13:** 2364.

The Infrared Limit of the Superstring

D. V. NANOPOULOS

Department of Physics
University of Wisconsin–Madison
Madison, Wisconsin 53706

A consistent and calculable theory of quantum gravity demands extended fundamental objects because point field theories of gravity are incurably unrenormalizable. The simplest extended objects—strings—contain spin-2 particles among their massless modes that are identifiable with the graviton. In general, string field theory leads to several inconsistencies, except if and only if we consider strings containing both fermions and bosons in their particle spectrum—superstrings[1]—in $D = 10$ space-time dimensions with $E_8 \times E_8'$ or $O(32)$ as the appropriate gauge group.[2] Such types of superstrings seem to satisfy all the usual rules of field theory, contain no anomalies[2] (Yang-Mills, gravitational, or mixed), and seem to be at least one-loop finite.[3] The best candidate for a superstring theory is the heterotic string,[4] where not only the number of space-time dimensions ($D = 10$) and $N = 1$ supergravity are derived dynamically, but where also the existence of the gauge group $E_8 \times E_8'$ or $O(32)$ is explained.[4] Interestingly enough, the point field theory limit of the superstring is $N = 1$ supergravity in $D = 10$ dimensions with $E_8 \times E_8'$ or $O(32)$ as a gauge group, plus some extra extremely useful terms. These include the addition of a Lorentz-Chern-Simons term (ω_L) for the field strength $H_{\mu\nu\rho}$ of the antisymmetric tensor field $B_{\mu\nu}$ contained in the $N = 1$, $D = 10$ supermultiplet. The existence of ω_L is essential for canceling anomalies of all types.[2] In addition, the fact that its supersymmetrization[5] leads automatically to the right combination of ghost-free[6] R^2 terms is essential in compactifying from ten to four dimensions.[7]

The survival of $N = 1$ supersymmetry (SUSY) in four dimensions (essential for the solution of the gauge hierarchy problem) picks out uniquely (?) Calabi-Yau manifolds[8] as the prime candidates for the compactification manifolds (K_6) of the six surplus dimensions. Calabi-Yau manifolds[8] are manifolds with $SU(3)$ holonomy and are thus Kähler and Ricci flat. Choosing the ground state of the superstring to be of the admittable type $M_4 \times K_6$ (with M_4 being a four-dimensional maximally symmetric space and K_6 being a six-dimensional Calabi-Yau manifold), we automatically oblige[7] M_4 to be the highly desirable Minkowski space; that is, the cosmological constant (Λ_c) is automatically zero at the classical (tree) level. This result is rather intriguing and is full of far-reaching consequences.

The continuation of the absence of anomalies in four-dimensions implies[7,9] $F \times F \propto R \times R$. This, in the case of the heterotic string, is only satisfied if we identify[7] the $SU(3)$ holonomy group with an $SU(3) \subset E_8 \times E_8'$ or $O(32)$. Thus during compactification, $E_8 \times E_8' \rightarrow E_6 \times E_8'$ or $O(32) \rightarrow O(26) \times U(1)$. While E_6 is an acceptable gauge group,[10] $O(26)$ is not because it eventually leads to a vectorlike fermion spectrum.[7] Remarkably enough, the particle spectrum that emerges in four dimensions looks very similar to the one observed experimentally. Beyond the gauge sector that contains the $E_6 \times E_8'$ gauge bosons and gauginos, there are N_g chiral multiplets transforming as the

$\underline{27}$ of E_6 (see TABLE 1), with some possible relics from $\delta(27 + \overline{27})$, as well as some gauge-singlets. The numbers N_g and δ are given by certain topological numbers characterizing the Calabi-Yau manifold K_6. Namely, $N_g = |\chi(K_6)|/2$, with $\chi(K_6)$ being the Euler characteristic number of K_6, and $\delta = b_{1,1}$, which is the Hodge number of K_6. Simply connected, Calabi-Yau manifolds always have $b_{1,1} \geq 1$ and $|\chi(K_6)| \gg 1$. That is not very exciting because it gives $N_g \gg 1$; however, most of us believe that the number of generations (N_g) cannot be a large number. Actually, this fact is one of many reasons to consider not-simply connected Calabi-Yau manifolds.[7,11,12] In this case, N_g is reduced considerably ($N_g = 3$ or 4 is possible), δ can be effectively set to zero, and $E_6 \times E_8'$ may automatically be broken into smaller groups by the Wilson-loop

TABLE 1. Transformation Properties under $SU(3)_c \times SU(2)_L \times U(1)_Y \times U(1)_E$ of the Chiral Matter Superfields Contained in the $\underline{27}$ of $E_6{}^a$

Chiral Superfield	$SU(3)_c \times SU(2)_L \times U(1)_Y \times U(1)_E$ Quantum Numbers
$Q \equiv \begin{pmatrix} u \\ d \end{pmatrix}$	$(3, 2, \tfrac{1}{6}, \tfrac{1}{3})$
u^c	$(\overline{3}, 1, -\tfrac{2}{3}, \tfrac{1}{3})$
d^c	$(\overline{3}, 1, \tfrac{1}{3}, -\tfrac{1}{6})$
$L \equiv \begin{pmatrix} \nu \\ e \end{pmatrix}$	$(1, 2, -\tfrac{1}{2}, -\tfrac{1}{6})$
e^c	$(1, 1, 1, \tfrac{1}{3})$
ν^c	$(1, 1, 0, \tfrac{5}{6})$
$H \equiv \begin{pmatrix} H^+ \\ H^0 \end{pmatrix}$	$(1, 2, \tfrac{1}{2}, -\tfrac{2}{3})$
$\overline{H} \equiv \begin{pmatrix} \overline{H}^0 \\ \overline{H}^- \end{pmatrix}$	$(1, 2, -\tfrac{1}{2}, -\tfrac{1}{6})$
N	$(1, 1, 0, \tfrac{5}{6})$
D	$(3, 1, -\tfrac{1}{3}, -\tfrac{2}{3})$
D^C	$(\overline{3}, 1, \tfrac{1}{3}, -\tfrac{1}{6})$

^{a}Group and generation indices are understood. In terms of Y_L and Y_R (associated with the maximal $SU(3)_c \times SU(3)_L \times SU(3)_R$ subgroup of E_6), the conventional weak hypercharge Y and the new hypercharge Y_E are given by $Y = (Y_L + Y_R)/2$, and $Y_E = (Y_L - Y_R/4)$, respectively. The properly normalized quantities $\hat{Y}$ and $\hat{Y}_E$ (such that tr $T_3^2 = $ tr $\hat{Y}^2 = $ tr $\hat{Y}_E^2$) are given by $\hat{Y} = \sqrt{3/5}\, Y$ and $\hat{Y}_E = \sqrt{3/5}\, Y_E$, respectively. According to our conventions, ν^c is the $SU(3)_c \times SU(2)_L \times U(1)_Y$–singlet contained in the $\underline{16}$ of $SO(10) \subset E_6$, while N is the $SO(10)$-singlet.

mechanism[13] with the exciting possibility[11] that the gauge couplings obey the successful grand unification relations, while the Yukawa couplings need not. Clearly, primordial E_6-invariance entails the allowable Yukawa terms, but the Yukawa coupling constants need not obey the E_6 Clebsch-Gordan relations.

All these facts are of invaluable importance in constructing realistic models. A large number of generations, especially containing 27 particles per gluezation, will destroy the smooth run of the renormalization group equations (RGE) and will demand unification at unacceptably low values [$\leq O(10^{12}$ GeV$)$]; this thus ruins the successful predictions of $\sin^2\theta_W$ and the longevity of the proton.[14,15] The same is true for large δ's. The Wilson-loop mechanism[7,11–13] is the only one that is presently known for breaking $E_6 \times E_8'$ without a simultaneous unacceptable breaking of the highly desirable residual $N = 1$ SUSY in four dimensions.

The need for E_6 breaking at low energies is obvious, while the breaking of $E_8' \rightarrow SU(N)$ is demanded[16] by the formation of the gluino-condensate needed[16,17] for local SUSY (supergravity) breaking.[18] Furthermore, the absence of E_6 Clebsch-Gordan relations between Yukawa couplings may automatically resolve long-standing problems related to proton decay or neutrino masses. The possible breakings of E_6 through the Wilson-loop mechanism have been meticulously classified,[15] though a small theorem, recently derived,[19] suggests a unique solution to the E_6 breaking pattern[20,21]: $E_6 \rightarrow G_0 \equiv SU(3)_c \times SU(2)_L \times U(1)_Y \times U(1)_E$. The decomposition of the $\underline{27}$ under $SU(3)_c \times SU(2)_L \times U(1)_Y \times U(1)_E$ is given in TABLE 1. A crucial assumption in deriving such an exciting result is the absence of an intermediate energy scale. This follows from the no-scale philosophy[22] developed during the last few years[23] and naturally emerges from the superstring, as we will discuss later.

Notice the existence of a new gauge interaction $U(1)_E$ at low energies with a well-defined "charge" relative (see TABLE 1) and absolute (as determined[19–21] by grand unification):

$$\alpha_E (M_w) \equiv \frac{g_E^2}{4\pi} = 0.0164. \tag{1}$$

The interesting phenomenology of a new, low-mass (100–1000 GeV), neutral gauge boson Z_E has been discussed in considerable detail elsewhere.[19–21,24] Here, it suffices to remind the reader that presently available experimental data put a lower bound at

$$M_{Z_E} \geq O(200 \text{ GeV}), \tag{2}$$

which may push down to[21] 110 GeV or stretch up[25] to $O(300 \text{ GeV})$ if we try to satisfy cosmological constraints coming from primordial nucleosynthesis.[25]

In regard to the chiral multiplet content, for models with no intermediate scale, RGE analysis demands[14,15] the existence of only $N_g = 3$ generations of $\underline{27}$ (as presented in TABLE 1), while δ may be taken naturally and safely to be effectively zero. These facts are of great importance when one tries to build up the Yukawa sector or, more generally, the superpotential f. Given the particle content of TABLE 1, primordial E_6-invariance allows only the following sets of couplings[19,26]:

$$f_1 = hQHu^c + h_DQ\overline{H}d^c + h_LL\overline{H}e^c, \tag{3}$$

$$f_2 = \lambda H\overline{H}N + kDD^cN, \tag{4}$$

$$f_3 = \lambda_1 D^cLQ + \lambda_2 De^cu^c + \lambda_3 Dd^c\nu^c, \tag{5}$$

$$f_4 = \lambda_4 DQQ + \lambda_5 D^cu^cd^c, \tag{6}$$

and

$$f_5 = \lambda_6 HL\nu^c. \tag{7}$$

While f_1 contains the usual Yukawa couplings of the "old" standard model [$SU(3)_c \times$

$SU(2)_L \times U(1)_Y]$, $f_{2,3,4,5}$ contain new "good" and "bad" couplings. The couplings contained in f_2 are of prime importance[19]:

(a) They allow the N-field to get a vacuum expectation value (v.e.v.) $\langle N \rangle \equiv x$, and they thus break the $U(1)_E$ and provide mass to the Z_E and to the D-quark. Notice that both m_{Z_E} and $m_D \propto \langle N \rangle$ ($\equiv x$), and in principle, if experimentally desired, they may be pushed higher than the usual $M_{W^\pm, Z_0} \propto \langle \overline{H} \rangle$ ($\equiv \overline{v}$).

(b) They give masses to all Higgs or Higginos, Neons, and Ninos, thus avoiding embarrassment with unwanted massless scalars (axion-types) or massless fermions.

(c) In an essential way, they help the realization of the "no-scale" scenario (as we will shortly see). For example, the presence of the $H\overline{H}N$ coupling is unavoidable[20,21] if $m_{1/2}$ (the primordial gaugino mass) is to be different from zero because it will be the only source left of SUSY breaking in the observable sector.[19–21]

In sharp contrast with the above remarkable consequences of f_2 (equation 4), one has to be careful[19,26] with $f_{3,4,5}$ (equations 5–7). The longevity of protons demands the exclusive presence (for making the D-quark unstable) of f_3 (equation 5) or f_4 (equation 6). The absence of large neutrino masses entails the banishing of f_5 (equation 7). It is highly remarkable that we have to rely on the existence of discrete symmetries and/or "topological" arguments[27,28] to forbid only f_3 (equation 5) or f_4 (equation 6) and f_5 (equation 7). This is not a formidable task because, for example,[19,26] a simple $Z_2 \times Z_2'$ symmetry—Z_2: $(L, e^c, \nu^c) \rightarrow (-1) (L, e^c, \nu^c)$ and Z_2': $\nu^c \rightarrow (-1) \nu^c$—allows $f_{1,2,4}$, but forbids $f_{3,5}$, etc. In addition, it is encouraging that examples have been provided[28] where out of 8,436 allowable Yukawa couplings, only 111 survive for "topological" reasons.[28] The extra bonus[19,26] coming from the form of the superpotential $f = f_1 + f_2 + f_{3 \text{ or } 4}$ also should not escape our notice. Automatically, there are no problems with flavor-changing neutral currents (FCNC) at the tree level, as well as no weak universality problems because D and d do not mix.[19,26] This latter property shows the deep connection between the disallowance of dangerous B-violating couplings and the exact validity of weak universality.

Up to now, starting from the superstring, we have derived[19] a unique low-energy gauge group $G_0 \equiv SU(3)_c \times SU(2)_L \times U(1)_Y \times U(1)_E$, a unique particle spectrum (see TABLE 1), and an (essentially) unique superpotential $f = f_1 + f_2 + f_{3 \text{ or } 4}$ that suitably bypasses unscathed all the usual traps [FCNC, weak universality, proton stability, (almost?) massless neutrinos, etc.] of the well-established low-energy phenomenology. Next, we have to determine dynamically the values of the different gauge coupling constants and mass scales in our model. This formidable task has been realized only recently,[19] and I give here a brief résumé of the present situation.[19]

Compactification on Calabi-Yau manifolds does not only preserve[7] $N = 1$ supergravity in four dimensions, but it also determines,[29,30] at the classical level, the precise form of the two, otherwise arbitrary functions,[31] $G(z_i z_i^*)$ and $f_{ab}(z_i)$. The Kähler potential G (a real function of its arguments) determines, through its second derivatives G_{ij} ($\equiv$ metric), the Kähler manifold where the fields z_i "live"—or in other words, the form of the kinetic terms of the scalar fields z_i. The analytic function $f_{ab}(z_i)$ determines the kinetic terms of the gauge boson and the corresponding gauginos.

Using an "educated" dimensional reduction that is highly imitative of compactification on a Calabi-Yau manifold, Witten succeeded[29] in deriving the following forms for G and f_{ab} at the tree level:

$$G = -\ln (S + S^*) - 3 \ln (T + T^* - 2\phi_i\phi_i^*) + \ln |W(S) + W(\phi_i)|^2 \qquad (8)$$

and

$$f_{ab} = S\delta_{ab} \qquad (9)$$

In writing down equations 8 and 9, we have used the four-dimensional gauge-singlet chiral superfields,[29]

$$S = e^{3\sigma}\phi^{-3/4} + 3i\sqrt{2}\, D, \qquad (10)$$

$$T = e^{\sigma}\phi^{3/4} - i\sqrt{2}\, a. \qquad (11)$$

Here, the two scalar fields, ϕ and σ, are the dilaton (ϕ) of the ten-dimensional supergravity theory and the "compacton" (σ) that scales the compactification manifold $g_{ij} = e^{\sigma}g_{ij}^0$. Also, the two pseudoscalar fields (D, a) correspond to components of the antisymmetric tensor field $B_{\mu\nu}$ (with ϕ_i denoting, collectively, the $E_6 - \underline{27}$ chiral multiplet). It should be stressed that because the $d = 4$ unified gauge coupling is given by

$$g_u^2 = \frac{1}{\mathrm{Re}(f_{ab})} : g_u^2 = \frac{1}{S_R} \text{ in leading order} \qquad (12)$$

and the loop diagrams renormalize g_u, then the simple expression of equation 9 will be renormalized by the matter fields in the observable sector. The D and a fields have been identified with the axion (D) and plation[32] (a) fields that are responsible for solving the strong CP-problem and also, maybe, the cosmological constant problem[32] (to all orders), respectively. We note, though, that recent claims[33] that state that Wess-Zumino terms in the effective ten-dimensional point field theory destroy the simple structure of equations 8 and 9 have not been found to be valid.[34]

The amazing thing about the form of G as given by equation 8 with its remarkable $SU(1, 1)/U(1) \times SU(n, 1)/SU(n) \times U(1)$ structure is that it is identical to the one that defines the so-called no-scale supergravity.[22,23,35-38] No-scale supergravity has been invented[35] by the demand of a naturally vanishing cosmological constant (at least, at the classical level) and has been subsequently used[22,36-38] for a dynamical determination of all of the mass scales in terms of a primordial mass scale, the Planck Mass M $(\equiv M_{P\varrho}/\sqrt{8\pi} \equiv 1/\sqrt{8\pi}\, G_{\mathrm{Newton}}) \equiv 1$. Indeed, it is a simple exercise[35] to show that by inserting the (T) part of equation 8 into the scalar potential[31] $(G_i = \lfloor\partial G/\partial z_i^*\rfloor, \ldots)$ of

$$V = e^{G}[G_i(G^{-1})_j^i G^j - 3], \qquad (13)$$

we get $V \equiv 0$ for every value of the field T, while the gravitino mass $m_{3/2} = e^{G/2}$ is different from zero. In other words, we get a theory[35] with a naturally vanishing cosmological constant at the tree-level, local SUSY breaking $(m_{3/2} \neq 0)$, but of undetermined magnitude. However, that is exactly what we are after.[22] We are going to let quantum corrections dynamically determine the magnitude of local SUSY breaking ($\propto m_{3/2}$), as well as the magnitude of gauge symmetry and global SUSY breaking.[22,36-38] This is the no-scale scenario, which has been implemented successfully in the past.[23]

The highly desirable form of G as given by equation 8 should not come as such a big surprise. As we discussed in the beginning, Calabi-Yau compactification leads to a vanishing cosmological constant for a large class of Calabi-Yau manifolds.[7] Because the observed four-dimensional chiral fields are the zero-modes of the compactification manifold, thus parametrizing its distortions, it follows that we get $\Lambda_c = 0$ for a continuous range of the chiral-field values. It has been proven in the past[36] for one chiral field (and very recently,[30] for any number of chiral fields) that the (T, ϕ_i) part of equation 8 is the unique G-function that leads to $V = 0$ for a continuous range of small field values.[30] Actually, we have recently been able to derive,[30] more or less, the whole form of equation 8 directly from the superstring without invoking the highly question-able[39] intermediate step of $N = 1$, $D = 10$, $E_8 \times E_8'$ point field theory. Therefore, several criticisms of Witten's intuitive approach[29] are highly undeserved and unfounded.

Let us now show the importance of each and every term in G (equation 8) and f_{ab} (equation 9) in getting a realistic model with well-determined and calculable super-symmetry and gauge breaking.[19-21] Let us start with the S-part, which is the part directly connected with local SUSY breaking. Indeed, nonperturbative effects of some unbroken subgroup $Q \subset E_8'$ may lead to gaugino condensation,[16] which in compliance[18] with the nontrivial form of f_{ab} (equation 9) triggers local SUSY breaking.[16,17] Integrating out the E_8'-gauge-sector degrees of freedom, the seeds of local SUSY breaking are "transmitted" to the derived effective superpotential,[17]

$$W(S) \ni h \exp\left(\frac{-3S}{2b_Q}\right), \tag{14}$$

where h is an unknown constant $O(1)$, and b_Q is the first nontrivial term in the beta-function for the subgroup Q of E_8' that condenses (e.g., for $SU(N)$: $b_Q = [3N/16\pi^2]$). The effective scalar potential for the S-field is then (using equations 8, 13, and 14)

$$V = \frac{1}{(16S_R T_R^3)} |W(S)|^2 \, G_S G_{\bar{S}} \, G_{S\bar{S}}, \tag{15}$$

which gives a minimum of $V = 0$ at $S_R = \infty$. Thus, the dynamical determination of S and hence (through equation 12) of g_u^2 would require an additional contribution of $\delta W(S)$ to $W(S)$. The simplest possibility[17] is a constant $C \propto \langle H_{ijk} \rangle \neq 0$, but any other feature of string dynamics generating a $\delta W(S)$ would be equally satisfactory. Assuming this to be the case, generically there will be a value of $S \neq 0$ or ∞ for which $G_S = 0$ and hence $V = 0$ (equation 15). This would give a vanishing cosmological constant $\Lambda_c = 0$ at the tree level in four dimensions, while local SUSY is broken:

$$m_{3/2} = e^{G/2} = \left(\frac{1}{16S_R T_R^3}\right)^{1/2} |W(S)|. \tag{16}$$

Because $1/S_R = g_u^2 = O(1)$ is also fixed, the only uncertainty in equation 16 is the value of T_R, which will be determined dynamically later. Incidentally, contrary to opposite, persistent claims,[39] it has been shown recently[40] that $g_u = 0$ cannot be reached by the classical field equations. Instead, for a considerable range in initial field values, these drive S_R towards the region $S_R = O(1)$, where the minimum is located. It is important to notice that equation 15 implies $M_{S_I} \sim M$, that is, the axion ($\equiv D$) becomes invisible, not because it is superlight, but because it is superheavy. In this case, because it decays instantly, no bound on $\langle S_I \rangle$ can be derived from the cosmological relaxation constraint.[41]

We now face the problems of dynamical determination of T_R and of supersymmetry breaking in the observable sector. These, of course, are resolved following the standard no-scale strategy.[23] However, there are certain amusing new facts that deserve some discussion. In the standard $SU(n, 1)$ no-scale supergravity,[37] the only allowed seed for supersymmetry breaking in the observable sector is the nonvanishing primordial gaugino mass, $m_{1/2}$. It has been proven that soft supersymmetry breaking gauge-nonsinglet scalar masses are naturally zero not only at the tree level,[37] but at the one-loop level[42,43] as well. Furthermore, this result is easily extended[43] to all higher loop orders if there is no nontrivial superpotential. These results[37,42,43] support our claim[19,21] that the primordial gaugino mass is the largest global SUSY-breaking mass parameter in the effective no-scale supergravity model (equation 8) derived from the superstring.[29,30]

However, who provides us with a nonvanishing $m_{1/2}$?—The superstring. We simply observe that in $N = 1$ supergravity,[31]

$$m_{1/2} = \frac{1}{4 \, \mathrm{Re}(f_{ab})} \, e^{G/2} \, G^i (G^{-1})^k_i \, f_{ab,k},\tag{17}$$

which surely demands nontrivial f_{ab} and of the appropriate form. Clearly, equation 9 is not enough, but as discussed after equation 12, it is renormalized by nongravitational interactions.[20,21] The astonishing thing is that if a $H\bar{H}N$ term exists in the superpotential $W(\phi_i)$, as it does in our case $[W(\phi_i) \ni f_2 \text{ (equation 4)}]$, then[19–21]

$$m_{1/2} = \frac{b_{E_6} g_u^2}{4} \left[\frac{1}{16 \, S_R T_R^3} \right]^{1/2} \cdot \lambda \, \frac{[|H\bar{H}|^2 + |HN|^2 + |\bar{H}N|^2]}{H\bar{H}N},\tag{18}$$

with $b_{E_6} = 27/16\pi^2$. In other words, the superstring-derived no-scale model[19] not only provides us with a nonvanishing $m_{1/2}$, but it predicts a strong correlation between the magnitudes of global SUSY breaking, $m_{W^\pm, Z_0}$ and m_{Z_E}. It now becomes apparent why m_{Z_E} cannot be too large [say, above $O(1 \text{ TeV})$] because then $m_{1/2} \gtrsim O(1 \text{ TeV})$ and the gauge hierarchy is jeopardized. Thus, this is another reason for excluding intermediate mass scales.

The only remaining problem is who determines T_R? Here, we can give a twofold answer.[19,21] It is conceivable that superstring dynamics (such as loop effects) determine T_R dynamically.[21,42] In such a case, we have to follow[19] the standard dimensional transmutation strategy[44] in which one keeps $m_{1/2}$ fixed, uses standard nongravitational interactions to break radiatively $SU(2)_L \times U(1)_Y \times U(1)_E$—thus determining $\langle H \rangle \equiv v$, $\langle \bar{H} \rangle \equiv \bar{v}$, and $\langle N \rangle \equiv x$ dynamically—and then calculates, self-consistently, $m_{1/2}$ through equation 18. This we call[19] the Hybrid Dimensional Transmutation model (HYDTRA), in contrast to the pure no-scale scenario[19,21] where T_R (and thus $m_{1/2}$) is left as a dynamical variable to be determined by low energy radiative corrections in the usual way.[22,36–38] Notice that in both scenarios,[19,21] equation 18 implies $T_R \sim O(1)$, which through equation 16 gives $m_{3/2} \sim M$. Notice how catastrophic such a result would be if m_0, the soft-breaking scalar masses, were not protected[37,42,43] as discussed above to be naturally zero (up to nongravitational radiative corrections). If they were not protected, then $m_0 \sim m_{3/2} \sim M$, that is, the end of gauge hierachy and the beginning of gauge chaos. It should be emphasized that not only $m_{3/2}$, but all other relevant scales such as the compactification scale M_c, the superstring scale M_s, the grand unification scale M_u, and the gaugino condensation scale Λ_Q are dynamically determined to be

TABLE 2. Particle and s-Particle Spectrum of the "Hybrid" Dimensional Transmutation Model for $h_u = 0.025^a$ and Two Representative Choices of the Parameters (k_u, λ_u)

	Mass (GeV)[b]	
Particle	(a)	(b)
Top-quark	42	41
D-quark	530	180
$Z_1 \; (\sin^2 \theta_W = 0.206)$	91.8	90.5
$Z_1 \; (\sin^2 \theta_W = 0.220)$	92.6	91.3
$Z_2 \; (\sin^2 \theta_W = 0.206)$	650	250
$Z_2 \; (\sin^2 \theta_W = 0.220)$	680	260
$H_a^\pm$	450, 540	160, 210
$H^\pm$	620	260
$\tilde{t}_1, \tilde{t}_2$	1500, 1600	540, 630
$\tilde{u}_R, \tilde{c}_R; \tilde{u}_L, \tilde{c}_L$	1500, 1600	580, 610
$\tilde{b}_R, \tilde{b}_L$	1500, 1600	560, 610
$\tilde{d}_R, \tilde{s}_R; \tilde{d}_L, \tilde{s}_L$	1500, 1600	560, 610
$\tilde{D}_1, \tilde{D}_2$	970, 1900	380, 720
$\tilde{D}_a, \tilde{D}_a^c$	1500, 1500	550, 560
$\tilde{\ell}_R^\pm \tilde{\ell}_L^\pm$	430, 540	160, 210
$\tilde{\nu}_R, \tilde{\nu}_L$	520, 540	190, 200
$\tilde{\chi}^0$	140	57
$\tilde{\chi}^\pm$	210	110

^{a}This corresponds to $m_{\text{top}} \simeq 40$ GeV.

^{b}Case (a), corresponding to $k_u = 0.055$ and $\lambda_u = 0.070$, gives $x/v \simeq 9.9$ and $\bar{v}/v \simeq = 0.4$, and is in agreement both with the particle physics limits and with the cosmological constraints; however, it is characterized by a rather heavy s-particle spectrum. Case (b), corresponding to $k_u = 0.050$ and $\lambda_u = 0.100$, gives $x/v \simeq 3.9$ and $\bar{v}/v \simeq 0.5$, and is characterized by a lighter s-particle spectrum; however, it does not satisfy the cosmological limit of reference 25.

$O(M)$ because they are all expressible as simple functions of the S and T fields.[16,20] Furthermore, the dynamical determination of the M_u in conjunction with the dynamical determination (through equation 12) of g_u^2 imply the dynamical determination (through the standard lore of RGEs) of all the gauge coupling constants.[16,19,20] This is an unprecedented incident in particle physics.

There are some advantages of the HYDTRA over the pure no-scale scenario. In the HYDTRA scenario, the "hidden sector" (S, T), which is there for providing SUSY breaking seeds, decouples completely from the observable sector. Thus, there is no mystery anymore why global SUSY breaking and electroweak breaking are of the same magnitude (see equation 18). Furthermore, such a complete decoupling of the two sectors automatically resolves the "Polonyi problem".[45] In the context of conventional cosmology, it has been difficult to see how the T_R field (in all models where the SUSY breaking scale is determined through a v.e.v. in the hidden sector) can relax to its minimum and not leave behind an excess of vacuum energy in the form of coherent waves. In the HYDTRA-type models,[19] $M_{T_R} \sim M$ and thus T_R, as being superheavy, decays instantly. Therefore, just like the axion S_I ($\equiv D$) case, it poses no cosmological problems.[45]

A complete and detailed study of the superstring HYDTRA and pure no-scale models has been given in reference 19. Here, we discuss a few highlights of their phenomenological consequences. As TABLE 2 shows, the expected SUSY spectrum and the extra-neutral gauge boson Z_E are rather "heavy"; for example

$$m_{Z_E}; m_{\tilde{q},\tilde{\ell}} > O(150 \text{ GeV}). \tag{19}$$

This makes their discovery rather impossible for the CERN Sp$\bar{\text{p}}$S collider, but it is hopefully within reach for the FNAL Tevatron or the Superstring Collider (SSC). Furthermore, in this class of superstring models, it is possible to create radiatively naturally tiny Dirac neutrino masses[46] that are suitable for solving the solar neutrino problem. In conjunction with stable photinos, it may also shed light on the dark matter.[47]

It should be stressed that because of the existence of the extra low energy $U(1)_E$ gauge symmetry, dangerous dimension-5 operators (causing catastrophically rapid proton decay) do not appear.[26,48] In addition, because the characteristic mass scale or ordinary dimension-6 operators causing proton decay is $M_{P\ell}$, it becomes extremely difficult in this class of superstring-inspired models to get observable proton decay.[48]

In regard to certain of our assumptions that lead us "uniquely" to our class of models, recent detailed studies have shown us that they are well founded. Indeed, it has been argued[49] recently on general grounds that E_6, $SO(10)$, and $SU(5)$ are the only superstring-allowable gauge groups. Further detailed studies of $SO(10)$- and $SU(5)$-based models show that these models are mired into rather grave difficulties.[50] Finally, it has been shown[51] that if we concentrate on E_6, then the rank-6 subgroups with intermediate scales are unattainable.[51] Thus, we are left with only rank-5 groups and no intermediate scales; this uniquely singles out our $SU(3)_c \times SU(2)_L \times U(1)_Y \times U(1)_E$ model,[19] no-intermediate scales, and only 3-generations. Furthermore, recent studies[52] support our view that gaugino masses are indeed the dominant source of supersymmetry breaking in the observable sector.

In conclusion, we have definitely not yet seen the last page of model building, but it becomes more apparent as the time goes by that superstrings are here to stay.

ACKNOWLEDGMENTS

It is a great pleasure to thank B. Durand and L. Durand for their kind invitation and the warm hospitality extended to me during the Aspen meeting.

REFERENCES

1. GREEN, M. B. 1983. Surv. High Energy Phys. **3:** 127; SCHWARZ, J. H. 1982. Phys. Rep. **89:** 223.
2. GREEN, M. B. & J. H. SCHWARZ. 1984. Phys. Lett. **149B:** 117.
3. GREEN, M. B. & J. H. SCHWARZ. 1985. Phys. Lett. **151B:** 21.
4. GROSS, D., J. HARVEY, E. MARTINEC & R. ROHM. 1985. Phys. Rev. Lett. **54:** 502; 1985. Nucl. Phys. **B256:** 253; 1986. Nucl. Phys. **B267:** 75.
5. CECOTTI, S., S. FERRARA, L. GIRARDELLO & M. PORRATI. 1985. Phys. Lett. **164B:** 46; ROMANS, L. & N. WARNER. 1985. Caltech preprint no. 88-1291.
6. ZWIEBACH, B. 1985. Phys. Lett. **156B:** 315.
7. CANDELAS, P., G. T. HOROWITZ, A. STROMINGER & E. WITTEN. 1985. Nucl. Phys. **B258:** 46.
8. CALABI, E. 1957. *In* Algebraic Geometry and Topology: A Symposium in Honour of S. Lefschetz, p. 78. Princeton Univ. Press. Princeton, New Jersey; YAU, S. T. 1977. Proc. Natl. Acad. Sci. **74:** 1798.
9. WITTEN, E. 1984. Phys. Lett. **149B:** 351.
10. ACHIMAN, Y. & B. STECH. 1978. Phys. Lett. **77B:** 389; BARBIERI, R. & D. V. NANOPOULOS. 1980. Phys. Lett. **91B:** 369; BARBIERI, R., A. MASIERO & D. V. NANOPOULOS. 1981. Phys. Lett. **104B:** 194; BARBIERI, R., S. FERRARA & D. V. NANOPOULOS. 1982. Phys. Lett. **116B:** 16.
11. WITTEN, E. 1985. Nucl. Phys. **B258:** 75.
12. BREIT, J. D., B. A. OVRUT & G. SEGRE. 1985. Phys. Lett. **158B:** 33.
13. HOSOTANI, Y. 1983. Phys. Lett. **126B:** 309.
14. CECOTTI, S., J. P. DERENDINGER, S. FERRARA, L. GIRARDELLO & M. RONCADELLI. 1985. Phys. Lett. **156B:** 318.
15. DINE, M., V. KAPLUNOVSKY, M. MANGANO, C. NAPPI & N. SEIBERG. 1985. Nucl. Phys. **B259:** 519.
16. COHEN, E., J. ELLIS, C. GOMEZ & D. V. NANOPOULOS. 1985. Phys. Lett. **160B:** 62.
17. DINE, M., R. ROHM, N. SEIBERG & E. WITTEN. 1985. Phys. Lett. **156B:** 55; DERENDINGER, J. P., L. IBANEZ & H. P. NILLES. 1985. Phys. Lett. **155B:** 65.
18. FERRARA, S., L. GIRARDELLO & H. P. NILLES. 1983. Phys. Lett. **125B:** 457.
19. ELLIS, J., K. ENQVIST, D. V. NANOPOULOS & F. ZWIRNER. 1986. Nucl. Phys. **B276:** 14; Mod. Phys. Lett. **A1:** 57.
20. COHEN, E., J. ELLIS, K. ENQVIST & D. V. NANOPOULOS. 1985. Phys. Lett. **161B:** 85.
21. COHEN, E., J. ELLIS, K. ENQVIST & D. V. NANOPOULOS. 1985. Phys. Lett. **165B:** 76.
22. ELLIS, J., A. B. LAHANAS, D. V. NANOPOULOS & K. TAMVAKIS. 1984. Phys. Lett. **134B:** 429.
23. For reviews, see: NANOPOULOS, D. V. 1984. *In* The Proc. of the XXII International Conf. on High Energy Physics, Vol. II. A. Meyer & E. Wieczorek, Eds.: 36. Academie der Wissenschaft der DDR. Leipzig; KOUNNAS, C., A. MASIERO, D. V. NANOPOULOS & K. A. OLIVE. 1984. Grand Unification with and without Supersymmetry and Cosmological Implications. World Scientific. Singapore; ELLIS, J. 1984. Proc. 5th Workshop on Grand Unification, Brown University. K. Kang, H. Fried & P. Frampton, Eds.: 436. World Scientific. Singapore; LAHANAS, A. B. & D. V. NANOPOULOS. 1987. The road to no-scale supergravity. Phys. Rep. **145:** 1.
24. BARGER, V., N. G. DESHPANDE & K. WHISNANT. 1986. Phys. Rev. Lett. **56:** 30; DURKIN, L. S. & P. LANGACKER. 1986. Phys. Lett. **166B:** 436.
25. ELLIS, J., K. ENQVIST, D. V. NANOPOULOS & S. SARKAR. 1986. Phys. Lett. **167B:** 457.
26. CAMPBELL, B., J. ELLIS, K. ENQVIST, M. K. GAILLARD & D. V. NANOPOULOS. 1986. CERN preprint no. TH.4473/86.
27. STROMINGER, A. & E. WITTEN. 1985. Commun. Math. Phys. **101:** 341.
28. STROMINGER, A., 1985. Phys. Rev. Lett. **55:** 2547.
29. WITTEN, E. 1985. Phys. Lett. **155B:** 151.
30. ELLIS, J., C. GOMEZ & D. V. NANOPOULOS. 1986. Phys. Lett. **171B:** 203.
31. CREMMER, E. *et al.* 1983. Nucl. Phys. **B212:** 413.
32. ANTONIADIS, I., C. KOUNNAS & D. V. NANOPOULOS. 1985. Phys. Lett. **162B:** 309.

33. Choi, K. & J. E. Kim. 1985. Phys. Lett. **165B:** 71; Ibanez, L. E. & H. P. Nilles. 1986. Phys. Lett. **169B:** 354.
34. Ellis, J., C. Gomez & D. V. Nanopoulos. 1986. Phys. Lett. **168B:** 215.
35. Cremmer, E., S. Ferrara, C. Kounnas & D. V. Nanopoulos. 1983. Phys. Lett. **133B:** 61.
36. Ellis, J., C. Kounnas & D. V. Nanopoulos. 1984. Nucl. Phys. **B241:** 406.
37. Ellis, J., C. Kounnas & D. V. Nanopoulos. 1984. Nucl. Phys. **B247:** 373; Phys. Lett. **143B:** 410.
38. Ellis, J., K. Enqvist & D. V. Nanopoulos. 1984. Phys. Lett. **147B:** 94; 1985. **151B:** 357.
39. Dine, M. & N. Seiberg. 1985. Phys. Rev. Lett. **55:** 366; Kaplunovsky, V. 1985. Phys. Rev. Lett. **55:** 1036.
40. Ellis, J., K. Enqvist, D. V. Nanopoulos & M. Quiros. 1986. Nucl. Phys. **B277:** 231.
41. Preskil, J., M. B. Wise & F. Wilczek. 1983. Phys. Lett. **120B:** 127; Abbott, L. E. & P. Sikivie. 1983. Phys. Lett. **120B:** 133; Dine, M. & W. Fischler. 1983. Phys. Lett. **120B:** 137.
42. Binetruy, P. & M. K. Gaillard. 1986. Phys. Lett. **168B:** 347; Breit, J. D., B. A. Ovrut & G. Segre. 1985. Phys. Lett. **162B:** 303.
43. Diamandis, G., J. Ellis, A. B. Lahanas & D. V. Nanopoulos. 1986. Phys. Lett. **173B:** 303.
44. Ellis, J., J. Hagelin, D. V. Nanopoulos & K. Tamvakis. 1983. Phys. Lett. **125B:** 275; Kounnas, C., A. B. Lahanas, D. V. Nanopoulos & M. Quiros. 1983. Phys. Lett. **132B:** 95; 1984. Nucl. Phys. **B236:** 438.
45. Ellis, J., D. V. Nanopoulos & M. Quiros. 1986. Phys. Lett. **174:** 176.
46. Mariero, A., D. V. Nanopoulos & A. Sanda. 1986. Phys. Rev. Lett. **57:** 663.
47. Nanopoulos, D. V. & K. A. Olive. 1986. UW-Madison preprint no. MAD/TH/86-27.
48. Gaillard, M. K. & D. V. Nanopoulos. 1986. UW-Madison preprint no. MAD/TH/86-13.
49. Campbell, B. A., J. Ellis & D. V. Nanopoulos. 1986. Phys. Lett. **181B:** 283.
50. Ellis, J., K. Enqvist, D. V. Nanopoulos, K. A. Olive, M. Quiros & F. Zwirner. 1986. Phys. Lett. **176B:** 403.
51. Campbell, B. A., J. Ellis, M. K. Gaillard, D. V. Nanopoulos & K. A. Olive. 1986. Phys. Lett. **180B:** 77.
52. Ellis, J., D. V. Nanopoulos, M. Quiros & F. Zwirner. 1986. Phys. Lett. **180B:** 83.

Index of Contributors